The Audience and Its Landscape

CULTURAL STUDIES

Series Editor
Paul Smith, *George Mason University*

The Audience and Its Landscape *edited by James Hay, Lawrence Grossberg, and Ellen Wartella*

Eye on the Flesh: Fashions of Masculinity in the Early Twentieth Century *Maurizia Boscagli*

Art and the Committed Eye: The Cultural Functions of Imagery *Richard Leppert*

Youth, Murder, Spectacle: The Cultural Politics of "Youth in Crisis" *Charles R. Acland*

Viewing, Reading Listening: Audiences and Cultural Reception *edited by Jon Cruz and Justin Lewis*

The Madonna Connection: Representational Politics, Subcultural Identities, and Cultural Theory *edited by Cathy Schwichtenberg*

Dreaming Identities: Class, Gender, and Generation in 1980's Hollywood Movies *Elizabeth G. Traube*

Enlightened Racism: The Cosby Show, Audiences, and the Myth of the American Dream *Sut Jhally and Justin Lewis*

FORTHCOMING

Girls: The Representation of Femininity in Popular Culture *edited by Elizabeth Ann Kaplan*

Being Indian and the Politics of Indianness *Gail Valaskakis*

Boys: The Representation of Masculinity in Popular Culture *edited by Paul Smith*

The Audience and Its Landscape

edited by

James Hay
Lawrence Grossberg
Ellen Wartella

Routledge
Taylor & Francis Group
New York London

Cultural Studies

First published 1996 by Westview Press

Published 2019 by Routledge
52 Vanderbilt Avenue, New York, NY 10017
2 Park Square, Milton Park, Abingdon, Oxon OX14 4RN

First issued in hardback 2019

Routledge is an imprint of the Taylor & Francis Group, an informa business

Library of Congress Cataloging-in-Publication Data
The audience and its landscape / [edited by] James Hay, Lawrence
Grossberg, Ellen Wartella.
p. cm.–(Cultural studies)
Includes bibliographical references and index.
ISBN 0-8133-2284-7 (hbk : alk. paper). –ISBN 0-8133-2285-5 (pbk
: alk. paper)
1. Television viewers–Research. 2. Motion picture audiences–Research.
3. Television broadcasting–Social aspects. I. Hay, James, 1952– .
II. Grossberg, Lawrence. III. Wartella, Ellen.
IV. Series.
PN1992.55.A88 1996
302.23'4–dc20 96-14081
CIP

ISBN 13: 978-0-367-31830-7 (hbk)
ISBN 13: 978-0-8133-2285-8 (pbk)

Contents

Part IV
Locating Audiences

Tables and Figures

Tables

Figures

Introduction

James Hay, Lawrence Grossberg, and Ellen Wartella

The genesis of this book is long and worth reconstructing. It is an allegory of sorts that involves many of the debates that have shaped audience study in recent years. Audience study in the late 1980s was fraught with both a sense of mission and a sense that no one had yet got it right. Its sense of mission had to do with the proliferation of work about the audience across various fields and disciplines. The sense that no one had got it right, however, was just as much the result of its dispersal. This book grew out of the sense of excitement and frustration surrounding the convergence and realignment of multiple disciplines and sites where studies about audience were being done at that time.

One of these realignments involved the distinction between "qualitative" and "quantitative" communication research. Qualitative research had emerged out of efforts, largely within the tradition of mass communication research, to rethink its empirical tradition by engaging critical theories that it perceived as more directly given to problems of interpretation. By the late 1980s it had become a quite recognizable project that saw itself engaged with but methodologically different from quantitative research.

Qualitative research cultivated a dialogue with critical studies in the humanities. But there remained strong perceptual differences between these fields about each other. The tradition of audience research lay within mass communication studies, which often perceived work being done in the humanities as preoccupied with texts and a formalist methodology. Spectator theory (in film criticism) and reader-response theory were often viewed in U.S. mass communication research as too laden with jargon, and perhaps even too theoretical, European, and threatening to its own "turf."

Work in the humanities, on the other hand, seemed to have little need even to feel threatened by mass communication research of any stripe. If those most invested in the humanities even recognized a distinction between "quantitative" and "qualitative" research, they would have still perceived any kind of mass communication research as dealing with subjects alien to the humanities and through a "scientific" project and a professionalism that had traditionally distinguished the humanities from other disciplines of knowledge. On one side, qualitative research took on the problem of interpretation but seldom attempted to deconstruct its own scientificity. On the other, work in the humanities was not quite willing to recognize its own relation to other professions and its own aspiration for as critical science (either in New Criticism or imported, Europeanisms such as semiotics).

The proliferation during the 1980s of critical studies about television could be seen as one consequence of a cross-disciplinary awareness by some in mass communication research and the humanities—though one could argue whether television criticism bridged or exacerbated their disciplinary differences. Given the proliferation of qualitative and critical work on television, television studies generally encountered great difficulty in establishing themselves within academic departments from either discipline. But it was through television studies that the methodological and theoretical debates across these disciplines came to bear most directly in their respective treatments of the audience. The most well-known audience work centered on television, while critical studies of television became distinguished from literary and film criticism precisely for the former's interest in the issue of the audience.

The tale of television studies during the 1980s was intertwined with the growing impact and tenuous status of "cultural studies." British Cultural Studies had often turned to television as a means of demonstrating or arguing certain issues about contemporary culture and everyday life. But because cultural studies was most interested in understanding media as cultural forms and media's relation to everyday life, it devoted considerable attention to the issue of the audience. Indeed television audiences often became a means for cultural studies to rethink traditional models of power and consciousness perpetuated in mass communication studies and literary and film studies. Cultural studies emphasized audiences' implication in broader hegemonic formation and a complex and constantly shifting cultural politics in everyday life. Just as important, cultural studies' attention to television and television audiences was informed by its more general interest in bridging sociological and textual methods in media studies while interrogating their assumptions and traditional subjects. But the 1980s, cultural studies had become one of the most vocal critics both of traditional mass communication research and of the kind of film and literary studies driven by textual, deconstructive criticism. Consequently, its discussion of audiences became quite influential across academic disciplines and departments but also stymied its acceptance in fields with strong attachments to one or the other of these methods.

Admist this disciplinary convergence and realignment, not to mention efforts by many to defend their intellectual and methodological traditions against the tide, audience study became increasingly associated with ethnography. For qualitative research, ethnographic methods effectively mediated their interest in interpretive issues while still remaining engaged with the tradition of empirical and scientific research of mass communication subjects. For British Cultural Studies, ethnographic methods (since its studies of youth subcultures) had held promise a way of understanding media culture in everyday life. Cultural Studies found in ethnography a way to combine empirical and textual study while arguing that traditional audience research was too empirical and textual study too detached and disinterested in texts' relation to popular struggles or contexts.

This interest in ethnographic methods occurred at a time when the mission of ethnography was itself in question. The debate was located within academic disciplines such as anthropology and sociology that had traditionally claimed ethnography as their method. More generally, however, the history of ethnographic research in cultural anthropology had increasingly come under attack by postcolonial theory, and the exigencies of doing ethnographic research in and of a global media culture became as theme for those who saw themselves doing a revisionist, "postmodern ethnography." While the definition and history of ethnography had come under attack, cultural studies of media audiences that valorized ethnography were often criticized for not knowing what ethnographic research really involved. Or cultural studies of television audiences, interested in how meaning was generated through media texts, were challenged on other fronts for believing that "fieldwork" somehow positioned them more favorably to accomplish this than an "armchair" critic or archivist.

The one general area of consensus across this range of shifting, occasionally contradictory positions was their rejection of the "hypodermic needle" conception of communication which assumed (some time long ago) that audiences were passive receptors—tablets on which were written media messages. What significance should be attached to audiences' "activity" beyond this one point of agreement, however, was not so easily agreed upon. And given the questions and conflicts over definition and method, the indeterminacy associated with "active" audiences was in part the consequence of media subjects, having become discrete objects of study about whose practices there was little consensus. Lurking around the assumption of audience activity was also a perception—somewhere between nihilism and deconstructionism—that the audience was purely a fiction, the arbitrary result of conventions and "fashion."

One of my favorite local anecdotes abut the various disciplinary and epistemological convergences or realignments through which audience had become an issue was a seminar conceived by Larry Grossberg and Ellen Wartella in 1988 at the University of Illinois that explored the possibility of an "interdisciplinary" discussion of the audience and audience study. Both were respectably knowledgeable about the theoretical writing that had shaped each others' thinking on the subject. The best of friends, with the greatest optimism about overcoming their differences (and a moment of faith that individual class sessions might even be a dialogue on selected issues), they gradually came to the cruel realization that interdisciplinarity is very hard work indeed.

Yet remaining undaunted (still friends) and probably out of frustration, the three of us (and I make no claims as a mediator between them) decided to organize a conference that would bring together scholars with various interest in the issue of audience and whose work was recognized as having a relation to various methodologies and intellectual traditions. At one point, we toyed with the prospect of encouraging participants to imagine what a multidisciplinary study of audiences might involve. But as we began to receive outlines for conference papers, and certainly over the course of the conference itself, we and most of the participants became acutely aware of just how deep were some of the dispositions about audience study and yet how willing all of us were at least to run with the possibility of a cross-disciplinary project. The conference Toward a Comprehensive Theory of the Audience, was held at the University of Illinois in Urbana/Champaign, September 6–9, 1990. The essays collected here were presented at the conference.

Throughout the conference, therefore, the interdisciplinary ideal itself was interrogated. Interdisciplinary work is always situated and subject to disciplines of knowledge. Certainly it is important continually to rethink intellectual traditions upon which any discourse or field of study is predicated and to recognize that "fields" are arbitrary, not particularly coherent, and constantly crossed by interlopers, poachers, and "dilettantes." Instead of the loftier aims of interdisciplinary projects, however, what may be needed is something closer in spirit to *bricolage*, the constant invention of new tools for working a changing landscape and making do with what is available in a given context.

The context or landscape of audience study over the last seven years is certainly not limited to academic debates and revisionism. It would be shortsighted to assume that audience study is bound by a narrow set of academic institutions and not recognize how it has taken shape amidst emerging and residual media technologies, businesses, and practices in everyday life. In recent years, radio and television programming and advertising have reimaged and mythologized the audience through talk shows. Strategies of political campaigning in the United States have come to exploit a myth of the "town meeting" and the instantaneous response associated with these talk shows. Home shopping, sex lines, and televised gaming are predicted upon a myth of direct participation. A convergence of media technologies and realignments among media corporations have contributed to as new myth of "interactivity."

Amidst the gradual erosion of the three broadcasting companies in the United States or of the "public" companies in Europe, newer companies worked hard to maintain the myth of the (demographically) coherent audience. The same media technologies, such as the remote control, that placed these networks on a domestic menu with more traditional broadcasters also made the image and definition of audience more of an issue. Music Television, Christian Broadcasting, Black Entertainment Television all are examples of efforts by broadcasters and advertisers in the United States to convince consumers (and each other) that audience definition matters. At the same time, however, their appeal to consumer lifestyles (i.e., to a whole range of "entertainment" and leisure products besides television) underscores the very problem in their defining audiences in terms of a particular "network" or even, for that matter, a particular medium.

This intense practice of media and product "tie-ins" belies a changing understanding of environment and the sites of media engagement. The definition of audience is, therefore, an issue of location—the situated and/or dispersed features of media engagement. The concept of the "home entertainment system" and "home computer system" affirm that a media site can be intersected by multiple media and that concepts of domestic or private space have been reorganized in relation to a material and conceptual redesignation of the media site. The policy implications of locating audiences have "come home" full force in the U.S.

Congress's effort to regulate the flow of "pornography" on television and the internet into U.S. households, where (it was argued) parents stood to lose their prerogative as guardians of their children's private lives.

All of these transformations have become part of a backdrop of practices against which attempts have been made recently to define, study, and locate audiences. In order to emphasize some of the strategic implications of audience study in relation to these conditions, the essays in this volume have been arranged into four groups. The first concerns discussions about audience study's relation to intellectual and research traditions, given their convergencies and realignments. The second section considers some current issues in defining audiences. Like those in the first section, these essays recognize that audience study rest upon changing assumptions, but they are particularly interested in the ontology of its subjects. The third section acknowledges the conflicts among current forms of research in order to emphasize the strategic, and thus political, ramifications of audience study. Essays comprising the fourth section offer some ways of thinking about audience study as a process of locating subjects, definitions and politics.

Most of the essays in this collection do share a sense that audience study is a situated practice. But working with this realization points out the difficulty or impossibility of cross-disciplinary projects and calls into question the mission of audience study. We make no claim that these essays offer a comprehensive understanding of current audience research. But it is our hope that they will serve to outline (however partially) the contours of an historical dialogue about the subjects, traditions, and context of audience study.

PART I

Audience Studies and the Convergence of Research Traditions

Viewers Work

Elihu Katz

1

If you think, as I do, that communications research is dominated by social scientists (even if these are marginal within their own mainstreams) you cannot help but have noticed the steady influx of "guest workers" from literature, cinema, linguistics, and elsewhere. If you think, as I do, that communications research is worse off because resident humanists split from the social scientists in the 1950s over the latter's obsession with short-run effects, you will welcome the reunion now in progress. A material reason for this is that there are jobs. Thanks to the academic diplomacy of our pioneers—of whom Wilbur Schramm is surely foremost in this respect—the university teaching of the media has become increasingly respectable, prosperous and interdisciplinary—and just in time too, since it takes some watching to watch the whole world watching the whole world watching.

The other reason is that there is a great convergence of interest in the audience. Literature, after 2,000 years, is urgently interested in matching real readers with ideal readers. Students of cinema need to know where all the voyeurs have gone. Sociolinguists—and a host of fellow travelers from mainstream social science—have discovered that we hang out at the lampposts under which social reality is constructed.

The secret satisfaction in all this is that we were there first. Media research, for better and for worse, was founded to tell communicators whom they were talking to. We do know a lot about audiences. In fact, we know a lot about reception, although it will not quite stand still. And, of course, we have much yet to learn.

What we know most about audience behavior is who chooses what; the problem here is that we know too much, in that the mountains of ratings have only rarely been mined for their lawfulness (an important exception is Barwise and Ehrenberg, 1988; for a humanist perspective, see Ang, 1991 and Morley, 1989). We also know quite a lot about why people choose as they do, and what they get out of it, at least from their point of view. What we know least about is the psychology and sociology of the viewing experience itself or, to use some double jargon, how viewers position themselves before the screen. It is fashionable nowadays to say that, unless we know how viewers "read"—what they make of what they see—we cannot talk about effects very well.

In this matter of reception, we are not of one mind. The elementary issue of passivity versus activity is still with us. There are those who continue to research and defend the proposition that television viewing is typically regressive. The viewer, they say, operates at a low level of mental arousal, performing simpliciti-cally in cognitive terms, and at a pre-Oedipal stage in psychosexual terms. In this same camp I would include those who insist that the viewer hardly exists. Television is just a background, "moving wallpaper," and nobody is actually watching, this proposition is supported by much evidence based on viewing viewers (Bechtel, Achelpohl, and Akers 1972; Collett and Lamb, 1986). On the other hand, there is much current research to support the image of the overburdened viewer, who, seated before the television set after a hard day, must yet erect schemes, scripts, frames, and decoding rules before turning on the set. In the old days, viewing was a lot easier. You pushed a button, and there was Lucy.

The question I wish to address is, Do viewers work? And if so, what sort of work do they do? I wish to begin with a passage on the subject from Schramm, Lyle, and Parker (1961). I then will confront Schramm et al. with four very recent, very different, studies, to see if we have gotten anywhere since 1961. Finally, I will discuss three studies that have occupied me over the last decade, to see whether any of that helps.

I

In their pioneering study of the interaction of television and children, Wilbur Schramm and his associates divided television texts into two types, which they variously call cognitive and emotional, socializing and wish-fulfilling, reality and fantasy, or later, reality and entertainment.

Citing Freud's principles of pleasure and reality as inspiration, they put the matter this way:

fantasy content invites the viewer to take leave of his problems in the real world; invites surrender, relaxation, passivity; invites emotion; works chiefly through abrogation of the rules of the real world; acts to remove, at least temporarily, threat and anxiety, and often offers wish-fulfillment; offers pleasure.	whereas reality content constantly refers the viewer to the problems of the real world; invites alertness, effort, activity; invites cognition; works chiefly through realistic materials and situation; tends to make the viewer even more anxious, in return for a better view of problem, offers enlightenment.

The authors emphasize that "while either type of content may be used in social interaction, as a source of information, and for entertainment, reality content (news, public affairs, etc.) primarily serves the function of supplying information, and entertainment content (Westerns, crime programs, comedy, variety shows, sports, etc.) serves the function of ministering to immediate personality and social needs" (Roberts and Schramm, 1971).

The statement lends itself well to modern discussion of the role of the reader. The authors do not propose, as positivists might, that the text dictates the reading, nor do they propose that the text is only an inkblot. Although they are avowedly functionalist, they steer clear of what Blumler et al. (1986) call vulgar gratificationalism. Rather, they use the language of "invitation" whereby certain texts express preferences for certain types of viewers. They make room for aberrant readings whereby real readers may use fantasy texts for information and reality texts for escape. The work of fantasy, they suggest, involves a concurrence in the breaking of real-world rules in a mood of "surrender, relaxation and passivity"(Roberts and Schramm, 1971). The reward is pleasure and the postponement or denial of threat and anxiety. For its part, reality content asks viewers to stay awake and suffer, offering enlightenment in return.

Schramm finds that there are two corresponding types of children who differentially seek out the two kinds of texts. Reality-seeking children, influenced by middle-class values of achievement, switch away from television to print and enrich their television diet with reality materials, especially as they grow older. The ranks of fantasy seekers, the authors find, augmented by middle-class children who are unhappy at home, consist largely of lower-class children who do not learn to delay gratification.

While refusing to pronounce judgment, the authors believe that fantasy predominates in children's use of television and that the normal process of maturation implies leaving television behind. They do not explain why adults come back for more.

What interests us, however, is not what people take from television but what they put into it. What do Schramm et al. have to say about viewers' work? Ostensibly, it appears, reality seekers are hard workers and fantasists have an easier time of it. Reality texts require the reader to postpone immediate gratification, to be alert, to invest mental effort, face the facts, cope with anxiety, and learn. The fantasy reader, on the other hand, is supposed to relax, lie back, and experience pleasure—to be passive, say the authors. But what about "taking leave"? Isn't that an activity? The denial of reality may be a useful therapy, but it is not as easy on the psyche as it sounds. And what about suspending disbelief in the act of cooperating with the storyteller? That does not sound much easier than testifying against the evidence of one's senses, as in the Asch experiment, and we know that to do so is actually painful. By the same token, wish-fulfillment requires that I have a wish and that I be engaged in the work of daydreaming. Freud's message is that all of these undercover processes are seething with activity. Even positivistic social psychology teaches that it takes effort to maintain attitudes, not just to change them, no less than critical theory wants us to know how much work it takes to maintain the status quo. It is an inescapable conclusion that fantasy viewers work, too, even if their work may be different. Why, then, do we call it passive?

II

Robert Merton repeatedly reminds us that science abhors monuments and that the only true tribute is to stand on somebody's shoulders in order to get a better view. You will understand, therefore, that I want us to be critical of

Schramm because of his place of honor, not in spite of it. So let us approach four recent authors and ask them what they see from there. Do viewers work? And if so, what sort of work do they do?

If you ask Kubey and Csikszenmihalyi (1990), the answer is no. Viewing, they say, is indeed like daydreaming. People who responded that they were engaged in watching television—when they were electronically beeped at random intervals over a seven-day period—rated themselves high on relaxation and motivation to perform the activity, but low on activity, alertness, concentration, feeling challenge or employing skill, and, surprisingly, affect. Reading is surprisingly like television viewing in that it is also rated low on activity, affect, and challenge and high on relaxation, but readers feel they are investing considerably more concentration, skill, and control than viewers. Among in-home activities, talking rates higher than television on almost all of these dimensions. Love and sports take all the prizes.

Here, then, is Schramm's fantasy viewer, but minus emotion and minus the aesthetic investment that fantasy viewers are asked to make. Even while accepting the label "passive," Schramm tells us about how fantasy viewers must at least cooperate. Kubey and Csikszentmihalyi's viewers are nearly asleep, and if they have viewed heavily, they feel the worse for it. While the authors take pains to reassure us that their findings are not necessarily incompatible with the competing image of the overburdened viewer, busily decoding, we are not reassured. In fact, Kubey and Csikszentmihalyi sound much more convincing in proposing that oral regression may describe television viewing. The viewer assumes a womb-like position, engages in eating, talking, smoking, or thumbsucking, while fixated on the pleasures being dangled before him. Using the same imagery, Beverle Houston (1985) thinks of viewing as a nonstop tease, tantalizing and frustrating. Kubey and Csikszentmihalyi warn us about such behavior; it is dysfunctional for individual growth, they say, bad for survival under stress. This is "the doctrine of horrible consequences" (Schur, 1988) at its best: if you watch television, you lower your chances of survival in a concentration camp. We might say about Schramm et al. that their use of the word "passive" diverts us from a better understanding of the viewing experience, but what shall we say about Kubey and Csikszentmihalyi? Perhaps we should be satisfied with passive, and quit while we are ahead. Viewing with others, by the way, raises the level of perceived challenge and the cheerfulness of the experience. Program and genre make little difference in viewers' self-descriptions; only cultural differences—where television is defined differently than in the United States, and is more demanding—may make a difference.

No less pessimistic about the dire effects of television on society, Neil Postman (1986) does not doubt that television is engaging. In his view, the problem is that the processing of reality by producers and viewers blurs the distinction between reality and fantasy and thereby emasculates society's ability to cope with change. Implicitly, Postman should be questioning television's ability even to tell a story, based as it is, in his account, on the decontextualized telegraph and photograph. But it is not stories that worry Postman; his concern is that television is incapable of rational argument. It is without context, sequence, syntax, or emotional restraint and hence cannot hope to be the political forum that some mistake it to be. It stands outside public space. Postman wants to teach us all—children and adults—about the danger of looking to television in order to relate to reality. He would be perfectly satisfied if the medium kept its place in the palaces of fantasy and pleasure, if thinking people would only understand that that is all it can do. Television news is entertainment, says Postman, not because aberrant readers may sometimes use it as such, and not because it is designed to keep us out, but because its pictures distract us from serious business.

It is perhaps ironic that Dave Morley (1980) should have to step in to save television news for politics. Standing at the vanguard of a neo-Marxist tradition that has only lately come to be interested in the empirical study of audiences, Morley tells us that viewers of the British television program *Nationwide* are actually doing political work, and not just being worked upon. Indeed, some viewers—as revisionist theory suggests—even read the program oppositionally, exposing the establishment bias that goes into the making of programs. Morley and his viewers affirm that television can also serve reality seekers. If Kubey and Postman are arguing—each in his own way—that television, the medium, and its audience, are unequipped to transmit or to receive ideological statements, Morley would argue otherwise. Not only are the viewers awake; they are also capable of ideological decoding.

My fourth witness is Sonia Livingstone (1990). Applying multidimensional scaling to audience perceptions of the characters in American and British serials, she demonstrates—in an extremely inventive way—that viewers have aesthetic structures in their heads, at least latently. For example, if one maps the *Dallas* characters on a multidimensional space, one finds them arrayed according to dimensions of morality and power. That the good guys are not always weak and vice versa is the dramatic impetus of the *Dallas* narrative, argues Livingstone, and the viewers betray an awareness of this. Livingstone believes that viewers relate to fictional characters according to the everyday rules of person perception in social psychology. If so, she might propose to Schramm et al. that fantasy is an invitation to reality seekers, since the suspension of disbelief is make-believe reality. Indeed, Livingstone's viewers work very hard: asked to retell the story of a television office affair, viewers begin and end the story persuasively, but at different points, and assign motives for action that researchers do not find in the script. Unlike Morley's viewers who do ideological work, Livingstone's viewers do referential, ludic, and aesthetic work, four roles to which I shall soon return.

Can these differences be reconciled? Is it possible to reconcile Kubey's passive viewers with Morley's and Livingstone's, or with Schramm's? Do viewers work, and if so, what sort of work do they do? One way to solve the riddle is simply to say, as Kubey does, that the two processes can indeed coexist; daydreaming may be passive compared to swimming, but perhaps it gives free rein to the imagination. Careful reading of Kubey and Csikszentmihalyi, however, leaves us unconvinced. More convincing is the closely related methodological answer that "passive" is an agreed-upon definition of what television viewing is like. People and researchers stereotype actions, and the labels—certainly in this case—do not tell all. But Kubey and Csikszentmihalyi have evidence to the contrary. The most appealing explanation, of course, would be if different kinds of genres and different kinds of programs evoked different degrees of activity. The female viewer watching "Holocaust" in one of Kubey's cases, for example, does indeed report herself to have been emotionally affected. But the researchers are not much impressed by the evidence that they have in hand; variations in programs do not much show up in their statistical analysis, although it should be noted that the data are inadequate.

Kubey and Csikszentmihalyi would agree on cultural differences as a possible explanation. They, as does Postman, repeatedly cite Salomon's (1983) finding that the mental effort expended on television viewing varies in different societies. It then becomes convenient to say that Morley and Livingstone come from a culture where television is treated more seriously, as it is, say, in West Africa and Israel. There may well be something in this, but it is too easy.

If we take stock at this point, several tentative conclusions may be proposed. First, even if Kubey and Csikszentmihalyi have something important to tell us, the active/passive dimension serves us poorly; it does not contribute much to a depiction of the viewing experience. We expect too much of it. It is not good enough to argue, as Kubey does, that activity is a variable, not a dichotomy. The problem is that it varies in only one dimension, call it physical movement, and we need it to vary cognitively and emotionally. Unfortunately, that won't solve the problem either since variables such as alert, excited, and happy have also been measured and also register very low in television viewing. "Relaxed" is television's main achievement. Unless the ratings are dismissed as mere stereotypes, our problem persists.

A second conclusion must be that the reality/fantasy dimension is in trouble, too. The problem is not so much in characterizing the text as one or the other, or even in classifying viewers' motivations, but in the interaction between the two. So far, it seems that fantasy texts on television ask fantasy seekers to approach them as real, while reality texts approached as real may be altogether subversive of reality. Postman's thesis, in this sense, is a technological counterpart to Lazarsfeld and Merton's (1960) narcotized news viewer.

Thirdly, it seems that the extent and nature of the investment in viewing may indeed vary under different cultural circumstances or, more generally, as a function of the demand characteristics of the viewing or postviewing situation. The mere presence of others, and, a fortiori, the anticipation of being invited by associates or researchers to reflect on what one has seen can apparently transform the viewing experience. Reality seeking, then, may be thought of as the motivational component of an institutionalized role—such as citizen—which one brings into the living room. The semioticians and social psychologists are trying to tell us that there are other, more personal, frames that viewers bring with them as well.

The concept of role may be of some use. In sociology, roles are normative expectations directed at holders of particular statuses. Thus, the text defines a role for the reader, the reader assumes a role vis-a-vis the text, the reader plays roles in society. (Roles are the building blocks of social structure and have the advantage of being more fixed than schema, scripts, frames, etc.). I want to suggest that reality seeking may seem more active than fantasy seeking because one assumes a publicly acknowledged social role that implies the taking of action. Viewing the Watergate hearings as citizens, for example, demands a response.[1] Fantasy texts ask you to check your roles in the wardrobe and come incognito to a costume party where the most beautiful girl gets to dance with the prince. Even this is a message of social mobility; as Ruth Katz (1973) suggests, it feeds back on real roles only when the dance is over.

It follows that the concept of escape is also not as good as it sounds. Katz and Foulkes (1962) proposed long ago that all mass communication implies escape. No matter how reality-oriented, one is transported by symbols from one's immediate environment to another space, or time, or community—to Kuwait, for example, or to Southfork. What matters is not only the destination defined by the text. Equally important is the identity one assumes for the trip, what one takes along, and what one brings back. Returning from Coronation Street with the bad news about social mobility (Liebes and Livingstone, 1992) or newly equipped for the gender struggle (Allen, 1985) is not so different from the psychosexual and social functions of children's fairy tales (Bettleheim, 1976) and is not usefully described as escape. Let us reserve escape for those who do not bring anything back.

To sum up, all that we have accomplished, if anything, is to shake up the two ideal types. Leave-taking, effort, activity, emotion, even rule breaking, want to be liberated from their one-sidedness. We need a moratorium during which to recombine these elements in new profiles of viewing experiences for different kinds of viewers interacting with different kinds of texts in different kinds of contexts—not randomly or anecdotally, but systematically.

III

In this final portion of the chapter, I wish to discuss three research projects that, I think, have some bearing. They represent joint work with Daniel Dayan and Tamar Liebes, on and off over some ten years. Each involves observing the interaction of viewers with a major genre of television: media events, the evening news, and *Dallas*. Each has something to say about viewers' work.

The media events project (Dayan and Katz, 1992) has outworn a number of subtitles, but they give the idea: the live broadcasting of history; the high holidays of mass communications; festive viewing; the experience of not being there; Durkheim, Live; ceremonial politics. Media events are unrecognizable in Kubey's and Csikszentmihalyi's terms; they are virtually a different medium. Audiences dress up to watch; they prepare their hearts—for the moon landings, for Kennedy's funeral, for the pope in Poland, for Sadat in Jerusalem, for the Olympics. They leave their doors unlocked; they seek out others to share the experience. They are offered a ritual role in a ceremony of civil religion. They cheer, cry, applaud, say amen. They talk to their television sets. They experience a sense of occasion, of communities (Turner, 1985). There are three kinds of events in our corpus; we call them contests, conquests, and coronations. Unlike the daily news, which is about conflict, these events celebrate reconciliation or the resolution of conflict. The type we call conquests—Sadat, the pope, the moon, Wenceslas Square—have a transformative aspect, not just a Durkheimian affirmation of the social order. These transformations emerge in the subjunctive mood—dreams of the possible—engendered by such occasions: that peace between Israel and Egypt is thinkable, for example. The heroes of such ostensibly real occasions seem to "abrogate the rules of the real world" by flying off to the moon, or crossing an enemy border unarmed, or breaking a world sports record with all of the trappings of fantasy.

Yet the readers are surely reality seekers. They come as citizens to the Kennedy-Nixon debates, jurors to Watergate, sports fans to the Mondial, Polish Catholics to the pope. But they surrender themselves, for a moment, to the liminality of a rite of passage.

These events fulfill the technological potential of the electronic media. Television is capable of reaching everybody, everywhere, simultaneously and directly—total, immediate, unmediated—although it does so only very rarely. And soon, even a certain amount of interactivity will prove possible. On these occasions, the public space is reinvigorated, and, ironically, its locus is the home. There is the further irony that reality itself is transformed in the process, not just because it behaves like fantasy, but because it exists in the air, not on the ground. Kennedy debated Nixon, but in their third debate one was in California and the other in New York: where was the debate?

Postman might not like this expropriation of reality by television, and not altogether without justification.[2] Kubey and Csikszentmihalyi would love it, however, because of the flow. Schramm et al. would have trouble deciding whether viewers were being referred to the real world or being asked to take leave of it, whether they were surrendering and alert or just alert, whether they were emotional or cognitive or both, whether the text was within the rules of reality or outside them, whether anxiety is altogether incompatible with wish-fulfillment.

This kind of festive television is a holiday not only from daily life but also from the daily routine of viewing. Like holidays, media events are publicly sanctioned interruptions, focused on some transcendent social value. But there are certain genres of everyday television that are also evocative of viewer activity.

Consider the nightly television news in Israel, where there is a single television channel and one major news magazine at 9 P.M. Because of the culture and the context, everybody views the news—some 70 percent nightly—and with a sense of commitment and often excitement. People like to view the news with others and often discuss it—during and after. Some observers are shocked to hear of this case of monopolized news in a democracy, but it takes only a moment longer to realize that Israel is a highly divided and highly politicized society and that this one news program is viewed, and largely trusted, by Jewish hawks, Jewish doves, and Israeli Arabs.

These viewers work, but the pluralism of the audience suggests that they may be doing several different kinds of work. We think (Liebes, forthcoming), for example, that the doves may have to do harder work than the hawks. Thus, in television coverage of the *intifada* uprising of recent years, Arab stone-throwers and Jewish soldiers were both portrayed in the role of combatants, although the Arabs were presented more exclusively in this role. Jewish hawks among the viewers understood this quite readily. Indeed, attribution theory would explain that they saw the disliked Arab as they believed him to be dispositionally, while the portrayal of the Jewish soldier as combatant and even aggressor is understood situationally. We know our side, they say to themselves, and know that we behave this way only when forced to do so. Television itself shows the Jews in many other roles. In other words, Jewish hawks do not even have to know the code that television news is about deviance more than about normality. It even helps not to know, because then can one accept the portrait of the Arab as if it were the norm.

The Jewish dove who has not learned that journalism is about deviance has a more difficult time. To find reinforcement for his political views, he must find some good in the other side, but television does not help him much. Only if he knows the code can he assume that most offscreen Arabs behave differently, and that even onscreen Arabs may be acting situationally rather than dispositionally. East Germans watching West German television news had something of the same problem. They saw the usual deviance—accidents, violence, homelessness, corruption—and had to decide whether West Germany was a civilized place or not. Knowing the code made it easier (Hesse, 1990).[3]

Finally, consider *Dallas* viewers overseas. What in the world do they make of it? In comparing focus group discussions of viewers of different ethnic origin in Israel, the United States, and Japan, Liebes and Katz (1990) found several different forms of involvement. Rather than trying to decide which was more or less involving—for good empirical reason, we believe—we tried to understand the differences among them. We labeled the forms Real, Ludic, Ideological, and Constructional, generating them from the cross-tabulation of two variables, Referential/Rhetorical and Closed/Open.[4] Real involvement, then, is at the intersection of Referential and Closed, where the characters and their values are treated as "given" and, typically, argued with. Sometimes they are treated as real but negotiable, where the viewers respond playfully in role-playing games. The

Referential/Closed viewers—the more traditional groups, in particular—stayed dressed in their identities, fascinated by the gossip but aware of the threat. The Ludic viewers—Americans and kibbutzniks—tended to check their identities at the door. Newcomers from Russia, especially, saw an ideological threat in the program, and argued that the doctrine of the unhappy rich is an ideological manipulation. Aimed against the producer, not the characters, these ideological readings are Rhetorical/Closed, and are roundly denounced by the Americans who refused to see any message at all in the program. In Japan, where *Dallas* had failed, we found viewers bemused over seeming contradictions between episodes in the plot, between the Americans in the story and the Americans they know, and between the *Dallas* plots and their own happy-ending family dramas. They declared the producers incompetent. We called this Constructional involvement, at the intersection of Rhetorical and Open.

While different groups specialize in the different forms of involvement—even though the real is the most frequent of all—a lot of commuting went on among the categories. Indeed, the concept of openness—applied to these data by Liebes and Katz (unpublished)—suits the movement of the viewers themselves in and out of the different forms. While Livingstone's study of serials concentrates on viewer activity in the Real and Ludic realms, the Liebes-Katz exposition is equally concerned with the rhetorical realm, which we call Ideological and Constructional. What is impressive about the results of these three studies—you will forgive my skipping their methodology, which is not much to be proud of—are the viewers'. They are wide-awake, and, what is more, articulate and creative in their reactions. In the case of *Dallas*, viewers draw on their own wisdom and experience to discuss the characters, and on their sophistication to discuss the producers. In the news study, too, parents used the news to socialize their children politically, occasionally based on a sophisticated understanding of how the news is constructed. Referential involvement, not just ideological involvement, was often oppositional, not just hegemonic.

Media events and the daily news are situated in public space, and it is therefore not surprising that people come dressed appropriately. Even in *Dallas*, which goes on in private space, the viewers are invited into the story with their domestic problems and family roles. But note that they do not quite assume the role of

Table 1.1 Types of Viewer Involvement in *Dallas*

	Referential	Rhetorical
Closed	Real	Ideological
Open	Ludic	Constructional

voyeurs. Unlike the voyeur of cinema, who abandons himself and identifies with the characters, the *Dallas* viewer keeps the lights on and sees himself vis-à-vis the characters, often confrontationally. The idea of talking back—parasocial interaction is thought to characterize frontal television genres (Horton and Wohl, 1956)—may apply to television narrative as well. Of course, *Dallas* viewers identify, too, but often ludically—that is, at some distance. Equally, they critique the producers—not only the characters—aesthetically and politically, thus moving some of *Dallas* into the realm of public space, too.

Why don't these viewers identify themselves to Kubey and Csikszentmihalyi? Why don't they tell Postman that, given the right circumstances and the right text and somebody who wants to listen, they know how to handle themselves in public space? Why don't they tell Schramm et al. that we need a better scheme? Why don't they tell their governments that we need more public channels, not more segmented ones? Why don't they tell themselves that they need fewer channels and programs that are worth staying awake for?

If the answer is that viewers of this kind are too few, that they are unaware of each other, that they are too tired to bother, or that they feel powerless, we will have come full circle. We will still be unable to reconcile the two positions: the one that holds that television viewing is an alienating affair—relaxing, but lonely and disconnected—and the other that sees television as interwoven with life, private and public. It is easy enough to exorcise phantom concepts such as active/passive by confining them to their strictly operational definitions. But that is not enough. There is simply too much evidence—not just Kubey—that most people, most of the time, are lounging in front of their sets, watching television, not programs. Or, to put the question otherwise, if you hold that the elaborate negotiations that go on between viewer and text resemble talk, why are "television" and "talk" coded by the respondents themselves at such opposite poles?

Marx defines work as the transformation of the useless into the useful, or more generally, as equal to the surplus value of the object. New theories of the audience maintain that viewers add value to what they view (e.g., Morley, 1989: 23). Going far beyond functional theories of media use, they proclaim the competence and creativity of the individual. They suggest that viewers do work—not just by staying awake, but by investing effort, by being critical, by making "public."

I often find that academic issues are not difficult to reconcile, and that different sides are saying similar things. If that is true in the present case—and I am not certain—the problems are not just substantive but epistemological. The first thing that must be done, I am sorry to say, is to agree on some rules of evidence, and the second is to do some research that will stand up to these rules, reliably and validly.

It is a good time for this project, too, since there is a lot of methodological inventiveness abroad, as recent studies attest. All of these try empirically to get inside the viewers' heads, and some of them try to get inside the viewers' natural habitats, where they view and talk and act upon television. Obtrusiveness remains a major problem, but so does representativeness. If some progress has been made in naturalness and unobtrusiveness, it is usually at the expense of representativeness, and at the expense of the mapping we need of the alternation—or even the coexistence—of different viewing patterns in the same individuals, in different groups, and in different cultures.

Notes

1. Lang and Lang (1983) elaborate on the role of "bystander," or witness, to which they attribute political importance. Dayan and Katz (1992) extend this point.

2. Dayan and Katz (1992) consider the ostensible kinship between these events and the nationalistic traditions of political spectacle as well as their emotional loading.

3. It is somewhat ironic, however, that there was more migration to the West from areas in East Berlin that could not receive West German television than from those that could.

4. In earlier attempts, we have used Hot/Cool for Closed/Open. We have also tried "critical" and "metalinguistic" as labels for what we are now calling "rhetorical." The basic distinctions remain, however. The one (Referential/Rhetorical) has to do with awareness of the constructedness of the text. The other (Closed/Open) has to do with the reader's sense of being able to negotiate with the text.

References

Allen, R.C. (1985). *Speaking of Soap Operas.* Chapel Hill, N.C.: University of North Carolina Press.

Ang, Ien (1991). *Desperately Seeking the Audience.* London: Routledge.

Barwise, P., and A. Ehrenberg, (1988). London: Sage.

Bechtel, R. B., C. Achelpohl and R. Akers (1972). Correlates between observed behavior and questionnaire responses on television viewing. In E. A. Rubinstein, G. A. Constock, and J. P. Murray (eds.), *Television and Social Behavior,* Vol. 4, *Television in Day-to-Day Life: Patterns of Age,* pp. 274–344 Washington, D.C.: Government Printing Office.

Bettleheim, B. (1976). *The Uses of Enchantment: The Meaning and Importance of Fairy Tales.* New York: Knopf.

Blumler, J. G., M. Gurevitch and E. Katz (1986). Reaching out: A future for gratifications research. In K. E. Rosengren, L. A. Wenner, and P. Palmgreen (eds.), *Media Gratifications Research.* Beverly Hills, Calif.: Sage.

Dayan, D., and E. Katz (1992). *Media Events: The Live Broadcasting of History.* Cambridge, Mass.: Harvard University Press.

Hesse, K. (1990). Cross-border mass communication. *European Journal of Communication* 5: 355–371.

Horton, D., and R. Wohl (1956). Mass communications and para-social Interaction. *Psychiatry* 19: 215–229.

Houston, B. (1985). Viewing television: The metapsychology of endless consumption. *Quarterly Review of Film Studies* (Summer): 183–195.

Katz, E., and D. Foulkes (1962). On the use of the mass media as "escape": Clarification of a concept. *Public Opinion Quarterly* 26: 377–383.

Katz, Ruth (1973). The egalitarian waltz. *Comparative Studies in Society and History* Vol. 15.
Kubey, R., and M. Csikszentmihalyi (1990). *Television and the Quality of Life: How Viewing Shapes Everyday Experience.* Hillsdale, N.J.: Erlbaum.
Lang, K., Lang and G. E. (1983). *The Battle for Public Opinion.* New York: Columbia University Press.
Lazarsfeld, P. F., and R. K. Merton (1960). Mass Communication, Popular Taste and Organized Social Action. In W. Schramm (ed.), *Mass Communication.* Urbana: University of Illinois Press. (Originally published 1948.)
Liebes, T., and E. Katz (1991). Parents, television, and the political socialization of children. Paper presented at the Annual Meeting of the National Academy of Education, Stanford University, May.
——. (1990). *The Export of Meaning: Cross-Cultural Readings of "Dallas."* New York: Oxford University Press.
Liebes, T. (forthcoming). Decoding television news: The political discourse of Israeli hawks and doves. *Theory and Society.*
Liebes, T., and S. M. Livingstone (1992). Mothers and lovers: How British and American soap operas cope with the women's dilemma. In J. G. Blumler J. M. Mcleod, and K. E. Rosengren (eds.), *Comparatively Speaking.* Newbury Park, Calif.: Sage.
Livingstone, S. (1990). *Making Sense of Television.* London; Pergamon Press.
Morley, D. (1980). *The Nationwide Audience: Structure and Decoding.* British Film Institute Television Monograph No. 11. London: British Film Institute.
——. (1989). *Changing Paradigms in Audience Studies.* In Seiter et al., *Remote Control: Television, Audiences and Cultural Power.* London: Routledge.
Postman, N. (1986). *Amusing Ourselves to Death: Public Discourse in the Age of Show Business.* New York: Viking.
Roberts, D., and W. Schramm (1971). *Functions and Effects of Mass Communications.* Urbana: University of Illinois.
Salomon, G. (1983). Television watching and mental effort: A social psychological view. In J. Bryant and D. R. Anderson (eds.), *Children's Understanding of Television.* New York: Academic Press.
Schramm, W., J. Lyle and E. B. Parker (1961). *Television in the Lives of Our Children.* Stanford: Stanford University Press.
Schur, E. M. (1988). *Politics of Deviance: Stigma, Protest, and the Uses of Power.* Englewood Cliffs, N.J.: Prentice-Hall, 1980.
Turner, V. (1985). Liminality, Kabbala, and the Media. *Journal of Religion,* 15: 205–217.

Combinations, Comparisons, and Confrontations: Toward a Comprehensive Theory of Audience Research

2

Karl Erik Rosengren

Under the main entry of the term *Audience* in the recent *International Encyclopedia of Communications,* the reader is referred to some forty articles, ranging from "Consumer Research" to "Bandwagon Effects," and including references to articles on such classics as Adorno, Hovland, Lazarsfeld, Lewin, Lippmann, McLuhan, and the French nineteenth century sociologist Jean-Gabriel Tarde. In addition—had the *Encyclopedia's* editorial principles admitted—some modern names could have been mentioned, including one or two authors in this book (Barnouw, 1989; cf. Beniger, 1990).

It is obvious, then, that the title of the 1990 conference that led up to this book, Toward a Comprehensive Theory of the Audience, was a quite daring one, especially considering the fact that the term "comprehensive" may refer to *either* to an attempt to cover the myriad empirical phenomena covered by the concept of audience, *or* to an attempt to reconcile and unite all the various theoretical schools and traditions laboring in the area, *or* to both of these attempts.

Both tasks are intimidating. Nevertheless I share the assumed presupposition of the editors of this book that those tasks have to be approached sooner or later. Personally would prefer sooner than later, and I also think we should start at the theoretical and methodological end, and then move to a theoretically founded, methodologically sound analysis of empirical data. The march will be a very long one, and if we never reach the final goal, at least we'll be moving in the right direction.

Before moving too far, however, we need to take a look at the two dimensions of comprehensiveness mentioned above. In principle, a theory may be comprehensive with respect to the world of phenomena and/or with respect to the world of ideas. A truly comprehensive theory successfully covers all variation over time and space in the relevant slice of the world of phenomena. In the world of ideas, a truly comprehensive theory is able to subsume all related theories as special cases of that comprehensive theory.

The former demand on the comprehensive theory may be called complete generalizability; the latter, complete inclusiveness. Neither demand will ever be fulfilled, of course. They represent ideals never to be reached. But ideals are good for one thing. They give us a sense of direction. They show us where to go.

Generalizability

A theory that is comprehensive with respect to the world of phenomena is invariant over time and space. It has a very high degree of generalizability. In a comprehensive theory of audience, then, this is one of the two questions we should ask: What is the invariance over time and space of audience theory?

In principle, the answer is fairly simple. The more our theories are anchored in basic psychological, social psychological, sociological, and communicative generalizations, the more invariant they are. And the more dependent they are on institutional specificities, the less invariant they will be. Yet in order to be truly comprehensive, the theory must also heed institutional circumstances. We are thus confronted with a true dilemma.

The solution seems to be not to get bogged down in substantively specific institutional details, but always to look for the basic formal dimensions of relevant institutions. Substantive specificity is all good and well, but more important is the search for those formal and formative dimensions of our institutions that define truly basic parameters.[1] The parameters will show different values, of course—over time, as well as between societies. But if our audience theory is ever to become comprehensive, the dimensions and their parameters themselves will ultimately have to be the same.

This is the ideal, in principle and under *ceteris paribus* assumptions. In reality, of course, the successful audience scholar must seek to strike a balance between generality and specificity, not only with respect to psychological and sociological dimensions, but also between institutional specificity and generality. What is absolutely necessary is to combine knowledge from the level-oriented sciences of psychology and sociology with institutionally oriented scholarship focusing on the interplay between historically given institutions of socialization—for instance, family, school, and the mass media.

Only many years of painstaking theoretical, methodological, and empirical research guided by various aspects of audience theory will provide certified, precise answers to the question of how stable over time and space our theories really are. The important thing in this connection, however, is that in order to know how comprehensive our audience theory really is, we must have those answers.

Some examples of the stability over time of theoretically guided empirical generalizations will be presented below. Similar examples may be found, but not all that many. As of today, we just cannot tell, therefore, how generalizable our theories are. There are some reasons to assume, though, that a truly generalizable, fairly detailed, and rather precise theory of the audience is, if not within actual reach, at least not entirely utopian. A complementary way to reach it is to look for that other dimension of comprehensiveness: inclusiveness.

Inclusiveness

As a starting point for my discussion of theoretical inclusiveness, I shall use a typology originally developed by the Anglo-American team Gibson Burrell and Gareth Morgan. Burrell and Morgan are sociologists, but their typology is eminently suited also to explicating structures and processes characteristic of communication studies. As a matter of fact, it has been used for that purpose more than once (Burrell and Morgan, 1979; cf., for instance, Gilljam, 1985; Rosengren, 1983, 1989).

Burrell and Morgan first identify eleven dimensions along which schools of sociology may differ. They then collapse these dimensions into two: basic assumptions about the *nature of social science,* and basic assumptions about the *nature of society.* Each of the two dimensions has two values. Assumptions about the nature of social science may be primarily subjectivistically or objectivistically oriented; assumptions about the nature of society may be primarily conflict or consensus oriented. Burrell and Morgan thus end up with a typology containing four main types.

The typology is rendered in Figure 2.1, offering also some examples of various schools of sociology classified in terms of the typology. The typology is very broad, but it is also very powerful, and it works well even for rather specific areas within communication research—audience research, for instance.

There are at least five main traditions in the area of audience research (Jensen and Rosengren, 1990). In Figure 2.2 the five traditions are presented, located within Burrell and Morgan's typology. (For illustrative purposes, a sixth tradition of media research has also been included in the figure.)

The locations of the different research traditions, are of course, approximate, and they are valid only at the level of a rough ordinal scale. Yet it seems intuitively quite natural that British *Cultural Studies* should be located in the cell characterized by a subjectivistically oriented perspective on science, and a conflict perspective on society (Turner, 1990). More traditional *Literary Criticism* also has a subjectivistic perspective on science, but as a rule it is characterized by a more consensual perspective on society. Mainstream *Effects Research* was—and probably still is–characterized by a combination of consensus and objectivistic perspectives (Bradac, 1989), while the two traditions of *Reception Analysis* (Jensen, 1987) and *Uses and Gratifications Research* (Rosengren Wenner, and Palmgreen, 1985) try to combine an objectivistic and a subjectivistic perspective, the former being somewhat more conflict-oriented, the latter, somewhat more consensus-oriented. A fairly clear-cut example of the objectivistic/conflict combination, finally, is offered by the British school of *Political Economy* (cf., for instance, Garnham, 1990; Golding Murdock, and Schlesinger, 1986).

Curran (1990) very convincingly shows that certain techniques and ideas that have been launched with some bravado as path-breaking innovations within

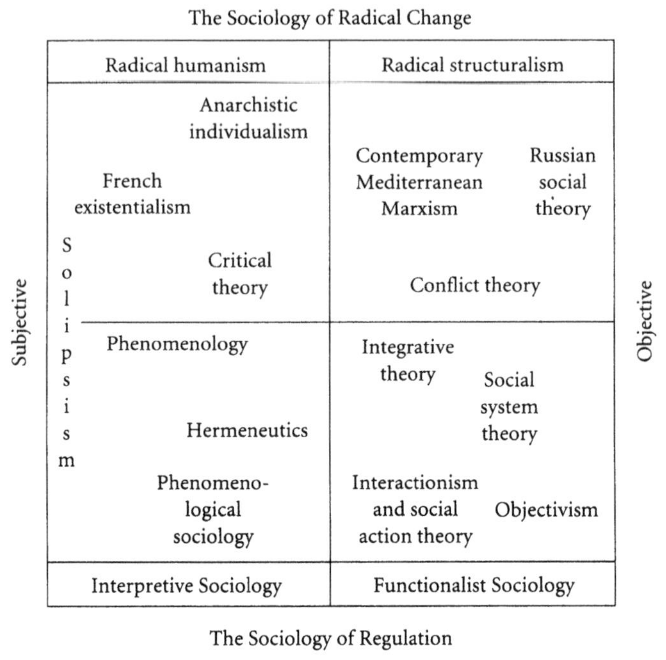

Figure 2.1 Typology of Sociological Theory

cultural studies and reception analysis oriented research were very sophisticatedly used about a half century ago—for instance, by the group of well-known mass communication scholars gathering around Lazarsfeld and Merton at Columbia. This means that basically the same techniques, approaches, and conceptualizations may be successfully applied in different cells of the typology, actually in diagonal—that is maximally dissimilar—cells.

All this provides support for the probably still rather controversial view that the dividing lines of the typology are not insurmountable. In somewhat more technical terms: what we have is not four *paradigms* by definition unable to communicate with each other, but rather a number of *schools* strategically competing for vital and scarce resources (above all, attention, recognition, status, and money; cf. Rosengren, 1989). In such competitions, the wheel does tend to be claimed as a new invention time and again.

To take another example: Out of traditional literary history and criticism there has gradually grown a rather specific tradition that has come to be known under the name of "Empirical Studies of Literature" and has established an international association of its own ("IGEL"), publishing its results preferentially in

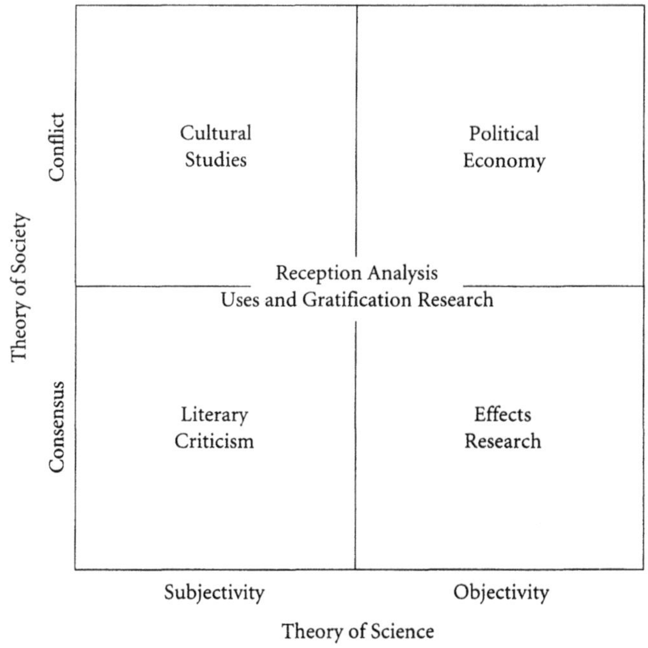

Figure 2.2 Typology of Communication Studies

internationally well-known journals such as *Poetics, Siegener Periodicum zur Internationalen Empirischen Literaturwissenschaft, Empirical Studies of the Arts.* The roots of this tradition go back at least to the late 1920s, but what the scholars of the tradition do is actually what in a more recent tradition has come to be called "Reception Analysis."

My point is that—to judge from their lists of references, at least—some recent reception analysts are not very well acquainted with this important research tradition (which in its turn, however, seems to be well acquainted with reception analysis as currently practiced in communication studies). To take one striking example: Stanley Fish—a pioneer and leader in the area—was mentioned in a footnote in Janice Radway's fine book *Reading the Romance* (1984), but there is no mention of, say, I. A. Richards, Gunnar Hansson, A. C. Purves, Siegfried J. Schmidt, or other American and European pioneers and leaders within this tradition, in spite of the fact that as a rule they have published some of their most important works in English.

What I would like to see in the broad area of audience research is much less intellectual parochialism between schools and traditions that are actually very similar to each other, and much more cooperation. This book will, I hope, help to make us take some steps in that direction.

Three Roads Toward a Comprehensive Theory of Audience

If we can reach it at all, we can reach the rather distant goal of a comprehensive theory of audience in a number of different ways. I would suggest that what we need in order to be able to approach that goal is more combinations, more comparisons, and more confrontations.

We need *combinations* of techniques and approaches traditionally used only within different research traditions with no or very few mutual contacts. We also need concrete temporal and spatial *comparisons* of results reached by scholars located within both similar and different research traditions. We finally need more *confrontations* between theories and methodologies presented by scholars from different traditions so that it will be clear what the various traditions have in common and what differentiates them from each other.

The rationale behind such combinations, comparisons, and confrontations may be explicated by means of a simple figure that was originally devised by the Swedish philosopher and sociologist, Haåkan Wiberg (1969). Scholars and scientists, having different backgrounds, may or may not produce basically the same results. Indeed, the same is true of scholars and scientists with the same background. This gives us Figure 2.3.

Obviously, the upper right-hand cell of the figure represents the triumph of science and scholarship: scholars and researchers arriving at the same results in spite of their different backgrounds. In the upper left-hand cell, all is fine and well—except the gnawing suspicion that the identity of the results may be the outcome of a common background rather than of sophisticated research. The lower left-hand cell represents failure: differing results in spite of a common background. The lower right-hand cell, finally, is often seen as failure, too. But nothing could be more wrong. That cell actually represents a promise. It offers an opportunity to take a close look at basic theories, concepts, and methods, to investigate the consequences of explicit stipulations and implicit assumptions. It offers the opportunity, that is, for potentially fruitful comparisons and confrontations between open-minded representatives of different schools—for instance, different schools of audience research.

Most comparisons of scholarly results are carried out in the left-hand column of the figure. My point in suggesting that more comparisons and confrontations be carried out in the right-hand column is that this column contains the most convincing cell of the figure, the one where scholars and researchers from different backgrounds reach the same results. If we are ever to reach the "comprehensive theory of the audience" toward which we are aiming, I believe that, at least to a considerable extent, it must be by way of a number of results from studies carried out in the right-hand column of Figure 2.3—that is, by means of combinations of different techniques and approaches, comparisons

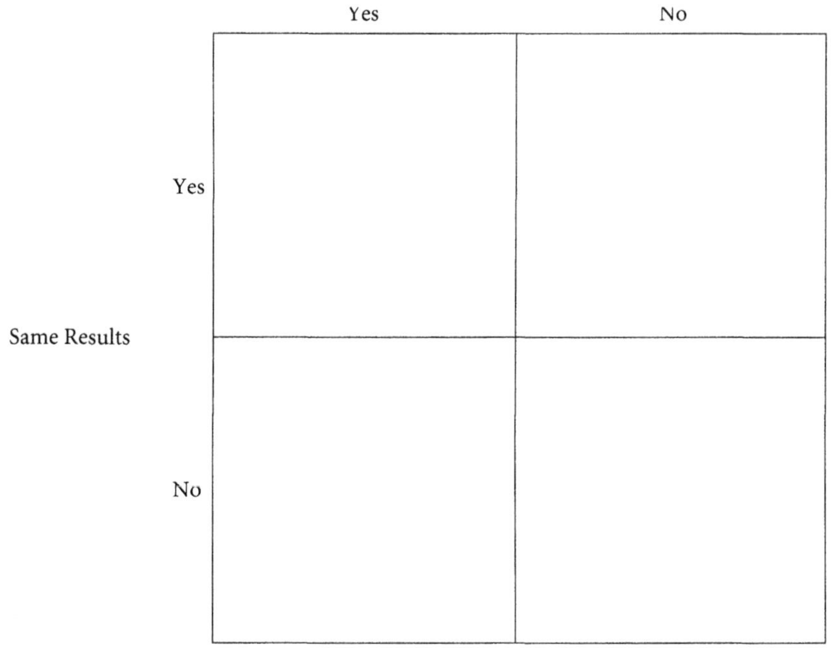

Figure 2.3 Wiberg's Figure

over time and space between results gained by scholars from different research traditions, and confrontations between related but different conceptualizations and methodologies by scholars from different research traditions (and sometimes from similar ones).[2]

In many cases we shall have to expect ending up in the lower right-hand cell of Figure 2.3. We shall then have to move from that cell to the upper right-hand cell. No doubt we will sometimes fail in those efforts. But in some few cases we will succeed. Those few cases will represent as many steps toward a comprehensive theory of audience research.

While waiting for all these combinations, comparisons, and confrontations to appear on the scene, I will content myself by presenting some examples of my own. In the rest of this chapter, I will thus present some combinations of techniques traditionally not very often combined, some comparisons over time and space of results in audience research, and some confrontations between conceptualizations and methodologies offered by scholars from different research traditions.

I will start by exemplifying some combinations of techniques traditionally not combined, then proceed to comparisons over time and space, and finally round off this section of the chapter by means of a confrontation between two related but different conceptualizations within a specific area of audience research. Parallel with this move from combinations to confrontations, we will move from the levels of biology and psychology to those of social psychology and sociology. First, however, the starting point for all these combinations, comparisons, and confrontations must be presented.

The Media Panel Program

The starting point for the confrontations, comparisons, and combinations alike is the *Media Panel Program* (MPP) carried out at the Department of Sociology, of the University of Lund in Sweden since 1975 or so. The MPP is focused on the character, causes, and consequences of media use among Swedish children, adolescents, and young adults. Results from the first ten years of the program were presented in the book *Media Matter* (Rosengren and Windahl, 1989), building on a number of doctoral theses that had been produced within the research program (Flodin, 1986; Hedinsson, 1981; Jarlbro, 1988; Johnsson-Smaragdi, 1983; Jönsson 1985; Roe, 1983; Sonesson, 1979). Additional results from the program have been—and will be—presented in more recent reports and chapters (cf., for instance, Hojerback, 1990; Johnsson-Smaragdi, forthcoming; Lööv and Miegel, 1989a, 1989b; Rosengren, 1991; Sonesson, 1989).

In terms of the audience research traditions of Figure 2.2, the MPP tries to merge the two traditions of uses and gratifications studies and effect studies into what has sometimes been called "uses and effects research" (cf. Rosengren and Windahl, 1972,

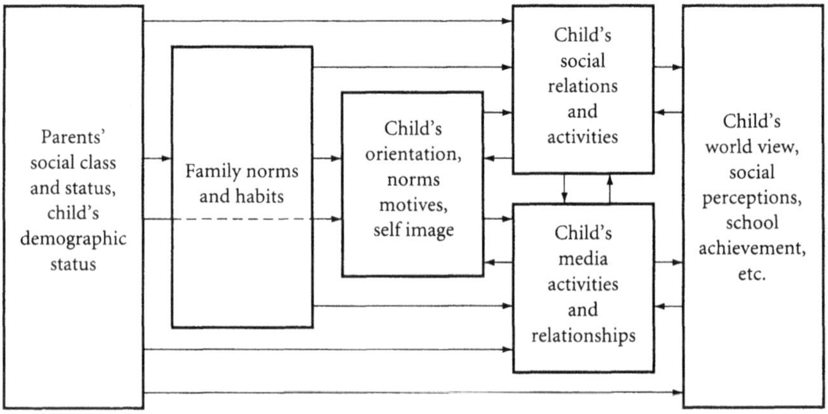

Figure 2.4 Conceptual Scheme of the Media Panel Program

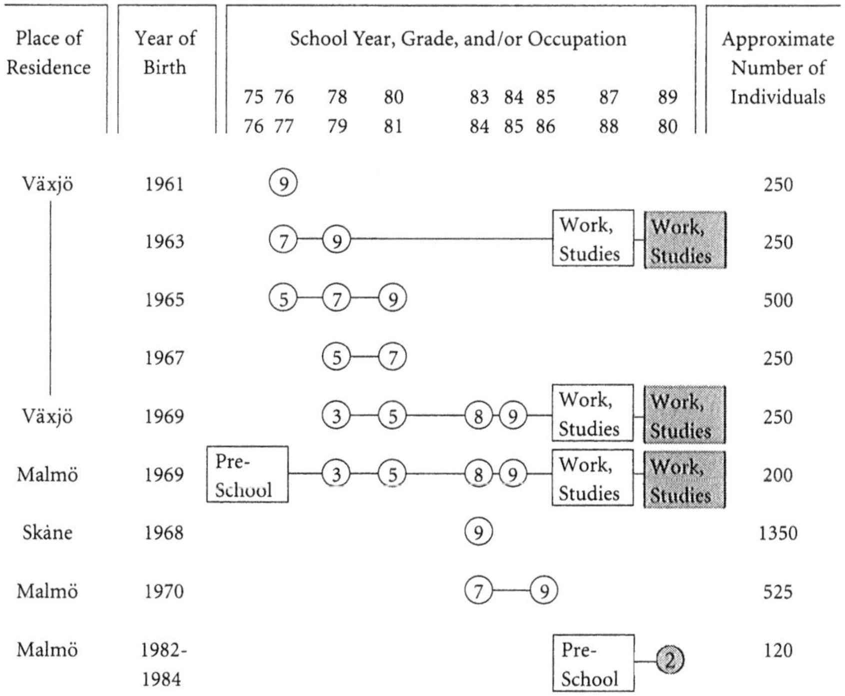

Figure 2.5 Design of the Media Panel Program

1989; Windahl, 1981). In so doing it regards audience members as willing and active subjects, and at the same time objects of strong forces within and outside themselves. It also recognizes that all societies are characterized by both conflict and consensus. The Media Panel Program is thus located at the very center of Figure 2.2.

In terms of a behavioral and social science at large, four main theoretical perspectives have been cultivated within the program: a developmental perspective, a social class perspective, a socialization perspective, and a lifestyle perspective. In order to admit the application of these four basic orientations, the overarching conceptual scheme of the program has deliberately been kept very simple, regarding media use as a bundle of variables affected by bundles of background and intervening variables and in its turn having effects of its own on knowledge, skills, attitudes, behavior, and so on—always allowing, of course, for the possibility of complex interactions and feedback mechanisms.

The basic conceptual scheme of the MPP is depicted in Figure 2.4. It will be seen that the media effects perspective is located toward the right-hand side of the theoretical model, and the uses and gratifications perspective is located in the middle. Especially the developmental perspective used within the program demands that the theoretical model be implemented within a longitudinal design.

The basic methodological characteristic of the Media Panel Program is that it builds on a combined longitudinal and cross-sectional design. It thus consists of a number of data collection waves that have been carried out on children, adolescents, and young adults in two towns in southern Sweden from 1975 onward. The design is shown in Figure 2.5, depicting a number of panels of children and adolescents passing through the school system and into work or continued studies during their early adulthood. (In order to confront the theoretical model of Figure 2.4 with the methodological design of Figure 2.5, it is necessary, of course, to repeat relevant sections of the theoretical model over time.)

So much for the theoretical and methodological characteristics of the Media Panel Program. Let us turn now to the empirical data.

The Bottom Line

A combined cross-sectional/longitudinal design is a powerful device. It admits, for instance, both cross-sectional, longitudinal, and diagonal analyses, making it possible to discern as clearly as possible the three basic types of time-related effects: cohort effects (controlled for in horizontal comparisons); situational effects (controlled for in vertical comparisons), and maturational (age) effects (controlled for in diagonal comparisons).

In addition, since the design is built on panels admitting coupling of data at the individual level, it admits strong causal modeling of causes and effects of mass media use, including long-term effects. By means of interpanel comparisons, it also admits the study of the temporal invariance of such models. Finally, when spatial variation is added to temporal variation—as is the case in the MPP—comparisons of the structural invariance of such models over both time and space may be undertaken (see below).

As demonstrated in Figures 2.6 and 2.7, the combined cross-sectional/longitudinal design is capable of yielding interesting results even at the descriptive level.

Figure 2.6 depicts graphically the amount of television viewing by children, adolescents and young adults in six panels and seven data collection waves of the MPP, covering the period 1976–1988, and ages 10–11 to 24–25. Figure 2.7 depicts the amount of listening to popular music during the same period of time, by the same panels, except—for technical reasons—panel M70. (For further technical details, see. Rosengren and Windahl 1989; Rosengren, 1991).

The great variation showing up within age spans that in current media statistics are often indiscriminately lumped together does provide some food for thought. The variation may be interpreted in terms of age, cohort, and situational effects. Although detailed analysis reveals the existence of both cohort and situational effects, it is clear that the greater part of variation stems from age effects. In this connection, age effects are best understood in terms of developmental theory.

Based on unweighted, standardized means of the curves of Figures 2.6 and 2.7, Figure 2.8 depicts the central tendencies in amount of television viewing

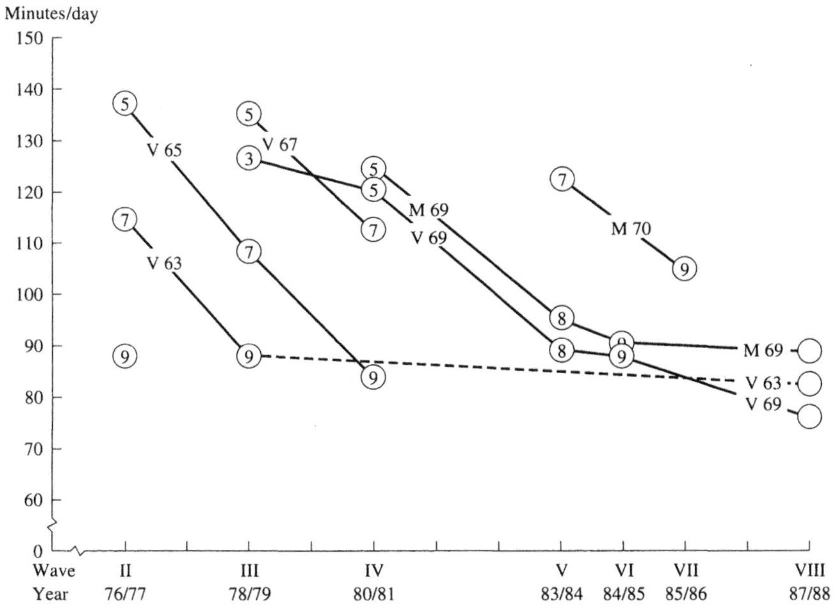

Figure 2.6 Amount of Television Consumption in Different Panels and Waves

and music listening among Swedish youths and young adults during the past fifteen years or so. The former curve shows strong decline; the latter, strong increase. As our children and adolescents grow into early adulthood, they leave the family medium of television for the peer-oriented medium of popular music (cf. Roe, 1983). The increase in music listening—as well as the decrease in television viewing—follow's curves strongly reminiscent of the classical S-curves summarizing so many well-known maturational phenomena. What we see is biological, cognitive, emotional, and social development expressed in terms of media habits.

Introducing also the variables of gender and social class, we could have made our results relevant not only to developmental theory, but also say, to, feminist and class-oriented theory. Nevertheless, as they stand these simple descriptive data are of interest. They represent the basics of audience research: the amount of media use. *They tell us the extent to which there is an audience at all among different segments of the population.*

Although only few, if any, directly comparable data seem to be available, even scholars from the most different traditions would probably have come to essentially the same results. And even in countries radically different from

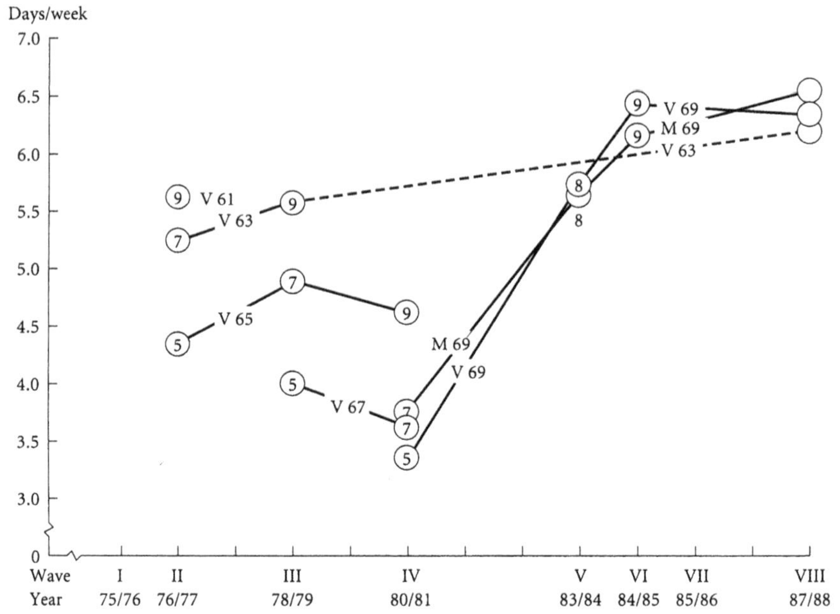

Figure 2.7 Amount of Listening to Music in Different Panels and Waves

Sweden, the results would have been comparable in the sense that the dimensions would have been exactly the same, and the main result probably much the same. Clearly, however, the parameters themselves (level of viewing, as well as rate of change) certainly would be different in different countries and times. This is a situation characteristic of all comparative research (Rosengren, McLeod and Blumler, 1991).

The details remain to be charted in future comparative research. For our present purposes, the important thing is to note that there are theories, methods, and techniques capable of producing nontrivial audience research results that probably have to be recognized as valid and relevant by representatives of even the most widely different schools of audience research. This is the bottom line, then. It actually implies the existence of widely accepted theories (commonsense theories, or more sophisticated ones, and more often than not, implicit rather than explicit ones), and at least roughly parallel techniques of data collection.

Our next step, therefore, must be to see whether, using different techniques of data collection we will be able to reach comparable (preferably identical) results. At the same time we will move from biologically rooted development theory to a higher level of complexity: the social psychology of that most basic of all small

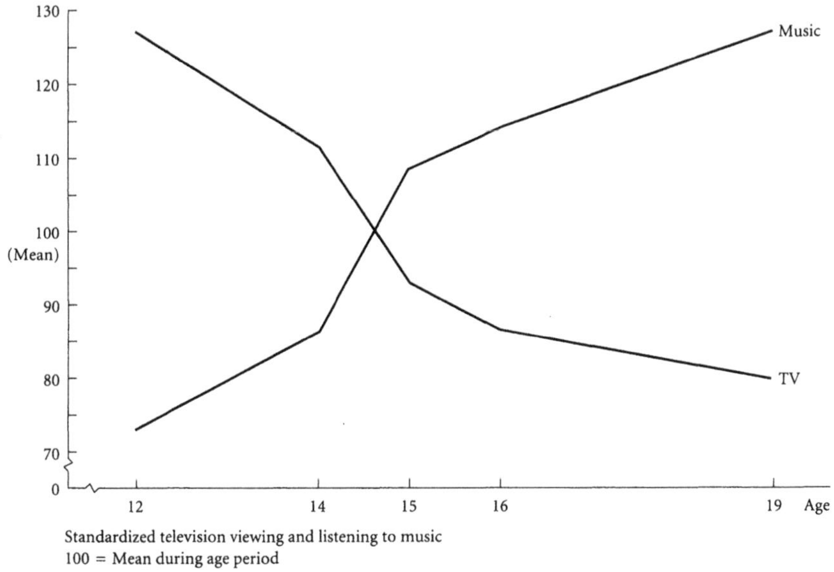

Figure 2.8 Development of Television Viewing and Listening to Music

groups, the family. In so doing, we will take yet another step in the direction of a comprehensive theory of audience research.

Different Techniques, Same Results

Validity is the very first requisite of a theory, and not least so for a theory claiming to be comprehensive. One way to test the validity of our results is to replicate our measurements, using different techniques of measurement—in short, using triangulation, a recipe that (especially after the publication of Denzin, 1970) has been often prescribed but unfortunately less often followed. Let us apply it to a rather complex phenomenon which has been given some attention in the MPP: the communication climate in the family.

Differential family communication patterns are among the most important of the many forces behind variations in the quantity and quality of children's, adolescents', and young adults' media use. As a matter of fact, the idea of family communication patterns represents such a basic social psychological phenomenon that it seems to have been independently discovered or invented time and again within different research traditions without much mutual contact (Rosengren and Windahl, 1989, 171; cf. Barnes and Olson, 1985; Tims and

Masland, 1985). Actually, a comparison between the various traditions in the area would be an undertaking well worth its while, and very much in the spirit of this chapter. Behind the substantive dimensions of family communication patterns such as, for instance, concept orientation and social orientation, one may find even more basic formal and formative dimensions such as amount of resources and amount of control (dimensions, incidentally, that could also be used to characterize in a very efficient way various types of audience research).

Family communication pattern variables were included among the central variables of the MPP from its very beginning, and they have remained so up until the very last of the many waves of data collection. In the first wave of the MPP we measured the parents of our children and adolescents by means of the well-known scale for family communication patterns developed by Chaffee and McLeod and their associates at the University of Wisconsin at Madison (Chaffee, McLeod, and Atkin 1971; Tims and Masland, 1985). Six years after the first wave a member of our team, Dr. Gunilla Jarlbro, returned to some of these adolescents (who by then had become young adults, 21 years old) and carried out long, informal, focused interviews with them about their childhood, their present situation, and their plans for the future. Combining her qualitative data with previously collected quantitative survey data, Dr. Jarlbro was able in a double-blind test, without any error at all, to locate sixteen out of sixteen young adults into the correct cell of the fourfold family communication pattern topology—the cell to which six years earlier, by means of the formalized, quantitative Likert scale questionnaire to which the parents had responded in a mail survey, their family had been shown to belong (Jarlbro, 1986, 1988).

Formalized quantitative techniques and informal "qualitative" techniques thus proved able to bridge a gap stretching over six years and between two generations. The theory behind the measurements was strong enough to do so, and in that sense it was comprehensive. The different methodologies, too, proved to be strong enough. Also, we should not forget that the theory and methodology, originally developed in the United States, is now found to be valid also in Europe. We know, then, *first,* that similar family communication patterns do exist in societies of different types, *second,* that they can be validly measured by means of very different techniques, *third,* that they seem to have a certain stability over time, and *fourth,* that they are relevant for the understanding of audience behavior.

All four results are important characteristics of a comprehensive theory of audience, and all of them demand comparisons over time and space, and/or combinations of different techniques to be satisfactorily demonstrated. In the next section of the chapter, we will turn to another type of comparison over time, at the same time moving to the next level of complexity: from the social psychology of the family to the sociology of culture and lifestyles.

Different Techniques, Different Results: Reconceptualization

Within communication research there is right now an increased interest in the notion of lifestyle. Lifestyle research, of course, has also become increasingly popular in other fields of research, for instance, marketing research, the sociology of medicine, and the sociology of culture (Lööv, 1990a, 1990b; Lööv and Miegel, 1989a 1989b; Miegel, 1990). Although they have much in common, and also show some interesting differences, the many different research traditions in the area can hardly be said to be in very close contact with each other (see, for instance, Donohew, Palmgreen, and Payburn 1987; Featherstone, 1987; Zablocki and Kanter, 1976).

As the children and adolescents of the different MPP panels have grown into young adults, they have been manifesting a number of different lifestyles, and their mass media use seems to be related to those lifestyles. The research program has developed parallel with the subjects under study. It is only natural, then, that the notion of lifestyle has gradually become quite central within the MPP.

Lifestyle is one among a whole family of terms and concepts covering a bewildering variety of phenomena that may be preliminarily thought of as "patterns of ideas, actions, and artifacts characterizing the way different categories and groupings of individuals lead their lives." Drawing upon previous research–from classics such as Marx and Weber onwards–Lööv and Miegel (1989a) of the Media Panel group made an attempt at clarifying the conceptualization of this rather diffuse family of concepts. Following Thunberg et al. (1982: 61), they did so by distinguishing between patterns of life determined mainly by societal structure, by the individual's position in that structure, *and* by the individual's own, more or less conscious, more or less idiosyncratic, choice.

As a result of this simple distinction, they arrived at a threefold typology, offering theoretical definitions of three different patterns of life, as follows:

- *Form of life:* patterns of life determined primarily by the societal structure
- *Way of life:* patterns of life determined not only by the societal structure but also by the individual's position in that structure
- *Lifestyle:* patterns of life determined not only by the societal structure and the individual's position therein, but also by the individual's more or less conscious, more or less idiosyncratic, choice

The advantage with this conceptualization it that it ties in neatly with generally acknowledged social theory. There is a societal structure, and people living in that structure tend to demonstrate an overall *form of life* common to most of them. There are positions in that societal structure, positions defined by their values along a number of basic dimensions such as age, gender, and social class. These positions define a number of roles, and individuals occupying a given position in the social structure—say, male, married, middle-aged office workers—tend to demonstrate an overall *way of life* common to most of them. Even identical roles, however, are differentially enacted by different individuals possessing different identities (Burke and Reitzes, 1988). Given a certain form of life and a certain way of life, then, there is always some leeway for individual choice of pattern of life, defining a specific, more or less consciously chosen *style of life.*

Forms of life may be studied by means of comparisons between different forms of societies (agrarian, industrialized, postindustrial, etc.). Ways of life may be studied by means of comparisons between categories of individuals located at different positions within societies (members of different social classes and class fractions, etc.). Lifestyles may be studied by means of comparisons between people with different values, norms, attitudes, tastes, and the like, after proper controls for type of society and societal position.

It is often maintained that the space available for differential lifestyles within given forms and ways of life is on the increase in industrial and postindustrial societies, but there is no agreement on the matter; much less, of course, is there any generally recognized opinion about the size of the potential increase (Featherstone, 1987; Lööv and Miegel, 1989 a, 1989b; Zablocki and Kanter, 1976).

In modern societies lifestyles are closely related to media use. The type of relationship between mass media use and lifestyle is by no means self-evident. For instance, use of the mass media may be seen "as one way of expressing a specific lifestyle, as one of the causes behind a specific lifestyle, *or* as being caused by a specific lifestyle." The first of these especially is highly relevant to audience research, since in that case audience activities are part of the lifestyle itself. Also regarded in the light of the two other types of relationship, however, lifestyle and audience activities are closely related.

Lööv and Miegel studied the relationship between media use and lifestyles in two different ways, a generalizing approach and an individualizing approach (Lööv, 1990a, 1990b; Lööv and Miegel, 1989b; Miegel, 1990). Applying exploratory factor analyses to the media use, leisure habits, interests, film and musical tastes of panels within the MPP, they came up with a number of patterns, that could then be related to positional variables (gender, class of origin, education, etc.) and to individually held values and attitudes. Controlling for the influence from structural and positional variables, the strength of the relation between, on the one hand, individually held values and attitudes, and on the other, patterns of tastes and habits, may be regarded as a measure of the strength of lifestyles in shaping the general patterns of life exhibited by the individuals under study. This way of studying lifestyles may be called the generalizing approach.

In their individualizing approach, Lööv and Miegel chose a dozen individuals who, according to the quantitative measurements, belonged to two decidedly different lifestyles. They then carried out long, informal, and intensive interviews with those individuals, focusing on their present situation in life (including work, leisure habits, tastes, media use, etc.) as well as their plans for the future. The interviews were repeated two years later (Lööv and Miegel, 1989b; Lööv and Miegel, 1991).

The striking outcome of this second approach was that, although the individuals were chosen for their high factor scores on the given lifestyle, there was large variation between individuals of the same lifestyle with respect to the concrete patterns of life actually characterizing the single individuals. As a matter of fact, the difference was so great as to be quite puzzling. In Kuhnian theory of paradigms, it represented an anomaly (Kuhn, 1970; Rosengren, 1989). In our more straightforward terms, we had come up with "different methods, different results."

The anomaly was solved when we realized that we were actually dealing with two different types of phenomena. On the one hand, we were dealing with lifestyle as an abstract pattern characterizing categories of individuals and perceived by means of advanced statistical analyses. On the other hand, we were dealing with an actual, concrete pattern of life, in all its bewildering richness, characterizing a given individual at a given moment of time and to be perceived only in actual social interaction between human beings. Obviously, these are two entirely different types of phenomena, which consequently should be kept conceptually and terminologically apart. *The abstract pattern of a "lifestyle" is to be distinguished from the concrete pattern of an "individual lifestyle."*

The research discussed in the previous section of this chapter demonstrated the validity of a given conceptualization by means of the application of two different methodologies. In this case, however, the application of different methodologies resulted in different results, leading to a reconceptualization. General, abstract "lifestyles" must not be mistaken for concrete cases of "individual lifestyles." The application of different methods may thus have very different consequences for the development of research. Two important consequences seem to be *validation* and *reconceptualization.* Both cases represent strong arguments for the application of different methods to the same scholarly problem. If more often resorted to in audience research, this strategy would no doubt be able to offer substantive contributions to the overall attempt to reach a comprehensive theory of audience.

There is a more general consequence to be drawn, however. Our example shows that, although focusing on seemingly the same phenomenon—say, some aspect of a television audience—so-called qualitative studies of a small number of individuals may well study a phenomenon completely different from the one studied in a parallel quantitative study of a large, representative audience sample. It need not always be the case, but sometimes it no doubt is.

It is often recommended that we change systematically between qualitative and quantitative studies. The strategy is a good one, but it is to be recommended only on the condition that researchers heed the possibility that when moving between qualitative and quantitative methodology they may well be moving between different objects of study. After all, forests have characteristics other than trees.

In the never-ending debate between qualitatively and quantitatively oriented audience scholars, then, these are two central questions: "What characteristics of a forest can be inferred from the study of a single tree? What characteristics of a tree can be inferred from the study of a forest?" Waiting for some answers to those basic questions, let us turn from comparisons between levels and over space to comparisons over time.

Temporal Invariance of Causal Structures

A comprehensive theory of audience should be valid over both time and space, and preferably across different types of operationalizations as well. In a previous section we saw that an important component in audience research, the notion of family communication patterns, does fulfill those basic conditions. The more explicit and detailed the theory, of course, the harder the test.

The well-known LISREL technique of structural modeling calls for a very high degree of explicitness; it is thus eminently suited to testing complex causal theories (Joreskog and Sorbom, 1989; cf., for instance, Cuttance and Ecob, 1987). Good examples of structural models in communication research have been presented by, among others, Milawsky et al. (1982) and Huesman and Eron (1986). Because of the cost in time and money, however, only few models have been replicated over time within precisely the same theoretical, methodological, and empirical framework. Actually, we have been able to find only one such replication with at least some relation to one of the main theoretic areas dealt with in the MPP (socialization): two cross-sectional LISREL models of schooling in the first grade, ten years apart (Entwisle and Alexander, 1987).

In her book, *TV Use and Social Interaction in Adolescence,* Ulla Johnsson-Smaragdi (1983) presented a number of detailed and precisely specified longitudinal causal models of, inter alia, the longitudinal relationship between children's and parents' television use. She tested her models by means of the LISREL technique. The continued data collection within the MPP has produced rich opportunities to probe the replicability of her models on later cohorts, and Dr. Johnsson-Smaragdi is currently busy doing just that.

A first, rough comparison of some relatively simple models was offered in Johnsson-Smaragdi and Hojerback (1989; cf. Rosengren 1991); further comparisons, richer and more detailed, will be presented in Johnsson-Smaragdi (forthcoming). Within the time/space framework available so far (4–5 years, and two cities in southern Sweden), the results of these studies show that the basic character of the causal models seems to be fairly invariant over time and space, while minor quantitative and qualitative differences may occur (a specific example being the problem of "reversed modeling" by parents of children's television habits; see Johnsson-Smaragdi, 1983; Rosengren, 1991; Rosengren and Windahl, 1989).

Results such as these raise a host of truly intriguing questions, all relevant to that basic question touched upon in a previous section of this chapter: What is the invariance over time and space of audience theory?

Dr. Johnsson-Smaragdi's results so far do suggest that there is indeed some invariance over time and space to be found even in relatively complex causal models. There thus seem to be reasons to assume that a fairly detailed and rather precise theory of the audience is, if not within actual reach, at least not entirely utopian. One way to approach it is by means of confrontations between theories and methodologies that have been produced within widely different research traditions.

Confrontations

Confrontations between scholars and scientists from different research traditions working within the same broad area of research are no common phenomenon. The normal thing, it would seem, is to live and let live. This strategy of mutual tolerance—or even mutual ignorance—is functional in that it saves time and evades bloodletting. But since it may be assumed, in some cases at least, that the different traditions are really studying the same phenomena from only slightly different perspectives, the strategy is dysfunctional in that it leads to unnecessary duplication of some scholarly work. In some cases it may actually occur at the price of clarification and thus lead to unnecessarily frequent dead ends.

This is not to say, of course, that confrontations do not occur. Some good examples of clarification through confrontation may be found, for instance, in James Curran's overview, "The New Revisionism in Mass Communication Research" (Curran, 1990). As a rule, however, such confrontations are exceptional. A good case in point is the area of lifestyle research touched upon above.

Lifestyle Conceptualizations Confronted

There are a number of traditions in the area of lifestyle research, but very little cross-fertilization between those traditions. One of the most prominent member of the field, Pierre Bourdieu, however, has been critically analyzed by some eminent sociologists of economy, politics, and culture (Coleman, 1988; DiMaggio and Mohr, 1985; Lamont, 1990; Martin and Szelenyi, 1987; Peterson, n.d.). It has also been less critically propagated by a large number of other scholars.[3] While the critical analyses have been centered mainly on the notion of different types of capital, the more uncritical presentations have often concerned lifestyle and media use—that is, topics of direct relevance to those interested in audience research. (For a good discussion of Bourdieu's views on these and related matters, see, for instance, Featherstone, 1987.)

Bourdieu arrived at the notion of lifestyle by way of his anthropological studies in North Africa, and by way of his interest in the notions of class, mobility, and capital—especially, perhaps, cultural capital (Bourdieu and Passeron, 1979, 1977). To him, the phenomenon of lifestyle manifests itself in every minute detail of the lives we lead, not least in media-related activities, such as reading, listening to music, going to the movies. His theory of lifestyle is therefore highly relevant to audience research. For our purposes, it is best studied in that wonderful book, *Distinction* (Bourdieu, 1984).

While writing *Distinction,* Bourdieu had been working with the theory of lifestyle for quite some time. Stemming from a French variant of anthropologically oriented sociology, his notions of theory are rather different from the ones dominating Anglo-American sociology and communication studies. His type of theory, rather than being made explicit, systematized, and tested at an early stage, grows organically over the decades. Having worked with the theory of cultural capital for a decade or two, he is ready to promise his patient readers that "a fuller presentation of the fundamental principles of this construction . . . is reserved for another book" (Bourdieu 1984: 572, note 17). What we have got so far, however, is quite fascinating. The interested reader of *Distinction* leaves the book with a somewhat different perspective on *Vanity Fair,* much as the readers of, say, Bunyan or Thackeray must have had in their days.

What is Bourdieu's notion of lifestyle, then? How does he conceptualize it? According to Bourdieu (1984: 171), the concept of lifestyle is defined as "a system of classified and classifying practices, i.e., distinctive signs ("tastes"), which is conditioned by the "habitus," in its turn defined as "a structured and structuring structure," in its turn conditioned by "objectively classifiable conditions of existence (class of conditionings) and position in structure of conditions of existence (a structuring structure)."

Thus a *lifestyle* is a pattern of life jointly determined by societal structure and individual positions in that societal structure. In the terms presented in a previous section, that is, Bourdieu subsumes *form of life* and *way of life* under the common term of *lifestyle;* he has no correspondence to what has above been called lifestyle and defined as patterns of life determined not only by societal structure and social position, but to some extent also by personal choice.

Similarly, in more mundane terms, Bourdieu's conception of *habitus* might be called individually internalized culture (Rosengren and Reimer, 1985; Reimer and Rosengren, 1990). Explicating it, he seems to be referring to a set of values, attitudes, opinions, and beliefs (Meddin, 1975) manifesting themselves in a set of coherent patterns of life, ranging from societally determined forms of life, over positionally determined ways of life, to individually determined lifestyles.

Bourdieu seems to give rather short shrift to lifestyle as defined in terms of individual choice. When structure and individual position in that structure have had their say, nothing much is left for individual choice other than in terms of rationalization of necessity. Thus, taste (an important element of habitus) "is a virtue made of necessity which continuously transforms necessity into virtue" (Bourdieu 1984: 175). This goes for all aspects of lifestyle, including, for instance, media use (television viewing, newspaper reading, theater habits, musical and literary taste, etc.). To use a parallel example from the sociology of religion: Bourdieu's notion of structurally and positionally determined lifestyles (i.e., forms and ways of life) could be likened to traditional sociological notions of religions, churches, and congregations, while the more individually determined lifestyles might be likened to *cults*—loose groupings of individuals worshipping an eclectic bric-a-brac of deities, marking their faith by subtle details of behavior, clothing, and other paraphernalia, the symbolic value of which can be fully appreciated only by those initiated into the cult.

The factors behind Bourdieu's both narrow and wide definition of lifestyle are at least twofold. Conceived in the heyday of Marxism in the 1960s, it is only natural that his theories should tend to minimize the role of individual choice. Empirically illustrated by data also collected during the 1960s, and in France—that is, in a society strongly stratified, not least with respect to cultural habits—it is perhaps even more natural that his data should tend to support his conceptualization and hypotheses. While there are thus good historical explanations for Bourdieu's conceptualization and empirical results, there is no reason why his thinking should not be confronted with theories and data from other regions of time and space—data and theories, preferably, less than thirty years old and stemming from parts of the world other than Paris and Lille.

In more specific terms, relevant to the theme of this book, such confrontations are currently being undertaken within the MPP program. It is too early to present any definitive results, but what findings have been presented so far—on the basis of qualitative and quantitative studies—do suggest that in western Europe of the 1990s, societal structure and individual position is not everything. There seems indeed to be some space for individually determined lifestyles. The space for structurally determined variation in ways of life, on the other hand, seems to be surprisingly small (Lööv, 1990a; Anshelm, 1990). This seems to be true not least with respect to lifestyles as manifesting themselves among members of different media audiences.

When confronting in this way Bourdieu's seminal studies of lifestyle and media use with other studies of lifestyle and audience behavior, one must in the end heed not only the overall conceptualizations but also the more detailed theoretical and operational definitions of the central concepts. Although Bourdieu's central concept of "cultural capital" has been rather uncritically accepted by most communication scholars, it has been very fruitfully related by a number of sociologists to basic concepts of economic and political sociology: physical, financial, human, and social capital (see., for instance, Coleman, 1988; DiMaggio and Mohr, 1985; Martin and Szelenyi, 1987; Peterson, Jones, and Miller n.d.). No attempt will be made here to add to those sometimes brilliant analyses. Accepting for a short moment Bourdieu's conceptualization, we will instead turn to another important problem: the way he relates what he calls economic and cultural capital to what he calls lifestyle.

The Relations Between Capital and Lifestyle

Bourdieu's basic conceptualization of the relations between habitus, lifestyle, and capital, of course, is strikingly simple: habitus and lifestyle are determined by economic and cultural capital. Unfortunately, it is not always quite clear what he refers to by the term "*determine*": logical definition or causal determination? As a rule, however, he is probably thinking of causal determination. In more traditional methodology than Bourdieu's, this could have been expressed by means of a causal three-term model, as graphically represented as in Figure 2.9.

Bourdieu uses a somewhat more sophisticated means, however. Drawing on a number of so-called analyses of correspondence, a French variant of factor analysis and multidimensional scaling (Benzécri, 1973; Lebart Morineau, and Tabard 1977; cf. Rijckevorsel and Leeuw, 1988; Weller and Romney, 1990), he presents a freehand fourfold table corresponding to, but different from, the one depicted in Figure 2.9 (Bourdieu, 1984: 128–129). When thus expressing the relationship between economic and cultural capital (EC and CC) on the one hand and the dispositional concept of habitus and the behavioral concept of lifestyle on the other, he introduces two new concepts derived from the basic ones: total amount of capital (TAoC), and composition of capital (CoC). TAoC and CoC represent the two main dimensions of his fourfold table.

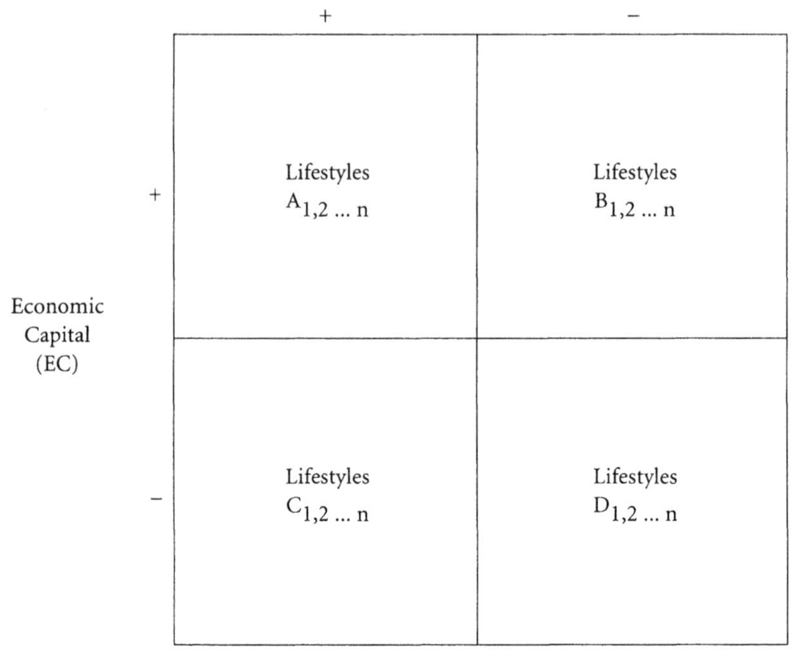

Figure 2.9 Correspondences Between Economic and Cultural Capital

Bourdieu does not specify the theoretical definition of these two new variables. There are a number of relatively simple options available and he probably had some of these in mind, at least implicitly:

$$\text{TAoC} = \text{EC} + \text{CC}, \textit{ or}\text{: TAoC} = \text{EC} \times \text{CC}$$

$$\text{CoC} = \text{EC} - \text{CC}, \textit{ or}\text{: CoC} = \text{EC} / \text{CC}.$$[4]

There are several interesting points that could be discussed in this connection, the most important of which is perhaps that the two dimensions of Bourdieu's fourfold table are not independent, built as they are on different combinations of the same basic variables of economic and cultural capital.

Now, let us combine Bourdieu's basic fourfold table with that of Figure 2.9. (The clue about how to do this lies in the fact that the upper-left/lower-right diagonal of Figure 2.9 represents the series of points where EC = CC.) Using squares of the same size, we then obtain the eight-cornered space depicted in Figure 2.10—the diamond shape of which may actually be traced in Bourdieu's original figure (Bourdieu 1984, 128–129).

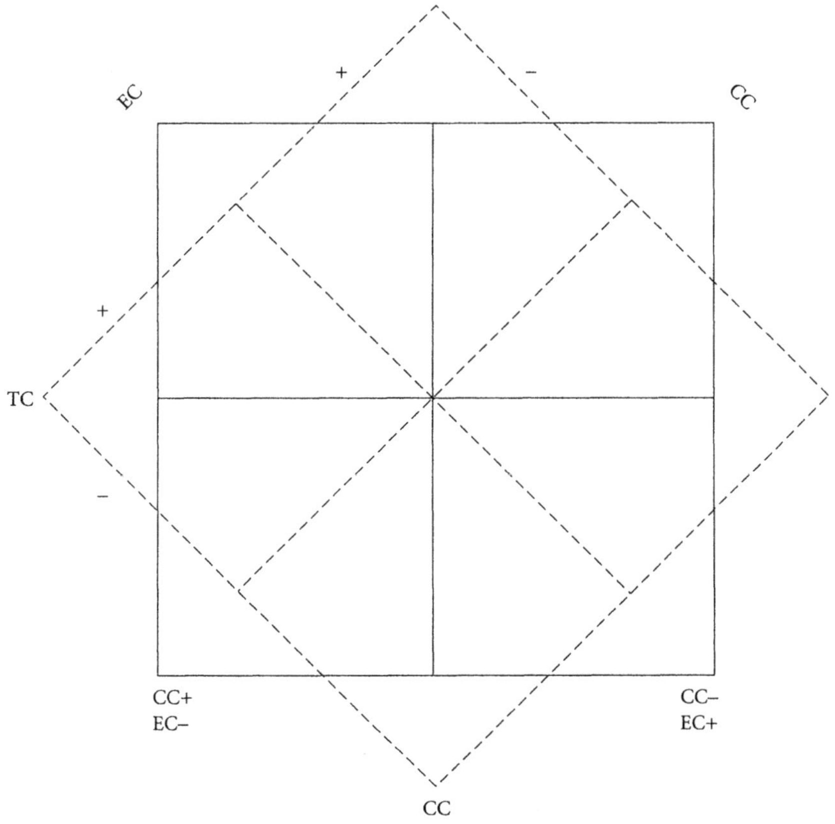

Figure 2.10 Bourdieu's Space of Lifestyles

It will be seen that the two fourfold tables are closely related. Although they do not cover each other completely, each one of them may be unambiguously expressed in terms of the other. Having thus related the two formal models to each other, the question inevitably announcing itself is: What, if anything, can be expressed in terms of one of those models that cannot be expressed in terms of the other?

This question might be used as a starting point for a serious confrontation between Bourdieu's theory of lifestyles and other theories of lifestyles—a confrontation that cannot, of course, be carried out within the framework of this article, as relevant it may be to any theory of audience activities.[5]

While Figure 2.10 to some extent clarifies the theoretical definitions implied in Bourdieu's thought, it does not tell us anything about the operational definitions used to establish the dimensions. By and large, however, cultural capital seems to have been operationalized by Bourdieu as quantity and quality of education received (by the respondents and/or their parents), and economic capital as income (from work or capital). In more traditional terms, cultural capital thus comes close to, but probably does not coincide with, human capital. To simplify, one might perhaps say that while human capital is related to a class dimension, cultural capital is related more to a status dimension. Social capital, in turn, is used to transform and enhance both economic and human capital (cf. Coleman, 1988).

A theory of audience aiming toward comprehensiveness can hardly afford to neglect these relationships. The precise relations between Bourdieu's conceptualization and what Bourdieu would probably call "traditional" theories of social class and culture remains to be chartered. Good foundations for that task have been offered by, for instance, Coleman (1988), Lamont (1988), Martin and Szelenyi (1987), and Peterson, Jones, and Miller, (n.d.).

Coda

Theoretical and methodological exercises such as these may appear trivial and somewhat narrow-minded to the reader more interested in the grand overall message of Bourdieu: the importance of inherited and achieved culture in shaping individual ways of life and lifestyles, including audience activities. But if we want to take Bourdieu as seriously as he certainly deserves to be taken, we must not only be his *lectores,* mechanically aping his terminology without questioning it the way we question all other terminology and related theoretical constructs. We must pay Bourdieu the compliment of confrontation.

Actually, I can think of few confrontations between scholars from different traditions that could be more rewarding than a series of large-scale and serious confrontations between Bourdieu's basic ideas and the ideas of prominent representatives of mainstream Anglo-Saxon audience research. A number of such confrontations will, I hope form a major component in future attempts to approach a comprehensive theory of the audience. The result will no doubt be an increased understanding of two concepts central to that theory: cultural capital and lifestyle. In the process some "allusions, gaps, and glissandos" (Lamont, 1990) will no doubt be replaced by more precise conceptualizations, and at the same time the productive force of the two concepts will have been retained. Only in so doing will we be able really to draw on Bourdieu's vast funds of scholarly learning, his rich theoretical imagination, and his many fascinating empirical results, so highly relevant for a future comprehensive theory of audience.

Conclusion: Comprehension Through Cumulation and Elimination

In this chapter I have suggested three roads presumably leading toward the distant goal of a comprehensive theory of audience:

- *combinations* of different techniques and approaches

- *comparisons* over time and space between results gained by scholars from different research traditions
- *confrontations* between related but different conceptualizations and methodologies developed by scholars from different research traditions (and sometimes from similar ones)

Actually, however, combination, comparison, and confrontation are just three different ways of achieving the more general goals of *cumulation* and *elimination.* In audience research—just as in all scholarly research—empirical data, research techniques, and methodologies as well as abstract concepts and theories must incessantly be combined, compared and mutually confronted so that less viable theories may be eliminated and more viable theories and data be cumulated until, by means of a finite and yet comprehensive theory, we may better comprehend some aspect of the infinitely rich world of phenomena.

This is certainly a rather abstract way of argumentation. However, some concrete examples, mainly from research within the Media Panel Program, have been offered in order to make the argumentation less abstract. Combinations, comparisons, and confrontations within the two fields of "audience research" and "lifestyle research" have been discussed at some length. Other combinations, comparisons, and confrontations could have been discussed, of course. In the end, however, such discussions must be much more detailed and thorough than is possible in a single book chapter.

Among promising areas for such combinations, comparisons, and confrontations, I would like especially to mention those conducted over time and space in order to arrive at an understanding of what results are related to specific structural conditions prevailing at a certain time and place, and what results are more independent of time and space. The strong structural changes currently under way in many media systems around the globe make this type of comparison especially timely. This avenue of research has recently been discussed at some length, and from a number of different perspectives, in the volume edited by Blumler, McLeod, and Rosengren (1991).

From a more general perspective, it would probably be beneficial if somewhat more active attempts at combinations, comparisons, and even confrontations could be substituted for the bland tolerance—sometimes, the dull ignorance—prevailing especially between humanistically and social science oriented audience scholars. Take as an example the two research traditions of "Reception Analysis" and "Empirical Studies of Literature" mentioned above. Although some more or less mutual awareness does exist between them, the two traditions have shown few if any cases of productive contact. Yet they have much in common—for instance, their humanistic background, which provides the valuable insight that in order to describe, explain, and understand audience activities, you must also describe, explain, and understand the "text" with which audiences incessantly interact.

Similarly, "reception analysis" and "uses and effects research" may be characterized by somewhat differential conceptualizations and methodologies (Jensen and Rosengren, 1990). At bottom, however, we are dealing with the same family of phenomena: the way individual audience members and/or different audience categories use media content, as well as the causes and consequences of that use for individuals, groups, organizations, and societies. By means of systematic combinations, comparisons, and confrontations, therefore, we should be able to arrive at a better understanding of the similarities and differences to be found in the results of the two traditions. We should gradually come to an agreement about what theoretical and empirical results may be cumulated into a joint fund of certified knowledge, and what about results, on closer inspection, we have to eliminate as not tenable in the strong light of two research traditions combined, compared, and confronted.

To return to a simile previously used in this chapter: Students of trees and students of forests should meet and enjoy the pleasure of the inevitable confrontation, so necessary in order to bring about both cumulation and elimination. This goes for the two traditions just discussed, and it may well be valid for the whole field of audience research.

Notes

1. The grid/group typology launched by Mary Douglas is a good example of creative use of such formal and formative dimensions (Douglas, 1970; cf. Gross and Rayner, 1985; Wildavsky, 1989). The task of confronting this very powerful typology with related typologies in social psychology and sociology offers a true challenge to the sociology of communication and culture.
2. Since different conceptualizations and methodologies may, of course, arise within the same or closely related research traditions.
3. The two types of reception, incidentally, correspond fairly well with Bourdieu's own distinction between *auctores* and *lectores*, the former being characterized as "producers," the latter as "reproducers, materialists without material, thoughts without instruments of thought;" cf. Bourdieu 1984: 510.
4. Having first carried out–in one way or another–the standardization necessary to combine the two basic concepts as just suggested, Bourdieu may, of course, also have introduced some sort of weighing after that.
5. Since Bourdieu's notion of "composition of capital" may be regarded as an attempt to catch aspects of interaction between economic and cultural capital, the notion of "interaction" would be highly pertinent to such a confrontation, both in formal and in substantive terms. While much simpler than Bourdieu's complex fourfold model, the fourfold model of Figure 2.9 is nevertheless able very clearly to express basic aspects of that interaction. It might be wise, therefore, to use Occam's razor.

References

Anshelm, M. (1990). Struktur och regional kultur. University of Lund Research Papers in the Sociology of Communication, No. 23.

Barnes, H. L., and, D. H. Olson (1985). Parent-adolescent communication and the circumplex model. *Child Development* 14: 149–170.

Barnouw, E. (ed.) (1989). *International Encyclopedia of Communications.* New York and Oxford: Oxford University Press.

Beniger, G. (1990). Identifying the important theorists of communication: Use of latent measures to test manifest assumptions in scholarly communication. In C. L. Borgman, (ed.), *Scholarly Communication and Bibliometrics.* Newbury Park, Calif.: Sage.

Benzécri, (1973). *L'analyse des données.* Paris: Dunod.

Blumler, J. G., J. M. McLeod, and K. E. Rosengren (eds.) (1991). *Comparatively speaking: Communication and Culture Across Space and Time.* Newbury Park, Calif.: Sage.

Bourdieu, P. (1984). *Distinction.* London: Routledge and Kegan Paul.

Bourdieu, P., and J.-C. Passeron (1979). *The Inheritors: French Students and Their Relation to Culture.* Chicago: Chicago University Press.

———. (1977). *Reproduction in Education, Society, and Culture.* Beverly Hills, Calif.: Sage.

Bradac, J. J. (ed.) (1989). *Message Effects in Communication Science.* Newbury Park, Calif.: Sage.

Burke, P. J., and C. D. Reitzes (1988). The link between identity and role performance. *Social Psychology Quarterly* 2:83–92.

Burrell, G., and G. Morgan (1979). *Sociological Paradigms and Organisational Analysis.* London: Heinemann.

Chaffee, S. H., J. M. McLeod, and C. K. Atkin (1971). Parental influences on adolescent media use. *American Behavioral Scientist* 14:323–340.

Coleman, J. S. (1988). Social capital in the creation of human capital. American Journal of Sociology 94 (Supplement):S95–S120.

Curran, J. (1990). The new revisionism in mass communication research: A reappraisal. *European Journal of Communication* 5:135–164.

Cuttance, P., and R. Ecob (eds.) (1987). *Structural Modeling by Example.* Cambridge: Cambridge University Press.

Denzin, N. K. (1970). *The Research Act: A Theoretical Introduction to Sociological Methods.* Chicago: Aldine Publishers.

DiMaggio, P., and J. Mohr (1985). Cultural capital, educational attainment, and marital selection. *American Journal of Soc iology* 90:1231–1261.

Donohew, L., P. Palmgreen, and J. D. Rayburn (1987). Social and psychological origins of media use: A lifestyle analysis. *Journal of Broadcasting and Electronic Media* 31:255–278.

Douglas, M. (1970). *Natural Symbols.* London: Barrie and Rockliff.

Entwisle, D., and K. Alexander (1987). The schooling process in the first grade: Two samples a decade apart. *American Educational Research* 23:587–613.

Featherstone, M. (1987). Lifestyle and consumer culture. *Theory, Culture and Society* 4:55–70.

Flodin, B. (1986). *TV och yrkesförväntan: En longitudinell studie av ungdomars yrkessocialisation.* Lund: Studentlitteratur. (With a summary in English.)

Garnham, N. (1990). *Capitalism and Communication: Global Culture and the Economics of Information.* London: Sage.

Gilljam, M. (1985). Pluralist and Marxist agenda-setting research. *Gazette* 34:77–90.

Golding, P., G. Murdoch, and P. Schlesinger (eds.) (1986). *Communicating Politics: Mass Communications and the Political Process.* Leicester: Leicester University Press.

Gross, J. L., and S. Rayner (1985). *Measuring Culture.* New York: Columbia University Press.

Hedinsson, E. (1981). *TV, Family, and Society: The Social Origins and Effects of Adolescents' TV Use.* Stockholm: Almqvist and Wiksell International.

Hojerback, I. (1990). Nya medier-nya klyftor? University of Lund Research Papers in the Sociology of Communication, No. 27.

Huesman, L. R., and L. D. Eron (eds.)(1986) *Television and the Aggressive Child: A Cross-National Comparison.* Hillsdale, N.J.: Erlbaum.

Jarlbro, G. (1986). Family communication patterns revisited: Reliability and validity. University of Lund Research Papers in the Sociology of Communication, No. 4.

———. (1988). *Familj, massmedier och politik.* Stockholm: Almqvist and Wiksell International. (With a summary in English.)

Jensen, K. B. (1987). Qualitative audience research: Toward an integrative approach to reception. *Critical Studies in Mass Communication,* 4:21–36.

Jensen, K. B., and K. E. Rosengren, (1990) Five traditions in search of the audience. *European Journal of Communication,* 5:207—239.

Johnsson-Smaragdi, U. (1983). *TV Use and Social Interaction in Adolescence: A Longitudinal Study.* Stockholm: Almqvist and Wiksell International.

——— (forthcoming). Structural invariance in media use: Some longitudinal LISREL models replicated. University of Lund Research Papers in the Sociology of Communication.

Johnsson-Smaragdi, U., and I. Hojerback (1989). Replikation av en LISREL-modell pa nytt urval. University of Lund Research Papers in the Sociology of Communication, No. 13.

Joreskog, K. G., and D. Sorbom (1989). *LISREL 7: A Guide to the Program and Applications.* 2nd ed. Chicago: SPSS Publications.

Jönsson, A. (1985). *TV—ett hot eller en resurs för barn?* Lund: CWK Gleerup. (With a summary in English.)

Kuhn, T. S. (1970). *The Structure of Scientific Revolutions.* 2nd ed., enlarged. Chicago: University of Chicago Press.

Lamont, M. (1990). Cultural capital: Allusions, gaps, and glissandos in recent theoretical developments. *Sociological Theory* 6:153–168.

Lebart, L., A. Morineau, and N. Tabard (1977). *Techniques de la description statistique.* Paris: Dunod.

Lööv, T. (1990a). Kulturell miljo och individuell stil. Strukturens och positionens betydelse for manniskors val av stil. University of Lund Research Papers in the Sociology of Communication, No. 22.

———. (1990b). Music video and lifestyle. Modern or post-modern aesthetics? Lund Research Papers in the Sociology of Communication, No. 24.

Lööv, T., and F. Miegel (1989a). The notion of lifestyle. Lund Research Papers in the Sociology of Communication, No. 15.

———. (1989b). Vardagsliv, livstilar och massmedieanvandning. En studie av 12 malmoungdomar. Lund Research Papers in the Sociology of Communication, No. 25.

———. (1991) Sju livsstilar. Om naågra malmöungdomars drömmar och längtan. Lund Research Papers in the Sociology of Communication, No. 15.

Martin, B., and I. Szelenyi (1987). Beyond cultural capital: Toward a theory of symbolic domination. In R., Eyerman G., Svensson and T. Soderquist (eds.), *Intellectuals, Universities, and the State in Western Societies.* Berkeley: University of California Press.

Meddin, J. (1975). Attitudes, values and related concepts: A system of classification. *Social Science Quarterly* 55:889–900.

Miegel, F. (1990). Om varden och livsstilar. Lund Research Papers in the Sociology of Communication, No. 25.

Milawskỳ, J. R., R. Kessler, H. H. Stipp, and W. S. Rubens (1982). *Television and aggression: A panel study.* New York: Academic Press.

Peterson, R. A., S. H. Jones, and O. Miller (N. d.). *Cultural capital in the status economy. A reconceptualization.* Mimeo. Vanderbilt University, Department of Sociology.

Radway, J. (1984). *Reading the Romance.* Chapel Hill, N.C.: University of North Carolina Press.
Reimer, B., and K. E. Rosengren, (1990). Cultivated viewers and readers: A life-style perspective. In: N. Signorielli, and M. Morgan, (eds.), *Cultivation Analysis,* pp. 181–206. Newbury Park, Calif.: Sage.
Rijckevorsel, L. A. van, and Jan de Leeuw (eds.) (1988). *Component and Correspondence Analysis.* New York: Wiley.
Roe, K. (1983). *Mass Media and Adolescent Schooling: Conflict or Co-existence?* Stockholm: Almqvist and Wiksell International.
Rosengren, K. E. (1983). Communication research: One paradigm or four? *Journal of Communication 33,* (3): 185–207.
———. (1989). Paradigms lost and regained. In B. Dervin, L. Grossberg, B. O'Keefe, and E. Wartella (eds.), *Paradigm Dialogues: Theories and Issues,* pp. 21–39. Beverly Hills, Calif.: Sage.
———. (1991). Media use in childhood and adolescence: Invariant change? *Communication Yearbook* 14:48–91.
Rosengren, K. E., J. McLeod, and J. G. Blumler (1991). Comparative communication research: From exploration to consolidation. In J.G. Blumler, J.M. McLeod, and K.E. Rosengren, (eds.), *Comparatively Speaking: Communication and Culture Across Space and Time.* Newbury Park, Calif.: Sage.
Rosengren, K. E., and B. Reimer (1985). Internaliserad kultur. Working Papers, No. 6. Gothenburg: Unit of Mass Communication Research.
Rosengren, K. E., L. A. Wenner, and P. Palmgreen (eds.) (1985). *Media Gratifications Research: Current Perspectives.* Beverly Hills, Calif.: Sage.
Rosengren, K. E., and S. Windahl (1972). Mass media consumption as a functional alternative. In D. McQuail (ed.), *Sociology of Mass Communications.* Harmondsworth: Penguin.
———. (1989). *Media Matter: TV Use in Childhood and Adolescence.* Norwood, N.J.: Ablex.
Sonesson, I. (1979). *Förskolebarn och TV.* Stockholm: Esselte Studium. (With a summary in English).
———. (1989). *Vem fostrar vaåra barn? Videon eller vi?* Stockholm: Esselte.
Thunberg, A. M., K. Nowak, K. E. Rosengren, and B. Sigurd (1982). *Communication and Equality. A Swedish Perspective.* Stockholm: Almqvist and Wiksell International.
Tims, A. R., and J. L. Masland (1985). Measurement of family communication patterns. *Communication Research* 12:35–57.
Turner, G. (1990). *British Cultural Studies.* Boston: Unwin Hyman.
Weller, S. C., and Romney A. K. (1990). *Metric Scaling: Correspondence Analysis.* Newbury Park, Calif.: Sage.
Wiberg, H. (1969). On the relationship between science and values. In L. Dencik (ed.), *Scientific Research and Politics.* Lund: Studentlitteratur.
Wildavsky, A. (1989). Choosing preferences by constructing institutions: A cultural theory of preference formation. In A. A. Berger (ed.), *Political Culture and Public Opinion,* pp. 21–46. New Brunswick, N.J.: Transaction Publishers.
Windahl, S. (1981). Uses and gratifications at the crossroads. *Mass Communication Review Yearbook,* vol. 2.
Zablocki, B. D., and R. M. Kanter (1976). The differentiation of life-styles. *Annual Review of Sociology* 2:269–298.

Audience Research: Antinomies, Intersections, and the Prospect of Comprehensive Theory

3

David L. Swanson

Intersections: A View of the Recent Past in Audience Research

These days, audiences are studied and theorized about by a bewildering array of disciplines, interdisciplinary fields, and particular approaches to research, and the relationships among these different views of the subject are quite complex. Limiting our concern to the field of communication research provides only a little help in sorting out the complexity because, even in this smaller arena, rapid development and changing research directions have broken down the conventional distinctions that formerly organized the field in ways that were analytically, if not always empirically, useful. Still, the idea underlying this volume encourages efforts to chart the territory or at least a few of its regions.

Revisiting, as so many others have done, the domain's natural history—some of the larger patterns and directions of its recent evolution—may shed some light on our subject. The problem with reconstructions of even the recent past, of course, is that they are too far removed from the details of the research that they describe. Yet so long as their limitations are kept in mind, analytic reconstructions of developmental patterns sometimes can reveal important features of the terrain.

James Carey (1990) has described the history of media and audience research in terms of a pattern of alternating cycles. In one cycle, the audience is seen as comparatively passive and weak in the face of powerful mass media. In the next cycle, the audience is seen as comparatively active and powerful, and it is the limits of media influence that are stressed. Then the field cycles back to the first view and the media are seen as sovereign once again. Carey suspects that these oscillations between opposing estimates of media versus audience power might well reflect the responses of audience research to a succession of external events, such as the Great Depression of the 1930s.

If Carey's reading is correct (and, of course, I have pushed his analysis farther than he would wish in order to make my own point), then it follows that the various approaches to audience research are unified by two common threads. First, in any given cycle, the different approaches share roughly the same view of the relative power of audiences and mass media. And this is because, second, the different approaches respond in roughly the same way to changes in the social and cultural milieux. In this reading, we perhaps may find an optimistic prospect for the project of a comprehensive theory that builds upon these common threads to integrate major research approaches in some way.

As a foil to Carey's view, I offer a somewhat different reading of the recent past. My reading arrives at approximately the same conclusion as Carey's does—that the diverse approaches to audience research share some common interests—but reaches that conclusion by a different path that points to some correspondingly different implications for the prospect of comprehensive theory.

My view is deeply indebted to James Curran's (1990) recent interpretation of some of the major developments in mass communication research since 1975 or so, primarily in Britain. In this section, I wish to support and extend his analysis and to turn it to the question of a comprehensive theory of the audience.

Taking as his starting point the mid-1970s and the then-conventional opposition between Marxist (or "critical" or "radical") and liberal/pluralist (or "effects") perspectives on mass communication, Curran concentrates on the subsequent development within the critical tradition of what he calls revisionist models of power and representation. Revisionist models have had their greatest impact in studies of audience reception, Curran believes, where, to oversimplify a bit, the evolution from the critical or radical to revisionist views have entailed two key moves. First, the "relatively unproblematic analysis of meaning" in the earlier critical tradition was replaced by a conception of meaning that stressed gaps, contradictions, inconsistencies, and internal oppositions within texts. As a result, in this conception's strongest form, audiences came to be seen as sites of the production of meaning and of possible resistance to dominant meanings inscribed in texts. Second, the audience was granted greater autonomy as it was reconceived as an "active producer of meaning," which was now seen as "constructed through the interaction of text and the social and discourse positions of audiences" (Curran, 1990: 145). Among the consequences of the movement from a radical to a revisionist view, according to Curran, are a more cautious and limited assessment of media influence and a shift from a political aesthetic (that is, "from whether media representations advanced or retarded political and cultural struggle") to a popular aesthetic (that is, to questions about the pleasures and popularity of mass media) (Curran, 1990: 146).

It seems to me as an interested tourist in the critical tradition that Curran has put his finger on something important in the movement from the study of ideology as imposed on audiences by univocal texts to the study of hegemony orchestrating conflicting interests within a dominant ideology, and then to the study of how active and creative audiences are able to construct from polysemic texts meanings that are pleasurable and useful in terms of the audiences' identities and experience. In much of this work, at least on my reading, connections have been attenuated between the political economy of mass media, the production of media content, and audiences' interpretation, consumption, and use of content. Increasingly these days, descriptions of audience attributes that influence the interpretation of media messages seem to be couched in the first instance in symbolic and cultural terms. The contexts of reception and interpretation seem to be defined in more microscopic ways that sometimes have less straightforward connections to society's gross political and economic structures. Devoting greater attention to the self-described perceptions and experiences of audience members has proved to be helpful, perhaps even essential, in navigating through the variability introduced by the multiplication of contexts and discourses of interpretation on the one hand and the comparatively greater independence and creativity accorded to audiences on the other.

While the power of media over audiences, according to Curran's analysis, has been progressively diluted and qualified within the revisionist critical view, the effects tradition has been moving in the other direction, arguing increasingly that in particular circumstances the media may exert considerable influence over their audiences. Among the components of the emerging view of more powerful media, I believe, is the movement from seeing audiences' interpretive and evaluative processes as almost completely sovereign, as in the old notion of the selective perception mechanisms, to a contemporary view of messages as capable of exerting some influence over audiences' interpretive processes, as in research on priming effects in the agenda-setting literature (e.g., Iyengar and Kinder, 1987) and in political information–processing research, where it has been suggested that messages may influence the contexts and dimensions of interpretation that audience members bring to bear in interpreting them (e.g., Lau and Sears, 1986, esp. 362–363). The same direction of development may be seen in the movement from conceiving of effects narrowly in terms of persuasion and attitude change to more elaborated conceptions of effects, in terms of emotion and cognition, for example (Lanzetta, Sullivan, Masters, and McHugo, 1985), occurring over longer periods of time, as in longitudinal agenda-setting and socialization studies (e.g., Neuman and Fryling, 1985), influencing a wider range of behaviors, such as willingness to express opinions in spiral-of-silence research (e.g., Noelle-Neumann, 1984), and resulting from the full range of media content (e.g., Swanson, 1990). Accompanying these developments has been a growing interest in exploring social, institutional, and cultural influences on audience response, as in some recent writing on media uses and gratifications (e.g., Blumler, Gurevitch, and Katz, 1985), and in the production of media content, as in some political communication studies that define national political communication systems chiefly in terms of relationships between political, governmental, media, and other institutions (e.g., Blumler and Gurevitch, 1975; Blumler and Thoveron, 1983). In such ways, the effects tradition may be seen as evolving toward conceptions of greater media influence over audiences within theoretical frameworks of increasing breadth and scope.

Those who are familiar with the conventional mythology of mass communication research will have anticipated Curran's conclusion:

> The most important and significant overall shift [during the last fifteen years] has been the steady advance of pluralist themes within the radical tradition: in particular, the repudiation of the totalizing, explanatory frameworks of Marxism, the reconceptualization of the audience as creative and active, and the

> shift from the political to a popular aesthetic. Because this revisionism has evolved in response to an internal debate within the radical tradition rather than as a direct response to pluralist texts, the extent of the movement towards the pluralist tradition has been partly obscured (Curran, 1990: 157–158).

And, further, that "by a curious irony, revisionist celebrants of semiotic democracy are thus moving towards a position that pluralists are abandoning" (Curran, 1990: 153).

All such reconstructions, including Curran's, invite researchers to object that the details of their views have been ignored or misrepresented in some way, or that vital distinctions have been overlooked. The pattern of development Curran describes is misleading in that it suggests that one view was abandoned and replaced by another, which in turn was abandoned as yet another view arose to take its place. In fact, it is rare in this research for any views to be abandoned altogether. As Curran reminds us, the views glossed as "revisionist" and "pluralist" are quite heterogeneous, and within each camp are some particular approaches that fit his characterizations fairly well, others that fit less well, and still others that fit not at all. We can agree, too, that ideas such as audience creativity and activity, for example, or the openness of media content to multiple meanings and interpretations are indeed embraced by opposing schools of research while we also recognize that these ideas take on quite different meanings within different approaches.

Nevertheless, with such objections admitted, I believe that Curran's diagnosis of the recent history of audience studies in mass communication offers a useful insight bearing on the prospect of a comprehensive theory of the audience. If the last fifteen years can be read in something like the way Curran suggests, then the intersections that have been crossed and recrossed by the major research traditions along their journeys in opposing directions may constitute one of the preconditions for a comprehensive audience theory. These intersections—audience activity and creativity, the processes through which media texts are experienced by audiences as meaningful, the reach and limits of audiences' interpretive autonomy, connections between the production and reception of media content and more general social and cultural processes and structures, and the rest—stand as points of contact and common interest among otherwise dissimilar views.

At this point, it is important to note some differences between the contrasting paths by which the present analysis and that offered by Carey both arrive at the conclusion that some common features intersect the various approaches to audience research. If, in fact, the critical and effects traditions have been moving in opposite directions during the last fifteen years, as the present analysis suggests, then these opposing approaches cannot be said to respond to external events in the same way. At least during the recent past, the theoretical impulses and content of contrasting views of audience research have shown more opposition and contradiction than resonance to some shared, underlying conception of the relative power of audiences versus media. That is, the threads that Carey found to unify audience research are, in the present analysis, unraveled.

Antinomies in Audience Research

Because contrasting approaches to audience research share some points of contact and common interests, as the preceding analysis suggests, it seems likely that the interpenetration of claims produced by competing approaches increasingly will become an issue to be dealt with. One way of dealing with common interests is to succumb to the temptation, to see points of contact as providing the basis for a comprehensive theory of audiences. A second course, advocated in this volume by David Morley, is to regard common interests as enabling us to triangulate observations and conclusions by employing multiple research methods. As Dan Nimmo and I explained in a commentary on roughly similar developmental patterns in political communication research,

> When practitioners of contrasting approaches find themselves making empirical claims about the same, or closely related, phenomena, there exists a basis for regarding their approaches as mutually relevant. Contrasting approaches can be seen . . . as contributing distinctive components to the most rounded understanding of the phenomena . . . in which various data sources, research methods, and theoretical conceptions are seen as best suited to studying complex social processes (Nimmo and Swanson, 1990: 22).

Multimethod triangulation of results produced by different theories and methods is one thing; a comprehensive theory of audiences that incorporates and, presumably, integrates some number of different approaches is something else again. Oppositions and contradictions between approaches are, in principle, no obstacle to multimethod triangulation; oppositions and contradictions are likely to be fatal to theory, however, comprehensive or otherwise. Surely one attribute of a comprehensive theory ought to be coherence, for an incoherent theory cannot help us better understand audiences. Attempting to integrate within a single overarching theory particular approaches, or elements within approaches, that are rooted in conflicting, contradictory conceptions must lead to an incoherent, confused result.

The terrain of audience research is riven by conflicts, contradictions, and oppositions. Competing approaches differ not merely in their theoretical commitments but also, and more importantly, in their conceptions of what an audience *is*, what constitutes authoritative information about an audience, and how such information may be interpreted to produce knowledge claims. Hence, major traditions have been described as "in some ways, fundamentally and irreconcilably opposed" (Curran, Gurevitch, and Woollacott, 1982).

Notwithstanding their lately discovered common interests, the relationship between some of the major opposing schools of research, such as, to use Curran's labels, the critical, revisionist, and pluralist, seems to me to be one of antinomy, authoritative contradiction between conclusions that, taken on their own terms, are equally logical, necessary, and reasonable. These opposing conceptions of what an audience is, where it is, and how we can know about it cannot be blended together to create some consistent and coherent general theory without radically revising the conceptions in ways that make them no longer recognizable, or so it seems to me. The authoritative evidence of one view may cut no ice with, and indeed may be seen as beside the point, by the other view. As a result, I can foresee no realistic possibility of constructing any integrative, comprehensive theory of the audience.

In a recent essay charting continuities and discontinuities across five major traditions of audience research, Jensen and Rosengren concluded: "It is unrealistic to hope to completely reconcile the differential legacies of arts and sciences which inform the five traditions of audience research. . . . Yet we do maintain that there are further possibilities of convergence at several levels of analysis" (1990: 229). I agree with both claims and turn now to some possibilities for capitalizing on common interests, even in the absence of a comprehensive theory.

Alliances and Convergence

There are several ways in which the common interests and mutual relevancies that link otherwise opposing approaches to research can be profitably exploited. And there is considerable incentive for trying to exploit our common interests. Processes of marginalization, domination, and resistance that are stressed by critical approaches and processes of empathy, shared experience, and cooperation by diverse groups to promote the common good that are stressed in effects research both are part of the social functions of mass communication in contemporary society. Neither approach profits by ignoring the concerns of the other.

The most obvious way we can exploit our common interests is simply to read and take seriously work produced by those with views different from our own—though this process is perhaps not so simple when some approaches traditionally have defined themselves in part by their opposition to other views. We have already been influenced by each other, and that should continue and increase. Consider, for example, the extent to which the future that Blumler, Gurevitch, and Katz sketched for gratifications research in 1985 was influenced by the work of critical theorists and reception analysts and, on the other side, the way in which some insights of uses and gratifications research have provided foils against which Morley has developed his own approach to family television and domestic media (e.g., Morley, 1986).

In particular, I would like to call attention to the way in which some of the major approaches can and have been read as identifying gaps and shortcomings in other views. For instance, there is little doubt, I think, that the richly textured analyses of media messages produced in critical studies have served to underscore to many effects researchers the inadequacy of their own resources for message analysis, or that some of the results of early ethnographic studies of audiences were very important in leading some critical researchers to conclude that their theories of texts needed to be complemented with theories of audiences. This process by which the emphases and accomplishments of one approach are seen as challenging other views to extend their reach and to elaborate gaps in their frameworks leads to an especially appropriate outcome for the conference that gave rise to this text: not a comprehensive theory of the audience, but more comprehensive theor*ies* of the audience.

As we strive to pay attention to each other and to construct more comprehensive theories within our respective approaches, it seems to me that the type of research that has come to be called reception analysis may have a special role to play, in something like the way Jensen (1987) has suggested. This approach encompasses the various styles of research we often describe as qualitative and (loosely) ethnographic, which involve interviewing audience members about their experience of media content, and which, as characterized by Jensen and Rosengren, "to different degrees, seek to integrate social-scientific and humanistic perspectives on reception" (1990: 213).

There are several distinctive features of reception analysis that are potentially important to the effort toward comprehensiveness. In my own experience, I often have found that theoretical ideas which seem obscure and arcane are more easily grasped when I see them used in reception analysis. This is so, I believe, because in the tradition of "thick description," theory hovers quite close to the ground in reception analysis. The researcher's constant movement back and forth between data and analytical characterizations makes the theoretical accomplishments of reception analysis accessible to the wide community of persons who are interested in audiences, not just to those who share the researcher's own theoretical and ideological preferences, and does so in a way that allows diverse readers to argue about the adequacy of the study's evidence and the appropriateness of its conclusions. That is, reception analysis allows dialogue among readers who themselves hold different theoretical views in a way that most other approaches do not.

Moreover, reception analysis provides an arena that is at least potentially hospitable to many of the concepts concerning reception processes that have been formulated within a number of different approaches to the subject. Reception analysis is committed to relating discourses produced by audience members to the discourses of media texts. In principle, notions of cognitive process and gratification-seeking may be deployed in reception analysis as comfortably as notions of subcultural resistance. So understood, reception analysis may serve as something like a bridge across a range of research traditions, as is attested to by the great variety of theoretical views that have been employed in reception analysis (for example, compare Radway, 1984, with Lindlof, 1988).

Reception analysis seems well suited to a bridging role because, in principle, it has no substantive theoretical content per se. Instead, reception analysis is committed to the sensitive interpretation of audiences' phenomenal experience and to relating that experience to media texts and to the pertinent features of the context within which audience experience occurs. That is, the defining commitment of reception analysis has to do more with the structure of explanation than with the substantive theoretical content of the explanation.

This is not to suggest that reception analysis is somehow unable to formulate concepts of its own or that reception analysts should see themselves only as engaged in translation or deploying concepts developed by others. On the contrary, the reverse side of the coin is that concepts formulated in reception analysis are certain to be relevant to the work of other approaches to audience research, providing yet another way of establishing connections between divergent views. But it is perhaps worth noting that in order to serve this bridging function as a basis for dialogue between approaches, critical reception analysis might profitably expand the range of concepts it employs beyond the hegemony of race, class, and gender.

Nor am I suggesting that all of mass communication research should be collapsed into reception analysis. With its limited focus, reception analysis falls short of addressing a whole range of important questions about mass communication, such as those concerning the production of media content and the political economy of mass media. Moreover, the interpretive methods and sampling procedures of reception analysis have better claims to heuristic value than to authority as judged by the commitments of a number of major research approaches.

But it does seem to me that its openness to explanatory concepts and its ability to analyze closely both phenomenal audience experience and media content can allow reception analysis to make a contribution of particular importance by creating space for discussion and argument across approaches. The result will not be a grand, master theory of the audience but perhaps a closer dialogue among conflicting views and a somewhat clearer understanding of how concepts developed in one view may contribute to other views, and this within a context in which each approach continues to pursue its own agenda, elaborate its own viewpoint, and increase its reach and comprehensiveness.

I realize that these few suggestions for a workable and productive modus vivendi in audience research are exceedingly modest steps and, to those who hope for a comprehensive theory of audiences, probably unsatisfying. Perhaps I can increase the appeal of what I am suggesting by describing a recent visit to something like this sort of future.

Beginning in 1987, I was a member of a group of American and French researchers who came together to conduct a comparative study of political communication in the 1988 presidential campaigns in France and the United States (Kaid, Gerstlé, and Sanders, 1991). This group was quite diverse and included representatives of a number of the major approaches to audience research, from media effects and public opinion researchers to narrative rhetorical critics and semioticians. We met early and often, arguing and negotiating what form the study should take. We quickly concluded that there was no single theoretical framework, either at hand or within reach, that could accommodate the conflicting views of all members of the group. We decided, then, to conduct not a single study but rather a group of almost two dozen separate but coordinated studies, each focusing on one or another aspect of the subject and reflecting the theoretical and methodological commitments of the particular researchers who designed and carried out that study. As a result, a wide range of approaches was represented in the results of the group's efforts.

Given the presence in the group of such opposing viewpoints, I would not have been surprised if we had been unable to reach agreement on any point. But things turned out to be just the opposite. Very many common interests quickly were identified between various of the approaches represented in the group. Practitioners of each approach willingly acknowledged limitations of their own views and deferred to other approaches that were better equipped to deal with particular questions. The range of viewpoints represented in the group brought to our discussions an unusual and, I think, stimulating breadth of perspective. I was struck by the similarity of the conclusions that were produced by researchers who employed different approaches, particularly in the domain of message analysis, and by the general tendency throughout the research group to explain the content of political communication in the same way, by pointing to the institutional interests that shape message production. In the end, no one had been converted to another viewpoint, at least as far as I know, but I think all of us came to a better understanding of the strengths and limitations of our own approach and its connections to other views.

It appears to me that the common interests and concerns of audience researchers offer a basis for dialogue of just this sort. There may come a time when attempts to formulate a comprehensive theory will be productive. For now, much can be gained by simply widening the circle of persons we relate to as colleagues.

References

Blumler, J. G., and M. Gurevitch (1975). Towards a comparative framework for political communication research. In S. H. Chaffee (ed.), *Political communication: Issues and Strategies for Research,* pp. 165–193. Beverly Hills, Calif.: Sage.

Blumler, J. G., M. Gurevitch, and E. Katz (1985). Reaching out: A future for gratifications research. In K. A. Rosengren, L. A. Wenner, and P. Palmgreen (eds.), *Media Gratifications Research: Current Perspectives,* pp. 255–273. Beverly Hills, Calif.: Sage.

Blumler, J. G., and G. Thoveron (1983). Analysing a unique election: Themes and concepts. In J. G. Blumler (ed.), *Communicating to Voters: Television in the First European Parliamentary Elections,* pp. 3–24. London: Sage.

Carey, James (1990). Paper presented at conference, Toward a Comprehensive Theory of the Audience, Champaign-Urbana, Illinois.

Curran, J. (1990). The new revisionism in mass communication research: A reappraisal. *European Journal of Communication* 5: 135–164.

Curran, J., M. Gurevitch, and J. Woollacott (1982). The study of the media: Theoretical approaches. In M. Gurevitch, T. Bennett, J. Curran, and J. Woollacott (eds.), *Culture, Society, and the Media,* pp. 11–29. London: Methuen.

Iyengar, S., and D. R. Kinder (1987). *News That Matters: Television and American Opinion.* Chicago: University of Chicago Press.

Jensen, K. B. (1987). Qualitative audience research: Toward an integrative approach to reception. *Critical Studies in Mass Communication* 4: 21–36.

Jensen, K. B., and K. E. Rosengren (1990). Five traditions in search of the audience. *European Journal of Communication* 5: 207–238.

Kaid, L. L., J. Gerstlé , and Sanders K. R. (eds.) (1991). *Mediated Politics in Two Cultures: Presidential Campaigning in the United States and France.* New York: Praeger.

Lanzetta, J. T., D. G. Sullivan, R. D. Masters and G. J. McHugo (1985). Emotional and cognitive responses to televised images of political leaders. In S. Kraus and R. M. Perloff (eds.), *Mass Media and Political Thought: An Information-Processing Approach,* pp. 85–116. Beverly Hills, Calif.: Sage.

Lau, R. R., and D. O. Sears (1986). Social cognition and political cognition: The past, the present, and the future. In R. R. Lau and D. O. Sears (eds.), *Political Cognition,* pp. 347–366. Hillsdale, N.J.: Erlbaum.

Lindlof, T. R. (1988). Media audiences as interpretive communities. In J. A. Anderson (ed.), *Communication Yearbook,* vol. 11, pp. 81–107. Newbury Park, Calif.: Sage.

Morley, D. (1986). *Family Television: Cultural Power and Domestic Leisure.* London: Comedia.

Neuman, W. R., and A. C. Fryling (1985). Patterns of political cognition: An exploration of the public mind. In S. Kraus and R. M. Perloff (eds.), *Mass Media and Political Thought: An Information-Processing Approach,* pp. 223–240. Beverly Hills, Calif.: Sage.

Nimmo, D., and D. L. Swanson (1990). The field of political communication: Beyond the voter persuasion paradigm. In D. L. Swanson and D. Nimmo (eds.), *New Directions in Political Communication,* pp. 7–47. Newbury Park, Calif.: Sage.

Noelle-Neumann, E. (1984). *The Spiral of Silence: Public Opinion—Our Social Skin.* Chicago: University of Chicago Press.

Radway, J. (1984). Interpretive communities and variable literacies: The functions of romance reading. *Daedalus* 113(3): 49–73.

Swanson, D. L. (1990). Popular art as political communication. In R. L. Savage and D. Nimmo (eds.), *Politics in Familiar Contexts: Projecting Politics Through Popular Media,* pp. 13–61. Norwood, N.J.: Ablex.

After Convergence: Constituents of a Social Semiotics of Mass Media Reception

4

Klaus Bruhn Jensen

If the 1970s marked a rise of cultural studies and critical theory within international mass communication research, the 1980s have witnessed the first stage in a process of convergence between the mainstream of the field and these "cultural" and "critical" undercurrents. Convergence has entailed a reassessment, to some degree, of the theoretical and methodological categories of communication studies, particularly in confronting and conjoining concepts and models from the humanities and social sciences. Nevertheless, convergence is still more evident in the rhetoric of much research than in its theoretical substance or empirical practice.

This chapter, accordingly, examines the possible terms of a further convergence between the humanistic and social-scientific traditions of mass communication research with special reference to reception studies. Starting from the premise that mass communication is simultaneously a social and a discursive phenomenon, I argue that a theory of social semiotics is needed in order to account for the social impact and uses of mass media. Mass communication is embedded in material social institutions and practices; it also works through language, pictorial signs, and other semiotic systems. While the two primary, qualitative and quantitative traditions of communication research each have a contribution to make to the development of a social semiotics, it is important to specify their explanatory value at different levels of analysis. Qualitative inquiry, as developed in humanistic scholarship, has a specific theoretical and methodological potential that is still waiting to be tapped within a framework of social semiotics.[1]

In the first section below, I briefly lay out the traditional conceptions of qualitative and quantitative inquiry as a prelude to the discussion of their convergence at a *theoretical* level of analysis. The second step of the argument introduces three concepts from discourse analysis that may help to integrate the different traditions of inquiry. The third and final part of the article brings together these two strands of the argument in a discussion of the constituents of social semiotics. This section also considers the prospects for the qualitative empirical reception analysis that has developed internationally over the last decade. In my conclusion, I suggest that the semiotics and pragmatism originated by Charles Sanders Peirce offers a promising avenue for a *social* theory of signs that may move mass communication research beyond the heritage of French semiology and of American content analysis.

Qualitative-Quantitative Distinctions Reconsidered

Two Modes of Inquiry

We may start to assess the respective contributions of qualitative and quantitative methodology by reconsidering the forms of knowledge that traditionally have been associated with each approach. The background to the two modes of inquiry is to be found respectively in the humanities and the natural sciences, or, in the classic German terms, *Geisteswissenschaften* and *Naturwissenschaften.* Culture and communication, from these perspectives, are conceived of as a source of either *meaning,* in phenomenological and contextualized terms, or *information,* in the sense of discrete vehicles transporting significance through mass media. Consequently, qualitative analysis tends to focus on the *occurrence* of its analytical objects in a particular context, as opposed to the *recurrence* of formally similar elements in different contexts. This suggests either an *internal* approach to understanding culture, interpreting and even immersing oneself in its concrete expressions, or an *external* approach that tries to establish a detached position outside of culture. Equally, media contents and other cultural forms can be said to give rise to a unique, indivisible *experience* through *exegesis* or, alternatively, to a set of stimuli that may be manipulated through *experiments,* thus producing variable effects that can be *measured.* Finally, whereas quantitative analysis will focus on the concrete, stable *products* of the media's meaning production, qualitative approaches examine meaning production as a *process* that is inextricably related to the wider social and cultural context. Table 4.1 sums up the two conceptions of communication that are normally associated with qualitative and quantitative methodology.

It must be emphasized that the dichotomies of Table 4.1 refer, first and foremost, to the self-conception of the two traditions. The social sciences, after an

Table 4.1 Two Modes of Inquiry

Qualitative	Quantitative
Geisteswissenchaften	*Naturwissenschaften*
meaning	information
internal	external
occurrence	recurrence
experience	experiment
exegesis	measurement
process	product

early qualitative phase, increasingly came to see the natural sciences as a standard also for social inquiry. (However, natural scientists themselves may perceive their research as more comparable, in several respects, to qualitative modes of inquiry.) In part as a response to this development, the humanities have come to advertise their aesthetic and historical perspectives on reality as being unique, thus contributing further to dichotomization. Today, the two sides of the dichotomies coexist uneasily in a number of social-scientific and humanistic disciplines and fields. Even though a unified science of communication may be neither possible nor desirable, at least in the short term, it seems worthwhile to explore in more concrete terms the complementarity of the analytical traditions. Their ends, means, and objects of analysis are hardly incompatible in an absolute sense; the question is to what degree and in what terms qualitative and quantitative modes of inquiry may be compatible.

Four Levels of Analysis

Although it may serve to confuse rather than clarify issues of scientific debate, the qualitative-quantitative distinction is a fact of research practice that has major epistemological and political implications that no researcher can ignore. For the purpose of examining these implications, it is useful to specify the analytical levels at which the distinctions of Table 4.1 may apply. Four such levels may be distinguished:

- the object of analysis (as identified and characterized with reference to the purpose and context of the inquiry)
- the analytical apparatus or methods (the concrete operations of inquiry, including the collecting, registering, and categorizing of data)
- the methodology (the overall design of the inquiry which serves to relate the constituent methods of data gathering and data analysis, further justifying their selection and the interpretation of the data with reference to the theoretical frameworks employed)
- theoretical framework(s) (the configuration of concepts that specifies the epistemological status of the other levels, and which hence assigns explanatory value to the specific rendition of the object of analysis that the methodology produces)

In the words of Anderson and Meyer (1988: 292), "it is method that generates the facts that become evidence within theory."

Whereas the four levels are interdependent, I want to suggest that, in principle, the labels of "qualitative" and "quantitative" refer to methodologies and, by implication, to the methods that constitute various methodologies. As it is the interchange between the concrete acts and tools of analysis (methods) and the overarching frames of interpretation (theory), a methodology makes up a heuristics, or a mode of inquiry. It is, then, at the level of methodologies that the distinctions of Table 4.1 apply.

The further relevance—the theoretical, explanatory value—of different methodologies depends on the particular purpose and area of inquiry (for arguments to that effect, see Lang and Lang, 1985, and Jensen and Rosengren, 1990). Very often in communication studies, it is evident that the methodological choices were made long before the issues and ends of inquiry had been posed, so that the methodologies become solutions in search of problems. One of the reasons that the use of qualitative methodologies in empirical communication studies is still relatively limited appears to be that these methodologies are not considered as an option, in part because students (and their professors) are still taught to think of survey and experimental designs as the standards of systematic science. Yet, the last few decades have produced systematic and professional conceptions of qualitative research (see the contributions in Jensen and Jankowski, 1991). Indeed, for purposes of theory development as well as for the social application of media studies, it is of the greatest importance that researchers assess the relevance of different methodologies with reference to the purposes and objects of analysis, asking *what* and *why* before asking *how*.

The role of methodology can be specified with reference to two of the other levels of analysis. First, no object of analysis is by nature quantitative or qualitative; it is thus framed by the medium or analytical apparatus employed. In preliminary terms, we could say that while the medium of quantitative analysis is numbers in a broad sense and their (numerical) correlations, the medium of qualitative analysis is human language expressing the concepts of everyday experience as they enter into a specific social context. The relevance of each of these media of research, to repeat, depends on the purpose of inquiry.

Second, the qualitative-quantitative distinction, in the traditional sense, loses its relevance at the level of theoretical frameworks. This is so despite the fact that qualitative and quantitative research traditions tend to rely on different types of theory. Theory is qualitative, insofar as it represents a configuration of interrelated concepts. At the theoretical level of analysis, statistics and geology are as qualitative enterprises as is art criticism. This is so even though much theory lends itself to formalization and numerical or graphic representation. Many, perhaps most, new insights depend on qualitative procedures that serve to relate the different levels of analysis, as witnessed also by examples from the natural sciences. In the postscript to the second edition of *The Structure of Scientific Revolutions,* Kuhn (1970: 182–184) refers to what he calls "symbolic generalizations"—the (qualitative) rearticulations of key concepts in a field which can open the field to new forms of empirical and mathematical analysis. In more abstract terms, various forms of qualitative analysis acquire general explanatory value, despite their "nonrepresentative" empirical samples because, as part of the analytical procedures, continuous cross-reference is made between the theoretical and other levels of analysis.

This last point is frequently missed in accounts of the foundations of communication studies, which tend to confuse analytical efficiency at the methodological level with explanatory value at the theoretical level, hence discounting qualitative analysis. One example is the handbook of Berger and Chaffee (1987), which proposes to set the standards for a comprehensive "communication science." While recognizing that "neither quantitative nor qualitative data have much meaning . . . in the absence of well-articulated theory," the authors nevertheless repeatedly imply, in their introductory sections and in their own chapters, that general (in the sense of predictive) theory grows from the quantitative measurement of the covariation of variables or operationally defined constructs, rather than from what they insist on calling "unspecified qualitative techniques" (Berger and Chaffee, 1987: 18). The one chapter in their handbook that considers qualitative approaches from the humanities, in a deferential tone, presents these contributions to the study of communication as 'nonscientific' (Farrell, 1987: 123).

Furthermore, Berger and Chaffee (1987: 144–145) neglect the fundamental theoretical problems that arise when "communication science" transforms the level of (verbal, visual, and other) *discourse* to the level of empirical, *numerical* analysis, stating simply that this "is not inherently problematic." What humanistic and other qualitative researchers have been demonstrating for some time now is that such a decontextualization of discursive meanings is one of the main obstacles to a better understanding of human communication. One may recall here the well-documented quantitative argument of Beniger (1988: 199) that, ironically, mainstream mass communication research, at least in the United States, may be the one field at the moment giving little attention to "theories of information, knowledge structures, communication, and the encoding and decoding of meaning." This state of affairs calls for more concretely exploratory theoretical as well as empirical work that would acknowledge the contributions of both qualitative and quantitative traditions.

To sum up, the qualitative-quantitative distinction should be taken to apply to methodologies—the structured sets of procedures and instruments by which empirical phenomena of mass communication are registered, documented, and interpreted. Different methodologies give rise to distinctive modes of understanding media and to specific applications of the findings in contexts of media production, education, and policy. Crucially, it is at the level of *theoretical frameworks* that the qualitative and quantitative analytical traditions might converge further, so as to increase both the explanatory value and the social relevance of the field as a whole.

Keywords of Qualitative Methodology

At the present stage of convergence, it may be useful to examine in a little more detail some specifically qualitative conceptions of methodology for the study of mass communication. My outline of certain central concepts below is meant to concretize the methodological level of qualitative research, which so far has often been left underdeveloped by the qualitative tradition. In addition to specifying the complementarity of qualitative and quantitative methodologies, the outline leads into a discussion of how the different methodologies might be conjoined within a theoretical framework of social semiotics. Qualitative methodology may contribute to a better understanding of the communicative process with particular reference to three concepts—discourse, subjectivity, and context. These concepts, while comparable to the standard social-scientific conceptions of the message, the communicators, and the embedding social structure, point toward a rearticulation of both the methodological and the theoretical levels of analysis.

The concept of *discourse*, first, is a legacy of the textual scholarship that, for more than 2,500 years, has been the foundation of most Western philosophy, theology, and other humanistic research. The underlying assumption here has been that language is the primary medium of interaction between humans and reality (in processes of perception, cognition, and action), and that consequently verbal texts may become vehicles of knowledge and truth. While originally this assumption applied to religious, scholarly, and literary texts, today much qualitative work employs the concept of discourse to refer to any use of language, or other semiotic systems, in a social context. Significantly, discourse now is said to include everyday interaction and its categories of consciousness, thus constituting the main medium of the social construction of reality (Berger and Luckmann, 1966). Through language, reality becomes social. Similarly, it is through language that reality becomes intersubjective and accessible for scientific analysis. In qualitative methodology, then, language and other semiotic systems represent both an analytical object and the central tool of analysis.

Subjectivity, equally, has come to be conceived in terms of language. In contrast to a philosophy of consciousness, in which subjects are understood as relatively autonomous agents that exercise moral and aesthetic judgment, recent theories of language and subjectivity have characterized the subject as a position *in* language (for a survey, see Coward and Ellis, 1977). Such a position, while negotiable, tends to carry a particular perspective on the world and on one's own identity and place in the world. In Althusser's (1971) terms, the subject is interpellated, or hailed, and as a result takes up particular positions. The mass media, clearly, are among the main sources of interpellation in the modern period. Moreover, the positioning of subjects in language implies their excommunication from certain other positions—the unconscious. Following Lacan's (1977) reformulation of Freud, it is this process of positioning that works to structure also the unconscious as a language. Thus, mass communication can be said to give voice to some discursive positions while silencing others, both of which lend themselves particularly well to qualitative inquiry.

Finally, humanistic communication theory has conceptualized the social structures in which mass communication is embedded as literally a *context*—a configuration of texts which must be "read" or interpreted, and which is the result of a process of historical change. This notion is in keeping with the classic understanding of history as, in one respect, a body of stories about the past. By shifting the analytical focus from specific stories as told by particular bards, to the deep structure or system of stories that dominates a given society or culture (Foucault, 1972), contemporary studies have shown how media and other agents of socialization serve to inscribe individuals in the culture. These stories give a sense of purpose to the social practices in which individuals, groups, and institutions engage, pervading everyday consciousness and action.

The keywords of qualitative methodology suggest a reformulation of dominant social-scientific approaches to mass communication, while simultaneously preserving and incorporating micro- and macrosocial perspectives on the communicative process. This is especially noteworthy at a time when both humanistic and social-scientific researchers have begun to reexamine classic issues of the reception and impact of mass media (Jensen and Rosengren, 1990). Whereas a convergence around the study of signs as used in social and cultural practices is perhaps particularly manifest in reception studies, it may be indicative of a larger turn of the field toward a theoretical framework of social semiotics.

Toward Social Semiotics

Mass communication is simultaneously a social and a discursive phenomenon. Signs, following the humanities, are a primary human mode of interacting with reality, establishing a continuous process of meaning production that serves to construct social reality as domains of political, economic, and cultural action. Further, both social-scientific and humanistic research traditions have looked beyond the aesthetic pleasures derived from signs and texts in private, thus hinting at a framework for studying the social uses of signs—a social semiotics which differs significantly from the semiology of Saussure and French structuralism (for the full argument and references, see Jensen, 1991a).

In contradistinction to the Saussurean dualism of signifier and signified, Charles Sanders Peirce developed a basic model with three elements: sign, object, and interpretant (for a collection of his works, see Peirce, 1958). A sign stands for an object, a phenomenon in the world, but only through reference to another sign in the mind of an interpreting subject, what is called the interpretant. The interpretant is neither identical with the interpreting subject nor an essence representing the content of that person's thoughts. Interpretation, hence, is seen as a continuous process, rather than a single act that internalizes external phenomena through a medium of signs. However, this does not imply acceptance of the solipsist reality of postmodernism, in which subjects are said to be caught in a web of signs, helplessly separated from social and material reality. In Peircean semiotics, signs are not *what* we know, but *how* we come to know what we can justify saying we know. Interpretants, in sum, are signs by which people orient themselves toward and interact with a reality of diverse objects, events, and discourses. The three-element model of sign use within social semiotics proposes to relate "the analysis of linguistic meaning to the idea of participants in communication coming to an understanding about something in the world" (Habermas, 1984: 397).

Peirce, moreover, suggested a concept of *difference* that implies a focus on the social uses of signs, not signs or texts in themselves. Though the sign remains the central explanatory concept in Peircean semiotics, meaning is here defined in relational rather than essential terms. The meaning of signs, accordingly, is determined not by any immanent features, but by their position—their relations of difference—within the total system of meaning production. While Saussurean semiology advances a similar argument (Culler, 1975: 11; see also Saussure, 1959), the emphasis of semiological studies in fact has been placed on the relations of difference *within* the language system, and less on the social uses of language and other signs. In summary, Peircean semiotics provides a framework for studying meaning production in its social context. When the discursive differences of media content are interpreted and enacted by social agents, hence serving to orient their cognition and action, the discourses of mass media may be said, in the terminology of pragmatism, to make a social difference. Meaning is a discursive difference that makes a social difference (Bateson, 1972: 242; Goodman, 1976: 227).

Certain forms of communication and interpretation make a special social difference and thus are particularly relevant for the understanding of society and culture. Whereas Peirce did not devote much attention to cultural practices, he characterized the scientific community as an institution conducting a form of interpretation that has important social consequences. Scientists, in one respect, are communities of knowers who produce a definition and legitimation of knowledge by some public, collective procedure, sometimes with major implications for social history (Kuhn, 1970; Lowe, 1982). Science, in a social perspective, represents an institution-to-think-with, by analogy to the anthropological, Levi-Straussian concept of objects-to-think-with (Schudson, 1987: 56).

The mass media constitute another, equally important institution-to-think-with. Science and mass communication, in different respects, serve to place reality on an agenda of public discussion; both institutions operate through social practices that presuppose a high degree of consensus concerning the relevant interpretive procedures. While their specific institutional hierarchies differ, certainly, both are important forces, and are increasingly so, in maintaining the political, cultural, as well as material structures of society (Galbraith, 1967). At the same time, of course, the interpretive communities of mass communication—the demographically but also culturally differentiated audience groups—are more diverse, complex, and, most important, inclusive than those of science. In principle at least, mass communication works as a cultural forum (Newcomb and Hirsch, 1984), which includes everybody, and which—again in principle—may take up any issue of power or social structure. Because they may, but frequently do not, fulfill this proclaimed mission, mass media institutions and their discursive representation of reality have become central sites of social conflict.

The concept of interpretive repertoires, or interpretive communities, has been introduced into empirical research by recent literary, cultural, and communication studies (Fish, 1979; Jensen, 1987; 1991b; Lindlof, 1988; Radway, 1984).

The interpretive perspective, by incorporating both social and discursive aspects of reception, may help to reestablish the link between social-scientific and discourse-analytical reception studies. (I prefer the concept of interpretive *repertoires,* because it indicates that audiences are not in fact formal groups, or communities, but dispersed and contextually defined agents who may employ a variety of repertoires to make sense of media. See the argument in Potter and Wetherell, 1987: 138–157.) The assumption behind the concept of interpretive repertoires is that media audience groups must be defined not simply by their formal social roles and demographic characteristics, but as significantly by the interpretive frames or repertoires by which they approach mass media content as well as other cultural forms.

This interpretive perspective begins to refocus research interest on the theoretical relationship between macrosocial structures, such as social classes and cultural institutions, and microsocial processes of, for example, media reception (see also the argument of Giddens, 1984, and the discussions thereof in Held and Thompson, 1989). In processes of mass communication, microsocial and discursive acts of interpretation enact what represents, at the macrosocial level, cultural practices. These cultural practices, similarly, serve to shape, and are shaped by, the different genres of mass communication. Genres, to reiterate, should be conceived as modes of address that imply specific social uses of communication in the context of particular political and cultural practices. By pointing to methodological interfaces between social-scientific and humanistic forms of inquiry, the categories of reception, genre, and interpretive repertoires are likely to prove especially constructive for the further development and convergence of communication studies.

Qualitative empirical reception studies, as developed in international research since the early 1980s (see, for example, Ang, 1991; Jensen, 1986; Liebes and Katz, 1990; Lindlof, 1987; Lull, 1988; 1991; Morley, 1980; 1986; Radway, 1984), may have a special contribution to make to social semiotics through in-depth analyses of the audience uses of mass media in specific cultural and historical contexts. Reception studies conduct a comparative audience-cum-content analysis of the discourses produced respectively by media and audiences in context, thus illuminating processes of the social production of meaning. After a period of focusing on theory development and basic research, some recent studies have begun to employ reception methodologies for examining specific issues and problems in the form of applied studies (see, for example, Corner, Richardson, and Fenton, 1990; Jensen, 1991b; Schlesinger, Dobash, Dobash, and Weaver, 1992). Reception studies, in sum, provide much evidence of a convergence that is still in progress.

In conclusion, social semiotics may contribute to an interdisciplinary reconceptualization of the theoretical framework of mass communication research. Discourse is approached as genres with specific uses in social and cultural practice; subjectivity is defined in collective rather than individual terms, as the expression of socially situated interpretive repertoires; and context is conceived of as a particular historical setting in which media institutions-to-think-with serve their various purposes. Within this *theoretical* framework, both qualitative and quantitative *methodologies* have important contributions to make. But further convergence will require much interdisciplinary research on the details of such a theoretical framework as well as more metatheoretical analyses of the complementary explanatory values of the two methodological traditions. If convergence has already happened in theoretical debates about the reception and audience uses of mass media, it is only now beginning to make its impact on the empirical practice of both audience studies and other mass communication research.

Notes

1. This chapter, originally presented to the 1990 Conference in Champaign-Urbana, Illinois, "Toward a Comprehensive Theory of the Audience," first appeared in C. von Feilitzen and K. Nowak (eds.), *Reception Reconsidered* (reprinted by permission). (The chapter draws on and summarizes certain of the principles and ideas that inform *A Handbook of Qualitative Methodologies for Mass Communication Research* (Jensen and Jankowski, 1991). For an elaboration of the history, systematics, and social uses of qualitative research, and for extensive references, the reader is referred to the relevant sections of the *Handbook.*)

References

Althusser, L. (1971). *Lenin and Philosophy.* London: New Left Books.

Anderson, J., and T. Meyer (1988). *Mediated Communication: A Social Action Perspective.* Newbury Park, Calif.: Sage.

Ang, I. (1991). *Desperately Seeking the Audience.* London: Routledge.

Bateson, G. (1972). *Steps to an Ecology of Mind.* London: Paladin.

Beniger, J. (1988). Information and communication: The new convergence. *Communication Research* 15(2): 198–218.

Berger, C., and S. Chaffee (eds.) (1987). *Handbook of Communication Science.* Newbury Park, Calif.: Sage.

Berger, P., and T. Luckmann (1966). *The Social Construction of Reality.* London: Allen Lane.

Corner, J., K. Richardson, and N. Fenton (1990). *Nuclear Reactions.* London: John Libbey.

Coward, R., and J. Ellis (1977). *Language and Materialism.* London: Routledge & Kegan Paul.

Culler, J. (1975). *Structuralist Poetics.* London: Routledge & Kegan Paul.

Farrell, T. (1987). Beyond science: Humanities contributions to communication theory. In C. Berger and S. Chaffee (eds.), *Handbook of Communication Science.* Newbury Park, Calif.: Sage.

Fish, S. (1979). *Is There a Text in This Class? The Authority of Interpretive Communities.* Cambridge, Mass.: Harvard University Press.

Foucault, M. (1972). *The Archaeology of Knowledge.* London: Tavistock.

Galbraith, J. (1967). *The New Industrial State.* London: Penguin.

Giddens, A. (1984). *The Constitution of Society.* Berkeley: University of California Press.

Goodman, N. (1976). *Languages of Art.* 2nd ed. Indianapolis, Ind.: Hackett.

Habermas, J. (1984). *The Theory of Communicative Action.* Vol. 1. Boston: Beacon Press.

Held, D., and J. Thompson (eds.) (1989). *Social Theory of Modern Societies: Anthony Giddens and His Critics.* Cambridge: Cambridge University Press.

Jensen, K. B. (1986). *Making Sense of the News.* Aarhus, Denmark: Aarhus University Press.

———. (1987). Qualitative audience research: Toward an integrative approach to reception. *Critical Studies in Mass Communication* 4(1): 21–36.

———. (1991a). When is meaning? Communication theory, pragmatism, and mass media reception. In J. Anderson (ed.), *Communication Yearbook,* vol. 14. Newbury Park, Calif.: Sage.

———. (1991b). *News of the World: The Reception and Social Uses of Television News Around the World.* Paris: UNESCO.

Jensen, K. B., and N. Jankowski (eds.) (1991). *A Handbook of Qualitative Methodologies for Mass Communication Research.* London: Routledge.

Jensen, K. B., and K. E. Rosengren (1990). Five traditions in search of the audience. *European Journal of Communication* 5(2–3): 207–238.

Kuhn, T. (1970). *The Structure of Scientific Revolutions.* Rev. ed.. Chicago: University of Chicago Press.

Lacan, J. (1977). *The Four Fundamental Concepts of Psychoanalysis.* Harmondsworth: Penguin.

Lang, K., and G. Lang (1985). Method as master, or mastery over method. In M. Gurevitch and M. Levy (eds.), *Mass Communication Review Yearbook,* vol. 5. Beverly Hills, Calif.: Sage.

Liebes, T., and E. Katz (1990). *The Export of Meaning.* Oxford: Oxford University Press.

Lindlof, T. (1988). Media audiences as interpretive communities. In J. Anderson (ed.), *Communication Yearbook,* vol. 11. Newbury Park, Calif.: Sage.

———. (ed.) (1987). *Natural Audiences.* Norwood, N.J.: Ablex.

Lowe, D. (1982). *History of Bourgeois Perception.* Chicago: University of Chicago Press.

Lull, J. (ed.) (1988). *World Families Watch Television.* Newbury Park, Calif.: Sage.

Lull, J. (1991). *China Turned On.* London: Routledge.

Morley, D. (1980). *The "Nationwide" Audience.* London: British Film Institute.

———. (1986). *Family Television.* London: Comedia.

Newcomb, H., and P. Hirsch (1984). Television as a cultural forum: Implications for research. In W. Rowland and B. Watkins (eds.), *Interpreting Television.* Beverly Hills, Calif.: Sage.

Peirce, C. S. (1958). *Selected Writings.* Garden City, N.J.: Doubleday.

Potter, J., and M. Wetherell (1987). *Discourse and Social Psychology.* London: Sage.

Radway, J. (1984). *Reading the Romance.* Chapel Hill, N.C.: University of North Carolina Press.

Saussure, F. de (1959). *Course in General Linguistics.* London: Peter Owen.

Schlesinger, P., R. E. Dobash, R. P. Dobash, and C. Weaver (1992). *Women Viewing Violence.* London: British Film Institute.

Schudson, M. (1987). The new validation of popular culture: Sense and sentimentality in academia. *Critical Studies in Mass Communication* 4(1): 51–68.

The Pragmatics of Audience in Research and Theory

5

James A. Anderson

The analysis presented in this chapter is based on the assumption that the concept of audience is indeed a discursive subject; too large to be apprehended directly in experience, it is a construction of our research and theorizing. Different research methods and different manners of theorizing produce different understandings of the term. An audience then is not a fact but a set of pragmatics invoked in our perspective. When the "audience" appears in our writings, it is as a working element of a given perspective. The work that it does, however, is usually submerged, making it appear as if writer and reader held to the same concept. What follows here is a partial attempt to distinguish and elaborate the term audience and then to consider the practical implications of these elaborations for our research programs. The effort begins first with a look at the classical heritage of the term derived from the audiences for public speaking and theater in the preindustrial eras of communication, and then at the constitution of the individual audience member (auditor) through which audiences are built up. It then moves to distinguish between two general classes of audiences that appear in contemporary writing, one of which I have designated as formal, the other as empirical.

The Heritage

The concept of audience is burdened with a classic heritage of individuals coming together both physically and socially to create the motive and the site for public presentations. According to at least one historic understanding, classic audiences were not aggregates of isolates but were interacting, interconnected social memberships. In the classic audience, the achievements of interpretation were located in the membership, which had the immediate and practical means of this accomplishment. Further, there was an established social connection between the members of the production enterprise and the members of the audience. The playwright was directly addressable by audience members, and audience members could be palpably identified by the playwright. Audiences for the products of industrial communication rarely meet the description of the classic audience (though they sometimes may, as discussed later). Nonetheless, we typically use this idea of the audience—or parts of it—as the default conceptualization when the term is invoked.

The result is some interesting prior assumptions that color discussions across a broad spectrum of inquiry, such as: (a) that there exists a communication contract between producers and auditors that renders the auditors responsible for the purposes and intended meanings of the producers and producers responsive to the interpretations and uses of their content by the auditors; (b) that audience members achieve some common understanding of the content presented; (c) that actions that follow exposure to content can be understood as emanating from those common understandings; and (d) that whatever feedback is received by producers can represent the audience in communication relevant ways. Assumptions such as these make some sense when the personal relationships between producers and auditors and within sets of auditors can be traced, but they are far less tenable when auditors have no contact with producers and share among themselves only some partial communality of exposure.

Consider the first of these, the assumption of a communication contract. Communication has been defined as the mutual accomplishment of meaning (Anderson and Meyer, 1988) and distinguished from interpretation and other semiotic activities such as appropriation, excorporation, poaching, and the like. As a communicant with you (as opposed to a producer of discursive material), you and I are bound into a contract that makes us mutually accountable for the communication performance. In conversation, my turn must make sense in relation to your turn. Even in the circumstance of the public presentation of a paper, such as the original presentation of this chapter, should a member of the audience fall publicly asleep or ask some irresponsible question, he or she would be censorable. As communicants, we are bound together in immediate ways. No such relation exists between producer and auditor in the industrial production of symbolic products. Producer and auditor are bound together in an economic relation that ordinarily precludes any opportunity for the mutual accomplishment of meaning. The purposes and intentions of the producer need mean nothing for the uses and interpretations of the auditor, and certainly auditor uses and interpretations are rarely supervised (though they may be observed and commented upon) by producers. The somewhat blasphemous question can be raised as to whether "communication" is the proper rubric for understanding audience practices. Perhaps interpretation or accommodation as the organizing concept would provide more fruitful analyses.

Recasting the vision of the mediated communication contract to one in which the social action determining the production of texts is mostly independent of the social action of their interpretation shifts the governance of interpretation from text to auditor practice. The claim for common meaning located in the second of our list of assumptions, therefore, becomes a claim of common practices of interpretation rather than a simple recognition of the delivery of common content. The question of common meaning ceases to be one answerable through methods of textual analysis and becomes one demanding methods more appropriate to practical action.

The move that privileges interpretive practices over delivery and exposure also calls into question both the claim of communality in the motives for subsequent audience behavior and the correspondence between producer interpretations of audiences and audience interpretations. The problematic status of both of these assumptions is heightened according to the increasing diversity of the interpretive practices postulated. The more freedom granted to audiences, the less understanding analyst and producer have as to why audiences behave as they do. That this disempowerment is not taken lightly is not surprising and perhaps explains the ferocity of some opposition to such theories.

The Auditor

Our constructions of audience are, quite naturally, inextricably bound up with our constructions of the individual. For our purposes here, these latter constructions divide nicely on the views of the individual auditor as site and the auditor as agent. In the Enlightenment's invention, the individual is the autonomous agent—the good person thinking well. This notion of free agency has several prior assumptions. It assumes (a) an accessible reality that exists independently of the powers that apprehend it; (b) that authentic choices are presented in that reality; and (c) that individuals select from those alternatives through processes of reasoning and choice under local, personal control. Clearly the particular individual is fully accountable and, I might add, needed for prediction. By the late nineteenth century, the twin forces of psychological science and material philosophies such as Marxism had mounted a concerted attack on Enlightenment rationality and had converted the individual to a site for the expression of internal and external forces. There are currently at least three strains of this Zeitgeist.

First, there are the cognitivists, who emphasize internal structures and can be seen as viewing the individual as an archive site—a repository of the processes of socialization in which cognitive structures such as values, beliefs, attitudes, scripts, schemata, and the like are stored. The analyst need not know the person because it is the value structure or the schema or whatever that is being played out. The individual is simply the particular place of its expression.

Second are the sociologists and culturalists, who emphasize external societal and cultural forces, seeing the individual as the site of the acts of society in which individual power is submerged, often overwhelmed and in radical formulations rendered useless in the face of cultural forces. The individual represents a particular point of intersection of these forces.

Neither group entirely cancels its subscription to agency, however. Cognitivists allow modification of cognitive structures once formed and changes in the initializing social processes through social reform. Sociologists/culturalists (both utopian and nonutopian) allow for emancipatory impulses, reform, and revolution. In these ways, agency and the reason for performing these analyses is maintained.

The third view is the structuralist perspective. Structuralists see the individual as an address where the outcomes of organism, community, and language are materialized in the fundamental structures of human action. It is these structures that are the object of scholarly interest. Radical structuralists have proclaimed the disappearance of the individual—"the death of man" as Foucault has put it. Radical structuralists reserve little sense of agency.

In our current phase of poststructuralism, actional theorists are recouping the idea of agency, but the agency is now socially directed rather than autonomous. Social action theorists (of whom I am one) see the individual as a socially bounded, knowledgeable agent who is a local and partial representative of societal memberships and whose actions are improvisations on cultural themes. The notion of agent used here recognizes the two senses of the term: the acting entity and the representative. As the acting agent, the individual recovers choice or "the ability to do otherwise." The local outcome of any improvisational performance, therefore, depends on the individual actor. As representative agent, however, the individual is immutably referenced to some "other" and loses the autonomy and independence characteristic of free agency. Actors and actions are both understood as situated in meaningful systems of memberships and behavior. The result is that the individual can neither be proclaimed dead nor decontextualized in global explanations. Explanation itself will always be local and partial.

In any extended discussion of audience, we will of course find all three conceptualizations of the individual auditor along with a wondrous multitude of variations. The particular assumptions concerning the nature of auditors move the analyses of audiences in particular directions. To briefly give three examples: The analyses of audiences "built" from free agency auditors will emphasize content (which is independently meaningful) and its "right" interpretation. Those of the "intersectionists" will partition audiences along race, gender, class, and similar lines, testing how these characteristics create different audiences for content that is viewed as both factually and semiotically the same. Social action theorists would find an audience (as opposed to a researcher-defined aggregation) arising only in some coherent social action. For them, there is not a singular "audience" of all those who watch a particular television program, there are many audiences involved in recognizable performances of being an audience of that program, and the program itself carries factually but not semiotically the same content for these audiences.

As different conceptualizations of the auditor give rise to different audiences, so too do different forms of analysis. The next several sections look at the audiences that arise in different sorts of formal and empirical analyses.

Formal Audiences

Formal audiences are those constructed in the discourse and practices of practitioners in the art and industry of content production, discourse analysis, and the like. Formal audiences are the creation of these practitioners as both a requirement for and an achievement of their occupational accomplishments. Of these, I would point out two: The first is the audience written to or encoded in the mediated product. The second is the foundation for or the prior assumption of any critical analysis.

The Encoded Audience

Eagleton has noted: "A writer may not have in mind a particular kind of reader at all, he may be superbly indifferent who reads his work, but a certain kind of reader is already included within the very act of writing itself, as an internal structure of the text" (1983: 84). I would call this formal audience the *encoded audience*. It is the audience that the community "knows in its bones" according to Gitlin (1983: 218). One finds this audience in the discourse that surrounds the necessary choices of production and their outcomes. This audience is fickle, gets bored, likes (or does not like) sex and violence, wants happy talk news, and all the rest. This audience is a construct of conflicting aphorisms, and its behavior is always understood on the terms of the industry. It is this star, that program, this schedule which explains audience behavior. The contradictions that are clearly apparent in this audience are a characteristic which in part helps practitioners to handle the uncertainty of a marketplace where worldwide and nationally nearly 80 percent of the new products of popular culture fail to meet standards of economic success (Anderson and Meyer, 1988; Fiske, 1987; Gitlin, 1985).

The behavioral patterns of an encoded audience are held in industry and organizational stories that are embellished by the personages of the working group. The repertoire of stories is diverse and sufficiently contradictory that it serves as an explanatory resource (Brown, 1990) for multiple positions on the same issue. Data from ratings, circulation studies, letters, personal contacts, close relatives, and the like are made sense of within these stories and within the standard practices of the industry and organizations for processing such information (Anderson and Meyer, 1988; Burgoon, Burgoon, Buller, and Atkin 1987; Turow, 1984). Representations of the audience are constituted from these explanatory resources for the internal and external political ends of the media organization.

In the industrial products of popular culture, then, multiple conceptions of the audience that come into play (Tunstall, 1991). Individual decisions affecting content are grounded in different visions of the audience to be. Writers, directors, actors, editors each participate in the final product with these different visions. This multitude of encoded audiences has been used as an explanation for the (argued) polysemic nature of popular culture products (Fiske, 1987).

Unless we are privy to hall and office talk, the encoded audiences of popular culture products appear to us primarily in promotional work as well as in the politicized statements about the audience that media people make to the press, watchdog committees, regulatory agencies, or buyers. There is at least some amusement in the clear contrast between the "intelligent audience" that appears in the discourse before legislative and social activist committees and the "seducible audience" that appears in the sales pitch to advertisers and affiliates. In fact, the industry—as might most of us working in audience research—believes in both.

The Analytic Audience

The second formal audience I would identify is the *analytic audience*. The characteristics of a particular analytic audience are established in the requirements of a critical or scientific claim. When Christopher Anderson writes, "in a medium that changes perpetually—even when the television set is switched off—nothing on television is precisely as we [the audience] imagine, remember, or hope. Even series television defined by repetition, forever plays havoc with our expectations" (1987: 114), he invokes an audience that behaves as it ought for his claim. Who are the members of this addled "we" who feel their dislocation at each power surge to the set? They are the clones of Anderson's argument.

The audience symposium headed by Martin Allor (1988) in *Critical Studies in Mass Communication* demonstrates the different sets of terms for audiences that are expected by the separate critical perspectives of political economy, post-structuralist film theory, feminist criticism, cultural studies, and postmodernism. Each of these critical approaches requires a somewhat different audience to propel its arguments. For example, feminist criticism requires an audience living the terms of a patriarchal society; cultural studies, an audience which re-presents the semes of narrative; and postmodernism a fractured and disconnected audience. The practitioners of all forms of criticism present their analytic audience to an encoded audience, a circumstance that complicates Hartley's (1988) claim that the critic is writing to improve, audience behavior. It is most often not the encoded audience one wishes to improve, for those members are typically our own, but the analytic audience—an audience safely removed from the writing.

It is not only critics who produce analytic audiences. Everett and Everett (1989) create an archetypal audience of newspaper readers when they consider "the benefits of competition" for such readers. If one is to argue benefits, one must presume readers all seeking certain ends. Finally, analytical audiences are created in survey and experimental studies where the requirements of a data collection protocol are founded on some set of assumptions of audience behavior. For example, the evidence in Helregel and Weaver (1989) that pregnant women select situation comedies in the first trimester and action drama in the last stages of pregnancy is based on the assumption that program choices in the home can be modeled by selections from a list on a questionnaire. This belief is an analytical creation of argument must occur prior to any empirical evidence.

Empirical Audiences

The remaining audiences fall into what I call the empirical set. This nomination is not to claim that such audiences are necessarily "actual audiences" or even populated by actual people. Empirical audiences are no less a discursive subject than formal audiences and always require some prior definitional work before they will appear, as the analysis of Helregel and Weaver above demonstrates. That definitional work establishes the boundary of the set. The question that can be reasonably asked is whether the set is empirically empty. Analysts who work inside the empirical rubric accept (though do not always fulfill) the responsibility of providing evidence for the answer to that question.

The discourse on empirical audiences divides over the issue of transcendence (which can also be read as generalizability or a historicity). Audiences constituted in the discourse of the social sciences are characteristically transcendent or somehow independent of the particular circumstances by which they become audiences. As a result the constituting discourse does not locate the audience in time and place. As we have seen, this discourse talks about cognitive or structural predispositions that are applicable to all humanity. Or it may talk about "children," for example, and the lessons learned about children in the United Kingdom are loosely presumed to apply to children elsewhere.

Transcendent audiences (generally conceptualizing auditors as independent actors or sites) are contrasted with situated audiences (auditors as social agents). The discourse constituting situated audiences takes direct account of the circumstances of being an auditor and accepts the responsibility (more or less) not to promote claims that transcend those circumstances. Audiences therefore appear in time, place, and situation and are uniquely marked by those conditions. This is not to say that there are no common requirements for being an auditor (e.g., the necessity of interpretation); it is to say that the performance of those requirements will be locally controlled.

In the sections that follow, I will consider more closely the discursive terms of transcendent and situated audiences and a few variations of each.

Transcendent Empirical Audiences

Traditional social science works to support categorical rather than situational claims. Categorical claims are presumed to be true for all elements that fall within the scope of the claim. Consequently discussion of an "adult audience" is assumed to be effectively true whenever an adult audience arises, with noted exceptions excluded.

The concept of the transcendent audience, then, depends on the principle of equivalent units. Within the scope of the claim, any adult or child or 18 to 49-year-old working woman or whatever is as good as any other and is functionally equivalent to any other. Adding one more increases the audience by just that much. In most theories and industry practices that depend on transcendence, each equivalent unit is complete, autonomous, and similar in its influences and determinants (e.g., Brabrow, 1988; Weaver and Newton, 1988). Other than their common categorical identification, there is no necessary connection among audience members—no "working" of the program material among members.

The transcendent audience is nearly always defined by discrete episodes of exposure (with their characteristics of place—where exposure happens, presence—who is there, frequency, and duration) to content that has a specifiable location (this newspaper article, that compact disc) whose meaning is fixed in that content, the interpretation of which is dependent upon the skills of literacy and an objective reading. The model in place in its simplest form is the exposure of an autonomous, more or less skillful, more or less objective auditor to a meaning delivery system of content. Some form of this model, of course, is necessary if the analyst is going to make claims based on content characteristics about why people are there and about predictable outcomes from exposure.

There are many variations on this model that emphasize the different components. Content emphases are found in so-called forms and conventions analyses, skills emphases are found in literacy programs, and the character of the auditor is found in uses and gratifications and lifestyle approaches. These variations could have allowed us to create several more categories of discursively constituted audiences—a temptation I resisted here, primarily because the basic framework remains quite visible in all the variations. Each of these variations, however, makes use of one of two approaches in which the audience is made visible: One is the method of aggregation by which audiences are "collected" by post hoc identification; the other is surrogation, a metonymic method by which selected individuals stand for all. Let us explore these audiences a bit.

The Aggregate Audience

In practice, most displays of an aggregate audience occur in the definitions and procedures of some sampling protocol. A. C. Nielsen's category "Working Women 18–49" is such an audience. As an audience, it exists in Nielsen's definitions of working women 18–49 years old. If respondents can be found who meet the requirements of the sampling protocol, then there are real people in that audience, although the audience itself is arbitrary. That is, it is an audience created by the set definitions and not necessarily by any theoretical understanding of what it means to be an audience or by any self-identification of the individuals selected. Some enlargement of these two points is instructive: The categories used in the most common audience studies have nothing to do with differentiated practices of audience members. They are in place because of the buying and selling practices of the industry. A woman 18–49 watching the news is equally a "member of that audience" regardless of her own nomination if she is there because someone else insisted on having the news on, if she watches intently, if she is more intent on reading the newspaper, and so on. And all the women in that category present all the possible reasons for being there.

It is these latter distinctions that create the fracture between the aggregated evidence and the claim of a functioning audience for that aggregation. To continue with this example, Nielsen's "Working Women, 18–49" is often used as a descriptor *for the reasons for attendance.* The facts of the set—that the elements are women, working outside the home, and in the age bracket 18–49—are used as the explanatory agency of attendance. The surrounding argument implies that there is some unified core (hence, the functional equivalence of the elements) of that category that is attracted by the content and explains why elements of the category are found in greater number when that content is present. The content produces the assembly of individuals. The effect is a *congregation* and not simply *a difference in distribution rates.* In the protocols of ratings and circulation studies, these two circumstances—whether a congregation or merely an aggregation—are actually indistinguishable. But the evidence is treated in industry production practices as if it were the former rather than the latter. In economic practices, it does not matter: If your property looks like a sand pit, sell sand. In short, the differences in distribution rates are used to sell the commodity audience to different buyers. (The commodity audience is an aggregated audience further activated by the economic practices of the industry.)

The problem of the difference between a congregation and differential distribution rates is engaged but typically submerged in every audience survey. In nearly every case, the question of who is there is accelerated into an answer of why they are there. They are there because they are black, male, higher educated, or whatever the defining terms of the categorical set (e.g., Brown and Schulze, 1990). Differences in distribution rates become audiences in our discourse about the content motives for these differences. Soap operas attract women, it is said as an example, because they are romantic stories about relationships, not because of the time of day they are on, the alternative programming available, or the cultural theme that women are supposed to watch soap operas and men are not. The larger aggregation of women, consequently, becomes an audience of women approaching the classical meaning with common motives and practices of interpretation. Ratings, of course, provide not a single shred of evidence for such a claim.

The inability of the method of aggregation to demonstrate the inner workings that would give credence to the claim of an audience also calls into question two other arguments typically advanced with this evidence: the arguments concerning audience trends in cross-sectional analysis and those concerning individual trends in panel analyses. Most straightforwardly, there is no basis for comparing an aggregated audience from, say, 1977 with a similarly formed audience in 1987 because even if the same sampling procedures are used, nothing else has remained the same. Distribution rates within the sample will either be different or the same, but the protocol provides no explanation for either circumstance. Our discussions of audience trends are opportunistic constructions that exploit whatever the findings.

The problem in panel studies is a bit less egregious because at least the material individual has remained the same. In understanding audiences, however, that difference is small comfort. (A useful discussion of these issues is found in a symposium of three articles in *Communication Yearbook,* volume 14 [1991]. The separate authors are Rosengren, von Feilitzen, and Murray.)

The facts generated by either procedure can, of course, be very useful—great sums of money are made in the selling of commodity audiences and some of us find solace in the precipitous decline of televiewing during the teen years. The evidence generated by the method of aggregation generally tells us what people are nominally exposed to what content, but the purposes and practices remain unrevealed. Situated audiences will talk of practices, but the most secure claim to purposes is in the next category of audiences, to which I now turn.

Surrogate Audiences

Surrogate audiences arise primarily in the discourse surrounding experimental studies when the focus of such studies is some purported universal human trait. Universality has been implicitly or explicitly claimed for (among many others) physiological (Lang, 1990), sexual (physiology rather than gender: Meadowcroft and Zillmann, 1987; as analogue to gender: Krendl, Brohier, and Fleetwood, 1989), cognitive (Abelman, 1989), psychoanalytical (Tamborini and Stiff, 1987), and even interpretive (Hoffner, Cantor, and Thorson, 1989) characteristics or practices. Because the trait under study is assumed to be held by all humans, it does not matter which humans get examined as long as "proper" experimental controls are put in place. The respondents in the study are the surrogates for us all.

Aside from the technical considerations of controls, the primary weakness of these claims is the evidence for universality. Universality is demonstrated by the replication of the study in different cultures, conditions, and so on. Without exception, the reported studies were single incident studies with no replication. Critics have become quite insistent that the protocol itself rather than a universal trait is the more likely explanation for any patterns or consistencies found within the respondent group.

While current evidence is poorly formed, there is also little doubt that, *on some level,* universals come into play in the practices of audiences. The more significant argument is centered on the level at which universals come into play. For some, the universals are completely accounted for when the content is produced within the boundaries of human sensory reception. For others, particularly structuralists, universals invade language and its narratives with common solutions and forms. For hardline objectivists and traditional critics alike, universal meaning can be expected. In this postmodern world, we have discourse from each of these positions and all the shadings in between.

Livening the discourse even more, we find universals coming in different orders. First-order universals hold for all; second-order universals hold for this culture or that language community; and so on until we reach the universal truths of audience symposium papers. Even though its universals can be bounded in scope, the surrogate audience remains transcendent as no element of local control is considered significant inside the scope of the claim. The next category of audience differs in exactly that regard, having founded its theory on the premise of local control.

The Situated Audience

The appearance of ethnographic methods in the study of media has given rise to descriptions of historic individuals performing the actions of media attendance (Alexander, Ryan, and Munoz, 1984; Anderson and Meyer, 1988; Lindlof, 1987; Lull, 1988; Morley, 1980, 1986; Traudt, Anderson, and Meyer, 1987; Wolf, Meyer, and White, 1982) situated in the social action of everyday life. Most (but certainly not all) of these studies use their evidence of actual practices to move to a set of generalized "rules" of situated media use. Consequently, audiences become distinguished by the social semiotic of the site of performance. Thus far, family television viewing has been the focus of most of the work.

The situated audience differs from the aggregated audience in that its members are identified as embedded in an ongoing social action that must be accounted for, thereby eroding the autonomy of the act of attendance and the functional equivalence of audience members. The behavior of any audience member becomes particularized by the actual circumstances of attendance.

The arguments here are often conflicted, however, as claim frequently widens to nomothetic vistas, so that evidence for, say, particular subordinated viewers in family settings (this child in this family) becomes evidence for all such members in all families. Such arguments are exactly another way of expressing categorical aggregates, the difference being solely in the method of collecting information about them. Nonetheless, those methods do tell us considerably more about the conditions of exposure to content and provide the basis for possible analysis of interpretive practices.

It is in this latter issue—the assumption of interpretive practices—that arguments from the situated audience differ most clearly from those using transcendent audience concepts. In its most sharply differentiated form, the concept of interpretive practices argues that content potentiates but does not activate meaning. The activation of meaning occurs in practices of interpretation that are performed by the individual but in response to the demands of the ongoing social action. The autonomous individual is denied and a collectively located individual invoked. Content is no longer a meaning delivery system that evokes meaning as a cognitive reflex, rather, content is a resource for the production of texts in the audience (Anderson and Meyer, 1988; de Certeau, 1984; Fiske, 1987).

Interpretation, then, for many actional theorists, is improvisational sense-making by a knowledgeable agent using communal tools of understanding. It is a socially embedded performance responsive to the demands of the social action in progress. It achieves a "made-meaningful" text of the content delivered. Its accomplishment is the local application of content resources in the service of some social action routine.

The perspective of interpretation clearly changes the traditional equation. The facts of content and audience are no longer secure. They can be poached in interpretation to accomplish different work (e.g., Livingstone, 1990). Consequently, we can no longer solve the equation by the careful study of content characteristics or of audience composition.

The granting of interpretive rights to the audience empowers the audience and moves to disempower content (for a study critical of this claim, see Budd, Entman, and Steinman, 1990). It also breaks the analyst's control of the audience that arose out of his or her analysis of content properties as predictors of audience behavior. The analyst can no longer claim certain audience outcomes because of the properties of content, as interpretive practices now intervene. Analysts have shown themselves reluctant to give up this power. Many media theorists grant interpretative rights to the audience while making claims as to what the interpretations ought to (or, in some cases, must) be.

Interpretive theorists engage the tension between content and audience power first by arguing that meanings are owned by the culture. They are not individual productions. An individual can reproduce, oppose, mock, metamorphize, and otherwise manipulate these meanings but is always condemned to their use. The struggle between content and audience is also engaged by posing audience texts (audience texts are the product of interpretation) that are collectively or separately (a) "overdetermined" by content, (b) the result of collective strategies of interpretation, and/or (c) tactically produced in social action.

Overdetermined texts are interpretations that are ideologically "naturalized," the default or re-presentational interpretation. Texts that affirm the dominant ideology are considered overdetermined. (The production of texts that repulse the overdetermined impulse are considered acts of resistance—a sort of semiotic guerrilla warfare.) The level of analysis engaged here is the most conventional and the least dependent on evidence of actual interpretive practice (Evans, 1990).

The concept of collective strategies of interpretation recognizes the multi-ideological nature of the postmodern world. Individuals typically have access to and must enter multiple domains of discourse (e.g., the classroom of a media literacy course and the family watching television). Such domains or "ways of speaking" are organized around strategic interpretations. Strategic interpretations produce the overdetermined texts of different memberships (discussed later).

Tactical texts are those put into action to accomplish some work. The evidence for tactical interpretations arises from the actual practices of situated individuals and is the least conventional. Tactical texts are those needed to buy this product now, to make a comment in an ongoing conversation, and the like. Tactical texts invoke, re-present, and play with overdetermined and strategic interpretations as an accomplishment in the current social action. Most analysts proceed on only one level and their analyses are often inflected with other theoretical intentions (which is why a lot of theoretical works simply talk past one another). Those who work at the full cultural level detailing ideologically overdetermined texts are the closest to critics using formal analysis and farthest from actual audiences (Altheide and Snow, 1988). Analysts at the strategic level, in my reading, have been most concerned with the discourse of organizations (Danowski, 1991; Goodall, 1990), although the ethnographers of our own societies are becoming more apparent (Conquergood, 1990). Those who work at the tactical level are the most taken by immediate variations granting great power to auditors in action and are the least sensitive to cultural determinations (Anderson and Meyer, 1988).

Situated audience models require a dependent self emerged in some collective, engaging the materialization of cultural semes within practices of interpretation that are themselves located in domains of discourse and quotidian social action in which those interpretations are made visible. Splitting that sentence into its elements, we have the concepts of the dependent self, the collective, the semiotic resources of cultural meanings in the repository of content, practices of interpretation, and the performances of social action that form the core of the discourse constituting situated, interpreting audiences.

While the situated audience provides the theoretically overarching concepts for the variations that follow, the temptation to multiply theoretical concepts was here irresistible because the elements in the model are often substantially restructured. Let us begin with the most similar and work our way to the most different.

The Strategic Audience

The strategic audience is one bounded by a set of interpretive strategies. I belong to an academic audience *if and when* I practice the interpretive strategies of that community. The strategic audience concept shifts the emphasis from the autonomous individual to a collective, most often called an interpretive community (Lindlof, 1988; Lindlof and Anderson, 1988; Radway, 1984). It is the community that develops the strategies, provides the means for dissemination and instruction, and supervises particular performances of them. The individual in any strategic situation is a local and partial representation of the interpretive community. Clearly, the members are not equivalent units, as the normal political processes of membership are presumed. Some members will be more important than others, directing the others to both what to "read" and how to read it.

As a side note, I might comment that our classrooms can easily be seen as the attempt of a collective to govern the interpretive practices of its individual members. As they are ordinarily self-contained sites, the power hierarchy and political processes show themselves with good clarity. Particularly helpful to the researcher is the explicit nature of the social action. Rules are formalized; interpretive performances are studied, practiced, and supervised; there is a clear entrance and exit to and from the purview of the collective. All of these characteristics make it relatively easy to identify the import of the community.

In most circumstances, however, the clarity of collective influence is not that apparent. In the ethnographic study of auditors, the researcher is presented not with a strategic audience but with individual tacticians of social action. The tactical auditor is concerned with the opportunistic progression of whatever is at hand. One can see only obscurely the strategies of, say, the middle class in the tactics of conversation, making dinner, or watching television. The successful analysis of the strategic audience must enlarge its scope well beyond the point of contact with mediated content. The formulation, representation, and performance of strategies of interpretation happen in the face-to-face interchanges of the many collectives with which an individual might identify. Anyone in education has seen the dutiful schoolchild repeat a classroom lesson well and then in stepping out into the hall, pass as easily into youthful resistance.

As you might suspect, the role of content in understanding audience outcomes becomes problematic. Different theorists describe this role differently, and their relative emphasis on content determinism is a useful distinguishing mark.

Some (Fiske, 1987; Hall, 1980) argue for content that is polysemic (i.e., offers meaning in excess); others (Anderson and Meyer, 1988) hold content to be a resource for (rather than a determinant of) interpretation. No theorist disputes the facts of the text (these words/images rather than those words/images) or that content is boundary setting, but the nature of the relationship between content and interpreter is debated with vigor.

Whatever the relationship, the strategic audience concept obviously lessens the emphasis on exposure to any particular content. It is not so much exposure but the practice of interpretation as governed by the community that is the agency of import. Meaning becomes a community production. Further, the community itself rematerializes in many different ways (in collective publications, public speech, conversations, etc.) the "active semiotic ingredients," if you will, of the industrially produced content. Consequently the connection to *an* exposure is not all that clear or necessary. Understanding the strategic audience, then, not only requires an extended study of content, but also an extended study of the collectives to which the interpretation of content refers.

The theory of strategic audiences is far better developed than the evidence for them (Carragee, 1990). Although there seems to be little doubt of their existence, the workings of interpretive communities remain to be revealed despite Radway's (1984) celebrated analysis of the romance reader. While offering an excellent beginning, Radway's analysis could not document the interconnections among readers upon which her argument depended. The only vehicle available for the dissemination of strategies in her analysis was the romance text itself. That lack forces either a return to content determinism or a recognition that more work needs to be done.

The Engaged Audience

The engaged audience is a genuine membership whose attendance to a text is directed as a sign of that membership. (Membership here refers to a recognizable relationship between an individual who seeks it and a collective that grants it.) The engaged audience, if you will, is a "declared" strategic audience openly practicing its strategies of interpretation in part to be recognized as a member. It is obvious, for example, that the political elite use both media and content, particularly that of news, in radically different ways from the nonelite. For most of us the news approaches something like the soap opera of life (Anderson, Chase, and Larson, 1990). For the political elite, news and its manipulation are stock-in-trade.

There are a number of useful distinctions that can be drawn between the strategic and the engaged audience concepts. Strategic audiences tend to be loosely coupled; the interpretive community less well developed, and community supervision less effective. Engaged audiences as memberships are tightly coupled in contracts of mutual dependencies. There are solid interpersonal networks and personal histories. Both audiences promote texts about their texts, but those of the interpretive community are less sophisticated—the fan magazine type. Engaged audiences often support costly, deeply insider newsletters that interpret the major texts of the membership. Engaged audiences appear much more likely to participate in the production process itself, supplying material or even content (press releases, planted stories, etc.). The symbiotic relationship extends well beyond the quid pro quo of purchase or subscription. In fact, the more the engaged audience constitutes the total audience for the text and its production enterprise, and the more homogeneous the audience itself, the more this audience approaches the classical form. In short, there are a substantially greater number of layers of connection (hence the term) with the engaged audience.

The Emergent Audience

If the engaged audience is the most deeply dependent on content, the emergent audience is simply opportunistic. Its connection to content is to take advantage of the common understandings that widely distributed content can make available. The emergent audience appears in the practice of some social action not related to audience membership. For example, if in the practice of gifting I determine that my gift qualifies as a gift by virtue of its *appearance in the media as a gift,* then I emerge as an audience member of the commercial or other content that does that work even though I may never have been exposed to the particular material. As another example, though most of us have not been to a Rambo film (unless I am seriously and sadly mistaken), most of us know the essence of the character and can both recognize and use a reference to it. So, in days past, when Reagan was presented as "Ronbo" it was a generally recognizable characterization.

The notion of the emergent audience extends the influence of media well beyond exposure. But it is an influence that is modified and perhaps entirely made over by social practices. This is not our predecessors' two-step flow in which the media delivered their content to opinion leaders who passed it on unaltered. Here we are talking of knowledgeable agents who reconstruct content for the political ends of the social action collectives in which they participate. The emergent audience lets us understand the ordinary practices of opposition, resistance, and inversion that populate the interpretations of any audience.

The emergent audience is clearly the most distant from our classical heritage, but it may be much more informative of the way most media are accommodated in our lives. I have not seen the Ninja Turtles (narratives widely available in live-action video and animated television) as of this writing, though I have read reviews and further know several folks who have seen them, and I have listened to their recountings and judgments. I could carry on an extended if not intelligent discussion concerning the film (much like American journalists on Marxism). Although I have not had the heart of the experience, the film is still present in my life and that presence has been accommodated. Given the small proportion of our populace who participate directly in the texts of national and international news, this dispersed contact must be, for most, the experience with what might be considered by journalism scholars as the significant affairs of the day. Most of the articles on the recent Balkan fractures probably have not been read by most of us, but that does not prevent us from knowing about and acting upon them.

Does this gauzy contact qualify those involved as an audience? Probably it does, as much as the generalized action of watching television or reading the newspaper where the act of attendance is the end itself. How often have we stepped away from the set with little memory trace save a recognition of the reproduction of industry forms and conventions? Exposure may or may not be significant.

Implications

I would not argue that the foregoing list is exhaustive and it certainly is not a set of exclusive categories. (In the empirical set, we real individuals wander through the categories willy-nilly, performing many of the requirements simultaneously.) Further, nearly none of this discussion is directed toward an understanding of why an individual auditor attends to a medium and its contents. It is concerned solely with understanding audiences appropriate to media products of sufficient scope to be mainly subject to market forces rather than personal interdependencies. Nonetheless, as a device to sharpen our own thinking about audience, it may have some utility. It is clear, for example, that the terms of any definition of audience have to account for exposure, content determinism, interpretation, the relational nature of the audience, the individual, and the collective. How that accounting is made will give shape to the audience that arises within our theorizing.

Causal lines of argument as well as many traditional critical and cultural arguments, for example, will nearly always see content as a determining agent the effects of which are dependent on exposure. Exposure, for its part, is the audience formulating act that divides auditor from nonauditor. Within the auditor set, frequency and duration of exposure will further define the audience. The audience will typically be an aggregation of individuals, sometimes divided by type. Collective action will rarely be treated. Functional variations of causal arguments usually just extend the explanation to include reasons for exposure. Can we make sense of audiences in this model? Clearly we have.

The sense we make of the audience, however, has certain practical consequences. As content is the effecting agent and exposure the vector of transmission, controls on effects are directed at these elements in the model. Censorship or suppression becomes a form of public policy along with regulations governing attendance. Exhortatory discourse about proper exposure is supported (see, for example, programs in media literacy such as Bazalgette, 1989, and Ontario Ministry of Education, 1989 or even children's books such as *The Berenstain Bears and Too Much TV* [Berenstain and Berenstain, 1985]. The auditor is clearly disempowered in the face of content as the explanation for effects. In the nutritional metaphor commonly used, the best one can do is to stay away from the bad and try to attend to the good.

Social action models, on the other hand, will find their audiences in common social action routines that in some way accommodate mediated content. The already meaningful performance of the routine (reading the newspaper, having dinner, working at the library) is the directing agent and the provenance of understanding the outcomes. Individual auditors are seen as knowledgeable actors competently using media to accomplish the ends of the social action. In these models, content becomes simply another resource to get the job done, though we may question the job itself. Scientific explanation arises in understanding the routine as a meaningful performative whole along with its uses in the tropes and figures of action (cf. Anderson, Chase, and Larson, 1990). Movements for social reform are directed toward the meaningful components of the routine: the use of content for informed purposes that serve the strategies of human emancipation.

So which is right? Can we move toward a comprehensive theory of the audience? The pragmatic answer is to ignore the question and to consider rather what the analyst is led to do from each model: What is the form of explanation that is supported? How is power distributed between the researcher and the researched in the performance of inquiry? Who is authorized to speak? And who is denied? How are lives touched or ignored? These questions take us back to the opening supposition of this argument: that the notion of audience is a discursive product—a social construction that makes meaningful our claims about it.

References

Abelman, R. (1989). From here to eternity: Children's acquisition of understanding of projective size on television. *Human Communication Research* 15:463–481.

Alexander, A., M. S. Ryan, and P. Munoz (1984). Creating a learning context: Investigations on the interaction of siblings during television viewing. *Critical Studies in Mass Communication* 1:345–364.

Allor, M. (1988). Relocating the site of the audience. *Critical Studies in Mass Communication* 5:217–233.

Altheide, D., and R. Snow (1988). Toward a theory of mediation. In J. A. Anderson (ed.), *Communication Yearbook*, vol.11, pp. 194–223. Newbury Park, Calif.: Sage.

Anderson, C. (1987). Reflections on *Magnum, P. I.* In H. Newcomb (ed.), *Television: The Critical View*, 4th ed., pp. 112–125. New York: Oxford University Press.

Anderson J. A., R. S. Chase, and T. Larson (1990). *Patterns of Viewing and Descriptions of Use for Broadcast Television News.* Dublin: International Communication Association.

Anderson, J. A., and T. P. Meyer (1988). *Mediated Communication: A Social Action Perspective.* Newbury Park, Calif.: Sage.

Bazalgette, C. (ed.) (1989). *Primary Media Education.* London: British Film Institute.

Berenstain, S., and J. Berenstain (1985). *The Berenstain Bears and Too Much TV.* New York: Random House.

Brabrow, A. S. (1988). Theory and method in research on audience motives. *Journal of Broadcasting and Electronic Media*, 32:471–487.

Brown, J. D., and L. Schulze, (1990). The effects of race, gender, and fandom on audience interpretations of Madonna's music videos. *Journal of Communication* 40:88–102.

Brown, M. H. (1990). Defining stories in organizations: Characteristics and functions. In J. A. Anderson (ed.), *Communication Yearbook*, vol. 13, pp. 162–190. Newbury Park, Calif.: Sage.

Burgoon, J. K., M. Burgoon, D. B. Buller, and C. K. Atkin (1987). Communication practices of journalists: Interaction with public, other journalists. *Journalism Quarterly* 64:125–132, and 275.

Budd, M., R. M. Entman, and C. Steinman (1990). The affirmative character of U.S. cultural studies. *Critical Studies in Mass Communication* 7:169–184.

Carragee, K. M. (1990). Interpretive media study and interpretive social science. *Critical Studies in Mass Communication* 7:81–96.

Certeau, M. de (1984). *The Practice of Everyday Life.* Berkeley: University of California Press.

Conquergood, D. (1990). *The Cultural Construction of Street Gangs.* Salt Lake City: University of Utah.

Danowski, J. (1991). Organizational media theory. In J. A. Anderson (ed.), *Communication Yearbook,* vol. 14, pp. 187–207. Newbury Park, Calif.: Sage.

Eagleton, T. (1983). *Literary Theory.* Minneapolis: University of Minnesota Press.

Evans, W. A. (1990). The interpretive turn in media research: Innovation, iteration, or illusion? *Critical Studies in Mass Communication* 7:147–168.

Everett, S. C., and S. E. Everett (1989). How readers and advertisers benefit from local newspaper competition. *Journalism Quarterly* 66:76–79, 147.

Feilitzen, C. von (1991). Children's and adolescents' media use: Some methodological reflections. In J. A. Anderson (ed.), *Communication Yearbook,* vol. 14, pp. 91–101. Newbury Park Calif.: Sage.

Fiske, J. (1987). *Television Culture.* London: Methuen.

Gitlin, T. (1985). *Inside Prime Time.* New York: Pantheon.

Goodall, H. L., Jr. (1990). A theater of motives and the "meaningful orders of persons and things." In J. A. Anderson (ed.), *Communication Yearbook,* vol. 13, pp. 69–94. Newbury Park, Calif.: Sage.

Hall, S. (1980). Encoding/decoding. In S. Hall, D. Hobson, A. Lowe, and P. Willis (eds.), *Culture, Media, Language,* pp. 128–139. London: Hutchinson.

Hartley, J. (1988). The real world of audiences. *Critical Studies in Mass Communication* 5: 234–238.

Helregel, B. K., and J. B. Weaver, (1989). Mood management during pregnancy through selective exposure to television. *Journal of Broadcasting and Electronic Media* 33:15–33.

Hoffner, C., J. Cantor, and E. Thorson (1989). *Children's responses to conflicting auditory and visual features of a televised narrative. Human Communication Research* 16:256–278.

Krendl, K. A., M. C. Brohier, and C. Fleetwood (1989). Children and computers: Do sex-related differences persist? *Journal of Communication* 39:85–93.

Lang, A. (1990). Involuntary attention and physiological arousal evoked by structural features and emotional content in TV commercials. *Communication Research* 17:275–299.

Lindlof, T. (ed.) (1987). *Natural Audiences.* Norwood, NJ: Ablex.

Lindlof, T. (1988). Media audiences as interpretive communities. In J. A. Anderson (ed.), *Communication Yearbook,* vol. 11, pp. 81–107. Newbury Park, Calif.: Sage.

Lindlof, T., and J. A. Anderson, (1988). *Problems in Decolonizing the Human Subject in Qualitative Audience Research.* Barcelona: International Association of Mass Communication Research.

Livingstone, S. M. (1990). Interpreting a television narrative: How different viewers see a story. *Journal of Communication* 40:72–85.

Lull, J. (ed.) (1988). *World Families Watch Television.* Newbury Park, Calif.: Sage.

Meadowcroft, J., and D. Zillmann (1987). Women's comedy preferences during the menstrual cycle. *Communication Research,* 14:204–218.

Morley, D. (1980). *The "Nationwide" Audience.* London: British Film Institute.

———. (1986). *Family Television: Cultural Power and Domestic Leisure.* London: Comedia.

Murray, J. P. (1991). Nothing lasts forever: Instability in longitudinal studies of media and society. In J. A. Anderson (ed.), *Communication Yearbook,* vol.14 pp. 102–110. Newbury Park, Calif.: Sage.

Ontario Ministry of Education (1989). *Media Literacy.* Ottawa, Ontario: Queen's Printer for Ontario.

Radway, J. (1984). *Reading the Romance: Feminism and the Representation of Women in Popular Culture.* Chapel Hill, N. C.: University of North Carolina Press.

Rosengren, K. E. (1991). Media use in childhood and adolescence: Invariant change? Some results from a Swedish research program. In J. A. Anderson (ed.), *Communication Yearbook,* vol. 14, pp.48–90. Newbury Park, Calif.: Sage.

Tamborini, R., and J. Stiff, (1987). Predictors of horror film attendance and appeal: An analysis of the audience for frightening films. *Communication Research* 14:414–436.

Traudt, P. J., J. A. Anderson, and T. P. Meyer (1987). Phenomenology, empiricism, and media experience. *Critical Studies in Mass Communication* 4:302–310.

Tunstall, J. (1991). A media industry perspective. In J. A. Anderson (ed.), *Communication Yearbook,* vol. 14, pp. 163–186. Newbury Park, Calif.: Sage.

Turow, J. (1984). *Media Industries.* New York: Longman.

Weaver, J. G., and G. D. Newton (1988). Structural determinants of the television news audience. *Journal of Broadcasting and Electronic Media* 32:381–389.

Wolf, M. A., T. P. Meyer, and C. White (1982). A rules-based study of television's role in the construction of social reality. *Journal of Broadcasting* 26:813–829.

PART II

Rethinking the Audience as an Object of Study

Recasting the Audience in the New Television Marketplace?

6

Jay G. Blumler

This chapter is not about audience behavior, effects, uses, or readings. It is about audience *imagery*—that is, potentially influential images, entertained by others, of what the audience for television programming may be like. It asks: Has the expansion and elaboration of a multichannel television system out of what was once a triopoly (dominated by three broadcast networks) significantly altered the way in which the U.S. viewing audience is conceived by those industry figures who cater to it?

This issue presupposes two more underlying questions. One concerns a facet of the process of media industry change that the literature has rarely raised or explored: Do prevalent notions of "the audience" change in response to changes of media structure, and if so, how? The other concerns the dramatic development of U.S. television that has taken place in recent years: How fundamentally different is the emerging system from the one that it is replacing? Is it quite transformed or relatively similar? Nobody would deny that this leopard, has changed some of its spots—but is it still a leopard, or has some new species been produced? An index of such change should be how providers of programming think about the audience for their offerings.

The source material for answers to these questions is a set of 150 interviews that were held with leading American broadcasting personnel in the course of an inquiry that was conducted in the period 1986–1988 with support from the John and Mary R. Markle Foundation. The interviewees included (among others) executives of the three mainstream networks (ABC, CBS, NBC), cable services, multisystem cable operators, Fox Broadcasting, local stations (network affiliates and independents), as well as many creative producers, writers, and directors of a wide range of programs for a wide range of markets. The common criterion for inclusion was that the informant should have held a position entailing experience of or likely to generate insights into the process of new television marketplace expansion—the forces driving it, the opportunities and constraints characteristic of it, and the consequences that might be flowing from it.

Two limitations of this body of material for clarifying the questions raised here should be borne in mind. First, since most of the interviews were held in informants' offices, they were at something of a distance from the working conditions and moments of actual program making. Second, images of the new viewing audience were not often an explicit focus of questioning, which dealt more centrally with the impact of multichannel abundance and competition on the conditions of television production, particularly on the kinds of programs that could (or could not) be made for sale and scheduling in a range of different outlets. Nevertheless, a lot was said in the natural course of the conversations about changes in and problems of holding the viewing audience in the new media landscape.

Before I present the resulting evidence, however, certain prior issues of conceptualization must be addressed.

Thinking About "the Audience"

Theories of the audience should take account of the fact that "it" is always a construction. As Fiske has bluntly declared, "There is no such thing as 'the television audience'" (1989: 56). If this is so, it does not seem to make sense to ask, "Does the *audience* influence media content?" (Cantor and Cantor, 1986: 214), what "the audience demands" of media materials (Schatz, 1981: 1), or "How can an audience directly cause artistic innovation?" (Billings, 1986: 201). Such formulations are inappropriate, because mediated audiences are no more than an aggregate of dispersed individuals, an "unstructured group" (Jarvie, 1970: 186), "a diffuse and anonymous 'body', lacking any pivotal focus of joint action or common expression" (Blumler, 1977: 241).

"The audience" by and large plays its part in mass communications, then, through constructions that emanate from and are lodged in *other* parts of the media system concerned. Notions of the audience deployed by industry leaders may therefore be regarded as systemic features of media arrangements, connecting "the audience" to the system and its principal actors through a construct, implying a part, desired and intended not necessarily by individual audience members (though they may absorb it as well), but by those seeking to mobilize it to their own institutional ends. As Ang (1991: 154) has argued: "Institutional knowledge is not interested in the social world of actual audiences; it is in 'television audience,' which it constructs as an objectified category of others to be controlled. This construction . . . enables television institutions to develop strategies to conquer the audience so as to reproduce their own mechanisms of survival."

Two important implications follow from this external constructedness of notions of "the audience." First, it implies a limited power of individual audience members to "assert [their] needs and compel respect for them" (Blumler, 1977: 241). This is not to deny the existence, sensitivity (in their own terms), and influence of feedback mechanisms in large-scale media industries, tapping certain audience member reactions and views (e.g., ratings and content-testing pilot research), measuring numerical success, and helping producers to "guess how markets might respond to their latest [program or] story" (Parker, 1986: 144). Nevertheless, most such mechanisms belong "to the financial economy of television, in which programs are made for sale to distributors," rather than to its "cultural economy," in which audience members make meanings out of what they have seen and heard (Fiske, 1989: 59).

Second, the power of other elements in the media industry to oblige individual audience members to conform to their assigned roles may vary according to certain system conditions that should be specifiable. Audience member autonomy in that sense might depend, for example, on:

1. The diversity of available program offerings—particularly with respect to the audience roles that are encouraged by the contents concerned (for discussion of how texts may facilitate or discourage the adoption of certain audience roles, see Blumler et al., 1985: 264–266).
2. The relative "openness" and "looseness" of program content structure (cf. Schlesinger et al, 1983: 32).
3. Differentiation of industry organization, including broadcasters with diverse aims, styles, and revenue sources (in contradistinction to what Billings terms a high level of industry institutionalization, "as measured by increased capitalization and consolidation of production, distribution and exhibition" [1986: 202].)
4. The forms of audience member response that the dominant feedback mechanisms are designed to tap.

Most empirical research into mass media communicators' audience images has been microscopic—more productive of close-ups than panoramas. The most common object of study has been a particular production team or individual producer, and the prime interest has been to learn whether and how the audience image, defined "as an external observer-judge against which the communicator unconsciously tests his product even while he is creating it" (Gans, 1957), has shaped a body of content. The role and influence of such images have accordingly been examined in the making of movies (Gans, 1957), the crafting of news stories (Pool and Shulman, 1959; Gans, 1979), the coverage of an election campaign (Blumler, 1969), the production of a documentary series (Elliott, 1972), the creation of a prime-time television drama (Espinosa, 1982), and the making of medical dramas for television (Turow, 1989).

Four features of research in this vein may be noticed. First, audience images emerge from it as undoubtedly functional for mass media communicators. They help producers to cope with uncertainty (McQuail, 1969). They yield a working psychology of expected audience reactions to certain features of content, in terms of what people should find appealing, credible, and understandable (or the opposite). They are often deployed as counters when production strategy is in dispute, injected to support the case for one approach over another. Most researchers have also concluded that content tends to be adjusted to communicators' impressions of audience requirements—though many other influences shape programming as well, and the possibility has been left open (see Blumler, 1969: 101–102) that particular images may be less reflective of producers' beliefs about audience likings (an "independent variable" in the production process, then) than of the kind of impact they would like to have *on* audience members (a "dependent variable").

Second, many audience images proffered by mass media communicators lack richness, complexity, or variety. They often come across in confident but stark terms as thumbnail sketches. An example is the statement of the network executive who declared, when taxed for the paucity of issue coverage on broadcast television during U.S. election campaigns, "All people are interested in is who wins" (Semetko et al., 1991: 45).

Third, more often than not, the audience images discovered by researchers have served a constraining function, as if bringing potentially straying content back into conventional line. Although on occasion more creatively enabling perceptions have been found (e.g., Blumler's [1969] observation of BBC election producers' concern to serve viewers wanting help in deciding how to vote, and Gray's [1986] study of community radio producers' relations to listeners with supposedly high levels of commitment and loyalty), audience images mainly come into play to prevent the inadvisable—curbing the strangeness of innovation, keeping debate within commonsense bounds, eschewing abstractions, and avoiding offense and divisiveness.

Finally, little empirical attention has been paid to those more macroscopic images of the target audience that may be thought appropriate for a medium as a whole, or some channel, outlet, or service within it, to address and cultivate. At this level scholars have mainly offered interpretative analyses, inferences, and classification schemes, as in the work of Ang (1991) for commercial and public television, Martell and McCall (1964) for mass periodical fiction, and McQuail (1987) for the mass media generally. That, however, was the level of audience conceptualization on which the informants interviewed for the study reported here mainly discoursed and with which the following analysis is concerned.

To explore whether the coming of multichannel television has affected orientations to the audience at this level, a typology of possible conceptualizations was required. Table 6.1 displays such a typology, based on modified versions of concepts presented in two passages (on "Alternative Conceptions of the Audience" and "A Typology of Audience Formation," respectively) of McQuail's *Mass Communication Theory: An Introduction* (1987: 218–224).

First of all, the audience may be construed in *market* terms. It is presumably fair to say that this is the notion by which U.S. commercial television was overwhelmingly guided in its three-network heyday. According to this notion, the audience is predominantly conceived:

- as a *commodity*, with viewing as an essentially commercial transaction, in which all that matters is a "sale" and program content matters only to the extent that it encourages a sale
- as *numbers*, yielding a "ratings discourse" about viewers (Ang, 1991: 50 ff.)
- as *individuals*, rather than as members of socially, culturally, or demographically formed collectivities[1]
- as *spectators*, whose attention-giving or withholding is the key feature of their

Table 6.1 Typology of Generalized Audience Images

Audience construed as:	Prime axis of orientation
Market	To sell via attention
Public	To serve a role
Fans of taste culture	To feed an identity
Social group	To satisfy shared interests
Involved audience	To provide an enriched communication experience

relationship to content (in contrast to any other forms of involvement or meanings they may derive from it)

Second, the audience may be conceived as a *public*, utilizing what is communicated to perform a valued social role. A characteristic example would be the audience member regarded as a citizen, served by information that enables him or her to play some part in civic affairs. As Ang explains: "In this context, the importance of radio and television programs lies in their potential to transfer meaningful messages rather than in their capacity as vehicles to deliver audiences to advertisers: programs and programming matter for their symbolic content rather than as agent for economic exchange value" (1991: 105).

Third, audience members may be construed as *fans of a taste culture*. In this conception, communicators may see themselves as catering to committed and absorbed followers of certain genres, content forms, authors, personalities, communication styles, and so on. The point of departure here is an assumed appreciation of content so dedicated that it has become a part of audience members' identities. This is reminiscent of Radway's (1984: 54) notion of an "interpretive community," which she applied to "certain readers" who were "particularly interested in a specific 'category' of books," choosing them over and over, "because essential features of their social life create needs and demands that are somehow addressed and fulfilled by these books."

Fourth, the audience may be conceived as a *social group*, the members of which have similar interests that communication can serve because of their shared "characteristics of place, social class, politics, culture, etc." (McQuail, 1987: 222). This is distinguishable from the previous category in having its point of origin in social structure rather than media materials.

Finally, there is the notion of the *involved audience member*. Not necessarily defined by particular social ties or tastes, this is someone who is disposed to follow media materials actively, to think about their possible meanings and significance, and to value the resulting deepening or broadening of experience. This image stands in contrast to the sufficiency of attention-giving, time-spending, and spectatorship for the market model.

Impact of Multichannel Television on Audience Imagery

The interview data support four propositions about the effects of the emergence of a multichannel system on how industry figures of U.S. television have come to think about the viewing audience.

I. Heightened Uncertainty

Uncertainty about the likely audience reaction to all individual program offerings has been heightened. Overall, the mood about generalized audience predilections was far removed from Carey's (1990) account of constructions of the listening audience held by many broadcasters in the early days of radio—of a family seated around the set, expectant, receptive, pleased to be fed and entertained by those who know how to do this well, accepting their offerings with almost a childlike credulity. Of course the audience for radio became a source of considerable anxiety as competition for its patronage increased, since it might or might not show up and might or might not respond positively. But among the 1986–1988 interviewees, there appeared to be a pervasive sense of the audience's having become even less controllable and predictable than before, a sense of having to feel one's way in trial-and-error terms hopefully toward a viable alignment with it, as well as a sense of this being a continuing and never settled or equilibrized process.

In part this stemmed from an image of the viewer as someone who prizes his or her autonomy and has been furnished with more means (e.g., remote control devices) and choices (more channels) to assert it. A production company executive referred, for example, to "the programmer in the home, which I believe drives most of television today, [and] is a consumer's desire to really control his programming decisions at any given point in time." In part it stemmed from the availability of so many viewing alternatives: "People are ready to search their dials for other types of programming" (independent producer). In part it stemmed from the dissolution of habit and channel loyalty as past props of audience share: "People are watching TV now for the program, not for the channel and not for those who made it" (producer-writer). This was most pronounced in the younger sector of the audience: "I have a ten-year-old daughter, who has zero channel loyalty: zero!" (cable executive). In part it stemmed from the accelerated pace of change in programming fashions and viewer tastes alike as more competitors more aggressively and ingeniously seek ways of attracting audiences: "Audiences are very fickle today. . . . I mean audience tastes change so rapidly from year to year and so vastly" (producer-writer-actress). As a business affairs reporter has painted the resulting predicament:

> The audience that you think you've got, that you thought you were talking to for the last ten years, may have in the last five or six wandered off somewhere. You

> are not exactly sure where they have gone, and you are not exactly sure who is left, leaving you in the disconcerting position of making television and asking at the same time: Who am I talking to (Krulwich, 1990)?

The audience uncertainties arising from all these influences were seemingly chronic and incurable: "Well, the question is how you reduce uncertainty. But the question that *creates* the uncertainty in the final analysis is whether the viewer will *watch.* And there is no deal you can make with the viewer. That's the problem" (production company executive).

II. New Spaces for Nonmarket Images

Openings have undoubtedly been created for all the nonmarket orientations to audiences, specified in the typology in Figure 6.1, to be cultivated and to move more to the forefront of providers' thinking, though often in somewhat impure, thinly stretched, or hybrid forms. The proliferation of channels has laid the foundations for this, enabling new entrepreneurs to challenge the hitherto established providers of products for large heterogeneous mass audiences by offering differentiated programming designed to appeal to targeted audiences. Remarks reflective of the development of nonmarket audience imagery were voiced mainly by two types of interviewees: (1) cable service executives, especially those responsible for commissioning programs for their channels; and (2) working program makers (producers, writers, directors), particularly the more self-consciously creative aspirants among them.

Among the former, certain threads of comment on audiences tended to recur. They saw themselves as catering to relatively bounded groups rather than to an undifferentiated mass audience, as in the case of the cable program commissioning officer who said (in line with the *social group* image): "We're not trying to be all things to all people. We're really focusing on that demographic, learning it, understanding it, caring about it, respecting them, and creating programs for them." Also expressive of this outlook was the Lifetime channel programmer who said: "I'm targeting a woman right now in her 30s who's better than average in education, above average in income, contemporary, urban, informed. . . . We're trying to get the aging baby boomer as our viewer right now. [So it's] gender targeting . . . but it's also generation targeting." They saw their services as catering to relatively focused viewing appetites, interests, and needs, as in the case of the cable executive who said (in line with the *taste-culture-fan* image): "We're for the videophile, for the person who is on the cutting edge in terms of his use of TV, use of video, and so on. We thought of ourselves as the yuppie channel when yuppie was used as a term of affection rather than one of disdain." They conceived their audiences as relating more personally and actively to their offerings than to old-style network fare. For example, a CNN editor claimed: "People feel that they have a greater say in our product than they had, say, in ABC's, feeling more free to write or phone because they pay for the service as opposed to receiving it free. . . . Also because Ted Turner is a known person, not just a corporation."

They also appeared to give program content per se a higher priority than in the schematicized market model. Content mattered more and provided the basis for some presumed mutuality of commitment between the channel and the viewer, linking, on the one side, a channel's identity—of being the place to go for a certain kind of content—with a presumed or hoped-for viewer loyalty to the channel, on the other side. There was, for example, the MTV executive who claimed that it catered to "the more active audience, an audience where music is really a part of their daily lives, especially teenagers and young adults." There was also the Nickelodeon executive who believed that: "In these past years we have built an audience that feels very strongly that Nickelodeon is their channel. . . . There's very much a sense of this is their channel, and we've worked hard through the packaging of the shows and the whole environment that we've created to give kids a very strong message, which is kids are great . . . this is a comfortable place to be . . . this is a full entertainment network just for you." For a C-SPAN executive, the audience as *public* seemed more salient: "Ninety-three percent of our viewers voted in the last election in 1984. Most of them are a little bit older, college educated. . . . More men than women. All walks of life naturally but a cut above. Many of them retired, all very interested in politics. Over time we have created what we call the C-SPAN junkie . . . , someone who is a devoted watcher of C-SPAN. There are clubs that have been formed."

Among the working program makers, an image of the *involved audience member* surfaced in two main themes of comment. One was the idea that with channel expansion, many viewers were learning to become more discriminating. They were exercising program choices and judgments more actively, and their tastes were becoming more discerning. According to a producer, for example: "People are not just putting on 2, 4, and 7 [the mainstream network channels]. They are now becoming more selective. This is very good news for the creative community." Another producer assumed: "This means that you have to satisfy them with more different varieties of food. You can't keep giving them hamburgers, particularly if somewhere they can enjoy gourmet food. In addition, they can pop in that video if they wish." As a writer concluded: "The viewer is less parochial than he was believed to be and less in need of the networks' old cookie-cutter approach to programming." Moreover, since for cable television the viewer makes a direct payment in the form of a subscription, he or she was presumed to want better material in return, worthy of the charge. An independent producer-executive, for example, explained, after characterizing the cable audience as a bit older, more upscale, and so open to more sophisticated material: "That is why to sell to network TV you need a hook, but to sell to cable you need the kind of material that will appeal to a more intellectual or an older audience. It's a bit more formularized when you think of what's appropriate for network material."

The other theme sounded by hopeful program makers concerned the possible release of a hitherto thwarted potential for audience involvement. As a producer-executive put it, "There is probably more appetite for quality in the viewing audience than we suspect." A director believed that: "When you lose the mass audience and you are catering for the demographics of pay cable, there is more freedom to rely on more complex subject matter." A producer-writer even proclaimed himself to be part of a crusade "to try to improve American television and to give the audience a chance to see that there is more, and there should be more than *Dallas* and *Wheel of Fortune* and so on. There should be all of that. But my feeling is that the time is ripe for something that the audience hasn't been given."

III. Forces for Dilution

Nevertheless, *the expression of such nonmarket perspectives on the audience was often qualified and diluted.* This is because catering to only one constituency is a less commercially remunerative prospect than is catering to more than one audience component; because twenty four hours is a very long time to program for just one constituency; because even to the degree that one aims to please a core constituency, one is bound to try to reach not only its most dedicated elements but also a wider and larger number of individuals who may seek material of the kind concerned only every now and then; and because prime time especially is still a period when people of different constituency types may be viewing together as a family or household.

As a Disney Channel executive explained:

> What we have been attempting to do is broaden this channel to make it a family, good taste, high-quality service. It's the idea that if you only bill this channel to satisfy pre-schoolers and preteens, in essence you should turn the channel off at 6:00 P.M. Why shouldn't we be filling that need of entertaining family in the evening? . . . Before I took this job, I met this lady while I was buying a car. She said, "Oh yes, we subscribed to the Disney channel for our child, but we quit subscribing because there wasn't anything for my husband and me." I said, "When would you watch it?," and she replied, "We'd watch it in prime time."

A similar logic was mentioned by an MTV executive:

> We have had traditionally a narrow appeal, and part of our success has come from that . . . the fact that we appeal to 12 to 34-year-olds . . . with a single focus on music. But when you think about a game show, it might have a fairly broad appeal with that demographic. And maybe we take a notion that's more palatable for a broader appeal on TV, like a *People Magazine* kind of thing . . . that has an MTV spin to it. I mean it isn't just rock stars, but it could be Steven Spielberg and Charles Manson in an interview that's maybe done with an MTV-type on-air personality. . . . Right now you get the hard core group out there, so our game is really to increase our core and to bring new people to the party. I mean it's a pretty fine steady business, and I don't want to kill the goose, but I think we can try and goose the goose until we see how high is up; then maybe we'll be satisfied.

Similar tendencies for cable services to expand outward from a homogeneous audience base toward a more heterogeneous appeal can be noticed in the cases of Lifetime, Nickelodeon, the Discovery Channel, and the Family Channel.

IV. Reassertion of the Audience as Market

Despite the emergence of such new images, in the end the market orientation to viewers hovered over all these relations like an enveloping and inescapable environment—penetrating and seeping through them, limiting their purview, compromising them, and bring them back into market-mode line. The interview material drew attention to five sources of this reversion to and overall predominance of the audience-as-market model.

One concerned the upsurge of yet another new image of the audience that multichannel television has spawned. This is the image of the *capricious* or *elusive* audience member—the viewer as *quicksilver,* as it were. This is someone who has grown up with multichannel television, become accustomed to its style, enjoys riffling through its many offerings, and incessantly grazes across the dial in search of more entrancing viewing pastures (*Channels,* 1988). In a network executive's words, "The audiences, because of all the alternatives that are available to them, keep looking around, bouncing around and hitting those buttons." As a comedy writer put it: "The switcher has changed TV. Now people do not have the patience to sit and wait and wait. That's why cable has had such a big rise, because while the networks go to those endless two minutes of commercials, you're flipping on and you've seen twenty other programs. And you may stay with one of them." A public affairs programmer commented that because "the public is now video educated, viewers expect a payoff every ten or twelve minutes, unlike the old-style documentaries, which expected the viewer to hang around for a payoff in the last ten minutes." A CNN reporter said: "We live in an environment where people are watching a channel for three minutes and then press the clicker. We've got to get them to watch in that environment." Even a Discovery Channel executive referred to its involvement in a "frame-by-frame struggle for viewers' attention" (Blumler, 1991).

All this has strongly reinforced the market model's primary emphasis on attention-gaining. As a writer explained, "Now the idea is to grab an audience quickly."

A second force behind the resurgence of a market mentality is the abiding incentive structure that plays on many leading producers, directors, and writers for television. After everything had been said by the program-making interviewees about the creative attractions of working for the new services, almost all, when asked to what outlet they would most prefer to take their *next* ideas, unhesitatingly mentioned the major broadcast networks. This was because with the networks the audiences are larger, reputations are made ("It's good for your career"), the resources for translating a concept into a good program are most likely to be on tap, and they offer the best gateway to lucrative ancillary markets. As a Hollywood studio executive explained: "The talent is still going to be drawn to those three networks. The networks can do some terrible programming; they can have a terrible season; but if you have a hit on network TV there is nothing comparable. There's absolutely nothing comparable to a show that consistently draws a forty-share-plus audience." Thus, despite the decline of the networks and the expansion of cable, much of the incentive structure is still skewed toward scale and grandeur of effort, a longer mass audience reach, bigger monetary rewards and prospects (through back-end earnings), and network-based credits and prestige (Blumler, 1989). As another producer-writer confirmed: "Believe it or not, we want to be in the prime-time network world, okay, more than anything. That is the big league. That's the New York Yankees. That's the big time."

Third, certain factors of economic and industrial organization have also favored the diffusion of market imagery in television sectors initially productive of other notions of the audience. In a multichannel television system, producers and distributors are under considerable pressure to tailor programs for acceptability in multiple markets (network, cable, syndication, home video) regardless of their initial scheduling destination. Partly this is because audience success for an individual program is such a chancy business that producers are continually striving to compensate for many failures by raking in all the available jackpots whenever an audience winner comes along. Partly it is because, as average audience and revenue levels fall, it is far less likely that a show's appearance in a single outlet of the splintered marketplace will suffice to cover the ever inflating costs and expected profits. As a producer-writer explained:

> The business has become more and more difficult with every year I have been in it. The marketplace is fragmented, network shares are being cut into by cable. On the one hand, there are more potential outlets for a supplier to sell his wares to, but, on the other hand, all of these outlets seem to have less money and the cost of production has not gone down.

Consequently, producers want programs "to travel sequentially" (as another put it). As an executive of a pay cable movie channel explained: "The prospect of a back-end afterlife is important; it's one factor of many that are important in making the deal because we're investing a significant amount of money. And in some cases even in our wildest dreams of success for the show on air, we need some back-end . . . to make the program economics work out for us." Another executive of a thematically targeted service even saw it partly as a test-bed for more lucrative rewards: "We've got the virtue of our own air, so we can put on a show, get it polished up, get it working, and then when it's ready, hand it off into syndication and so we kind of use it as a laboratory." Industrial trends of concentration and conglomeration have also furthered these processes, encouraging the production of content suitable for phased appearance in the diverse outlets controlled by large corporations.

Fourth, there is the sheer fact that all the enterprises that provide American television (except for public television and C-SPAN) are businesses and must ultimately satisfy commercial criteria of market success. As an independent producer put it: "Each of these channels has a president or a general manager who is being driven to deliver higher and higher revenues, ratings, and profits." Thus, the market-related "ratings discourse" about viewers (Ang, 1991) tends to become the common currency of audience success for all services. As the commissioner of programming for a thematically targeted cable channel explained: "It's still a ratings business . . . and if your shows are up in the ratings, sales are up and you get more subscribers. . . . So it's pretty sales driven." As another cable industry leader concluded: "You can never lose sight of the fact that you're in the entertainment business—and it's getting eyeballs to the set."

Finally, over and above such economic and organizational supports for market-based audience imagery, a specifically American cultural influence pulls in the same direction. This is a tendency to regard involvement in the competitive making, selling, and showing of programs as a game of chance, akin to gambling. Although games can be exciting and fun, in the American culture they are also deadly serious, of course, intended above all to be won. This justifies a vigorous and unsentimental form of play, ruling out uncalculated effort as unrealistic and unprofessional. "The audience" (the larger the better) is then conceptualized as a "prize" for winning the game. As a writer observed: "The whole point about American television is that each player is trying to beat the other players out. Who's going to be number one? Who's going to be 10.2 and the other guy is 10.1? That tenth of a point sometimes is a headline in *Variety.* I mean it's all competition for numbers." In short (according to a station manager): "You do wanna win. It's a very competitive business." In such a culture, then, market-type rewards compatible with the game metaphor are prioritized, while other possible communication goals and ways of serving the audience tend to be pushed to the sidelines—"benched," as it were—for the duration of the game.

Explanations and Implications

A mixed picture has emerged from this study of images of the viewing audience associated with the development of U.S. multichannel television. In terms of the theatrical metaphor of the chapter title, although it has been given more roles to play, most of them are little more than bit parts, still subservient to the dominant lead (the audience-as-market). In themselves, the changes are real enough, offering American viewers more choice among different role options, particularly for cable subscribers confronting fifty-channel menus. Yet the changes concerned have also been modest and limited and are perhaps even becoming peripheral. Although the palette of imagined audience roles has been stretched to cover new notions, these rarely stand out free and clear but have been absorbed and coopted instead back into the market model's embrace.

This pattern is probably best explained by the process of radical yet partial change through which the U.S. television industry passed in the 1980s. Technologically, it was transformed—by cabled abundance; satellite delivery of thematic services nationwide; proliferation of the remote control device; diffusion of videorecorder/players; and an increase in multiset households. Economically, however, it was little changed. Although the role of viewers' personal payments (through cable subscriptions and videocassette rentals and sales) increased greatly, most services continued to depend heavily on advertising revenue (excepting only a small number of pay television channels), and production had to be supported by what it could earn in the various program markets.[2] Consequently, the integrity of those additional ways of conceiving the audience that multichannel expansion had initially prompted was eventually threatened because, on their own, they could not work in commercial terms.

It is intriguing to relate these findings to current images of the audience held by scholars of reception analysis. How far do they correspond? In much recent academic work, the audience has been reconceptualized "as an active maker of meaning" (Curran, 1990: 145), and television is no longer "the dominant monster it [was] often thought to be; viewers have considerable control not only over its meanings but over the role that it plays in their lives" (Fiske, 1987: 74). In short, as Liebes and Katz have noted: "The reader/listener/viewer of communications theory has been granted critical ability. The legendary mental age of twelve which American broadcasters are said to have attributed to their viewer may, in fact, be wrong" (1989: 204).

Some of the findings presented in this chapter suggest there was a potential in new television marketplace developments for closing this gap. At least some creative producers, writers, and directors were encouraged to think of viewers as more active meaning makers. Nevertheless, they still had to place their programs in commercial networks and services, in high earnings from which they had enormous financial and psychic stakes. Regarded in this light, though it is a "plus" that the image of the involved audience member has at last been planted in the system, it is also something of a "minus" that its roots appear relatively shallow and vulnerable to hostile conditions.

Does this matter, when viewers can apparently perform a certain amount of creative work even on what Liebes and Katz have succinctly termed "dumb genres" (1989: 204)? If audience member orientations to mass media materials are qualitatively as rich as recent reception research implies, it is presumably preferable for content to be on a par with this in maturity and support for imagination and reflection. Ultimately at issue here, however, is how providers' images square with the value of "respect for the audience member." Arguably, this is more likely to be upheld when the audience member is conceptualized as a citizen, an avid and informed fan, a member of a cohesive social group, or as poised to engage actively with meanings of content. It is most in danger of being flouted when viewers are treated, along market lines, "as numbers not subjects; as pocketbooks without brains; as lambs for delivery to the advertiser" (Blumler, 1989: 34–35).

Notes

1. This can apply to audience segmentation, when certain categories of viewers defined, say, by age, income, or gender are preferred because of their greater value to advertisers as individuals with higher purchasing power.

2. I am indebted to Professor David Waterman of the Annenberg School for Communication, University of Southern California, for an analysis of these trends performed on behalf of the Markle Foundation–supported inquiry into the New Television Marketplace.

References

Ang, Ien (1991). *Desperately Seeking the Audience.* London: Routledge.

Billings, Victoria (1986). Culture by the millions: Audience as innovator. In Sandra J. Ball-Rokeach and Muriel G. Cantor (eds.), *Media, Audience, and Social Structure,* pp. 200–213. Beverly Hills, Calif.: Sage.

Blumler, Jay G. (1969). Producers' attitudes towards television coverage of an election campaign: A case study. In Paul Halmos (ed.), *The Sociology of the MassMedia Communicators,* pp. 85–115. Keele, Eng.: University of Keele.

———. (1977). 'The election audience: An unknown quantity? In *TV and Elections,* pp. 241–247. Turin: Radiotelevisione Italiana.

———. (1989). *The Role of Public Policy in the New Television Marketplace.* Washington, D.C.: Benton Foundation.

———. (1991). The new television marketplace: Imperatives, implications, issues. In James Curran and Michael Gurevitch (eds.), *Mass Media and Society.* London: Edward Arnold. 194–215.

Blumler, Jay G., Michael Gurevitch, and Elihu Katz (1985). Reaching out: A future for gratifications research. In Karl Erik Rosengren, Lawrence A. Wenner, and Philip Palmgreen (eds.), *Media Gratifications Research: Current Perspectives,* pp. 255–273. Beverly Hills, Calif.: Sage.

Cantor, Muriel G., and Joel M. Cantor (1986). Audience composition and television content: The mass audience revisited. In Sandra J. Ball-Rokeach and Muriel G. Cantor (eds.), *Media, Audience, and Social Structure,* pp. 214–225. Beverly Hills, Calif.: Sage.

Carey, James (1990). Paper presented at conference, Toward a Comprehensive Theory of the Audience, Champaign-Urbana, Illinois.

Channels (September 1988). The new TV viewer: Confronting a nation of grazers. 8(8): 53–62.

Curran, James (1990). The new revisionism in mass communication research: A reappraisal. *European Journal of Communication* 5(2–3): 135–164.

Elliott, Philip (1972). *The Making of a Television Series.* London: Constable.

Espinosa, Paul (1982). The audience in the text: Ethnographic observations of a Hollywood story conference. *Media, Culture, and Society* 4(1): 77–86.

Fiske, John (1987). *Television Culture.* London and New York: Methuen.

Fiske, John (1989). Moments of television: Neither the text nor the audience. In Ellen Seiter, Hans Borchers, Gabrielle Kreutzner, and Eva-Maria Warth (eds.), *Remote Control: Television, Audiences, and Cultural Power,* pp.56–78. London and New York: Routledge.

Gans, Herbert J. (1957). The creator-audience relationship in the mass media: An analysis of movie making. In Bernard Rosenberg and David Manning White (eds.), *Mass Culture: The Popular Arts in America,* pp. 315–324 Glencoe, ILL.: Free Press.

Gans, Herbert J. (1979). *Deciding What's News.* New York: Pantheon.

Gray, Herman (1986), Social constraints and the production of an alternative medium: The case of community radio. In Sandra J. Ball-Rokeach and Muriel G. Cantor (eds.), *Media, Audience, and Social Structure,* pp. 129–142. Beverly Hills, Calif.: Sage.

Jarvie, I. C. (1970). *Movies and Society.* New York: Basic Books.

Krulwich, Robert (1990). Presentation at PBS-CPB conference, Exploring Prime Time, Hilton Head, South Carolina.

Liebes, Tamar, and Elihu Katz (1989) On the critical abilities of television viewers. In Ellen Seiter et al. (eds.), *Remote Control: Television, Audiences, and Cultural Power,* pp. 204–222. London and New York: Methuen.

McQuail, Denis (1969). Uncertainty about the audience and the organization of mass communications. In Paul Halmos (ed.), *The Sociology of the Mass—Media Communicators,* pp.75–84. Keele, Eng.: University of Keele.

McQuail, Denis (1987). *Mass Communication Theory: An Introduction.* London, Sage.

Martell, M. U., and G. J. McCall (1964). Reality orientation and the pleasure principle: A study of American mass periodical fiction (1890–1955). In L. A. Dexter and D. M. White (eds.), *People, Society, and Mass Communications,* pp. 283–333. Glencoe, Ill.: Free Press.

Parker, James J. (1986). The organizational environment of the motion picture sector. In Sandra J. Ball-Rokeach and Muriel G. Cantor (eds.), *Media, Audience, and Social Structure,* pp. 143–160. Beverly Hills, Calif.: Sage.

Pool, Ithiel de Sola, and Irwin Shulman (1959). Newsmen's fantasies, audiences, and newswriting. *Public Opinion Quarterly* 23(1): 145–158.

Radway, Janice (1984). Interpretive communities and variable literacies: The functions of romance reading. *Daedalus* 113: 49–71.

Schatz, T. (1981). *Hollywood Genres: Formulas, Filmmaking, and the Studio System.* Philadelphia: Temple University Press.

Schlesinger, Philip, Graham Murdock, and Philip Elliott (1983). *Televising Terrorism: Political Violence in Popular Culture.* London: Comedia.

Semetko, Holli A., Jay G. Blumler, Michael Gurevitch, and David H. Weaver (1991). *The Formation of Campaign Agendas: A Comparative Analysis of Party and Media Roles in Recent American and British Elections.* Hillsdale, N.J.: Erlbaum.

Turow, Joseph (1989). *Playing Doctor: Television, Storytelling, and Medical Power.* New York and Oxford: Oxford University Press.

Toward a Qualitative Methodology of Audience Study: Using Ethnography to Study the Popular Culture Audience

7

Andrea L. Press

In critical communication circles, there has been much recent discussion of the issue of using qualitative methods for studying the popular culture audience (Jensen, 1987; Lindlof, and Meyer, 1987; Fiske, 1987; Allor, 1988). Qualitative methodologies—in-depth interviews and participant observation, primarily—have been proposed in opposition to the quantitative methodologies that dominate mainstream audience research—primarily sample survey results and manipulations thereof, and experimental techniques. But why this interest by communication researchers in using qualitative methods? What is it researchers seek from qualitative methods that they find lacking in the quantitative approaches?

There is a stock answer to this query. Communication researchers have become increasingly dissatisfied with a purely demographic approach to audience study and more interested in investigating the ways in which subjectivity is constructed at the site at which the audience constitutes itself—a realm somewhat resistant to quantitative methodology. This has led us to challenge the term "audience" altogether for the way it emphasizes the passivity of this group and does not seem to allow an opening for studying the subjectivity of the audience (Fiske, 1987; Radway, 1988; Grossberg, 1988; Lindlof and Meyer, 1987). These challenges have opened the way for a much broader interpretation of what audience study is: we have to borrow from the human sciences—anthropology, sociology, psychology, history, and philosophy—a very general definition of the human community/communities, and of subjectivity itself, to begin to address these new interests.

As many have already argued, we in communication research need a broader definition of the "audience" as a human community (or one group from among many human communities). Focusing on the constitution of this community as popular culture/mass media audience per se will yield us only one facet of the whole. Of greater interest is the integration of this audience-aspect into the whole lives of individuals and communities, the interplay between people's dimension "as audience" and the meanings, rituals, practices, struggles, and structural roles and realities that make up the rest of their lives.

Ethnography has been widely proposed in recent works on the subject as the most promising qualitative method for audience studies, given our new definition of their domain (Morley, 1986; Radway, 1988; Grossberg, 1988; Fiske, 1988).[1] The way this term is used, however, unlike its very specific meaning in the discipline of its origin (anthropology), has been as a sort of catch-all phrase to include *all* qualitative methodologies, rather than to refer solely to ethnographic method per se—it has been used to refer to in-depth interviews and short-term observation as well as the longer-term participant observation that the term classically connotes. Ethnography in the classical sense has been widely used, particularly in anthropology but in sociology as well,[2] for studying the cultural beliefs, rituals, and practices of groups of people. But what is the classical ethnographic method? Can it be productively adapted to the goals of audience study? Will the ethnographic method yield the richer, more contextual understanding of the popular cultural "audience," however we want to define it, that we seek?

Ethnography is the method that defines anthropological research in the Western academic world. As a method, it has evolved within the context of the anthropological tradition. Ethnography began as an attempt to scientize the type of writing that characterized personal memoirs, travel books, journalism, and accounts by missionaries, settlers, and colonial officials that preceded it. In his introduction to *Argonauts of the Western Pacific,* first published in 1922 and regarded in anthropological circles as perhaps the first truly successful ethnography, the early British anthropologist Malinowski offers this instructive, basic definition of ethnography and its method:

> The first and basic ideal of ethnographic field-work is to give a clear and firm outline of the social constitution, and disentangle the laws and regularities of all cultural phenomena from the irrelevances. . . . This ideal imposes in the first place the fundamental obligation of giving a complete survey of the phenomena, and not of picking out the sensational, the singular, still less the funny and quaint. The time when we could tolerate accounts presenting us the native as a distorted, childish caricature of a human being are gone. This picture is false, and like many other falsehoods, it has been killed by Science . . . the whole area of tribal culture *in all its aspects* has to be gone over in research . . . for joining them into one coherent whole. . . . An Ethnographer who sets out to study only religion, or only technology, or only social organization cuts out an artificial field of inquiry, and he will be seriously handicapped in his work (Malinowski, 1922: 10–11; emphasis in original; partially quoted in Pratt, 1986: 27).

Malinowski turns to ethnography for reasons similar to our own: like Malinowski, we seek a method capable of recognizing the other as subject, of acknowledging the reciprocally human qualities of those we study. But Malinowski points out two main features of ethnography in this passage which highlight the differences between anthropological uses of ethnography and our own, and with which we must come to terms if we are legitimately to adapt this method to the communication field. First, ethnography as Malinowski defines it requires a sense of a "culture as a whole;" the ethnographer's goal is to operate as a cultural outsider to join a culture's pieces together into this coherent whole; ethnographer's who attempt artificially to carve out a piece of culture as their main object of study will find it difficult to achieve this goal. This is best done from an outsider's perspective; insiders are incapable, so anthropological theory asserts, of sensing these wholes that constitute culture, as this passage from the French anthropologist Levi-Strauss well illustrates:

> The ethnographer, while in no wise abdicating his own humanity, strives to know and estimate his fellowmen from a lofty and distant point of vantage: only thus can he abstract them from the contingencies particular to this or that civilization. The conditions of his life and work cut him off from his own group for long periods together; and he himself acquires a kind of chronic uprootedness from the sheer brutality of the environmental changes to which he is exposed. Never can he feel himself "at home" anywhere: he will always be, psychologically speaking, an amputated man (Levi-Strauss, 1974: 58; quoted in Geertz, 1988: 36).[3]

Second, ethnography, at least as it was originally conceived, was meant to replace—to "kill" with true scientific record—the long and varied traditions of nonscientific writing about alien cultures that preceded it. It is meant to provide us with a *record* of culture and the individuals within it. Of course, more current theories of ethnography (Geertz, 1988, 1973; Clifford and Marcus, 1986; Marcus and Fischer, 1986) have moved beyond Malinowski's emphasis on "objective" science to a more interpretive, semiotic theory of what ethnography is. Thus Geertz tells us that ethnography is an attempt to converse with others, to "enlarge the universe of human discourse" (1973: 14). But even Geertz is careful to retain the sense that ethnography is actually a species of scientific practice; aside from observing and participating in social discourse, an ethnographer by definition "inscribes social discourse; *he writes it down*" (1973: 19), creating what in effect becomes a scientific document used to further the ends of anthropological practice.

Culture as a Whole

Malinowski's definition of ethnographic method presumes that culture forms a coherent whole. In fact, the discipline of anthropology itself arose from the belief in this notion—that culture coalesced in each separate society into unique "patterns" that the observer could, through meticulous, long, and detailed field observations, describe, as in Benedict's programmatic saying that "a culture, like an individual, is a more or less consistent pattern of thought and action" (1934: 46).

Certainly the extent to which this can be true of complex modern societies is a matter of continuing debate (although Benedict thought that it was);[4] and certainly this idea has been modified when anthropologists have turned to complex societies as fieldwork sites, which they—like us—have increasingly done. The concept of the outsider coming into an alien culture and being able to sense its distinctive patterns and characteristics has certainly been modified as it has become more acceptable among anthropologists to attempt to study some aspect or aspects of their own, complex cultures, or of complex cultures generally. Yet what is striking about recent discussions of ethnography in communication studies is the virtual absence of any consideration of the idea of the cultural whole (Fiske, 1987; Morley, 1986; Lindlof and Meyer, 1987).[5]

Recent communication literature also shows an almost complete lack of attention paid to the problem of the communication researcher's status as an "insider" to the culture he or she studies. In our field, we need more discussion of the meaning of doing ethnography in our own cultures, which after all is what we are talking about when we debate using ethnographic techniques to study audiences in our own societies. The situation we normally construct in audience research is one in which the observer sets out to learn fundamental cultural patterns about his or her own culture, or about a subculture not as far removed as the truly alien culture posited by the anthropological model. This is a significant alteration of the traditional ethnographic pattern, one that at the very least deserves some attention if we are to construct a workable ethnographic method for audience study.

The Sociological Context of Audience Ethnography: Interpretive Method Meets "Science"

Neglect of both the "culture as a whole" issue and the "insider-outsider debate" resulting from it adds up to the somewhat simplistic assumption, common in communication writing, that we can do an "ethnography of," for example, family television-watching or women's reception of romance novels, and that somehow this sort of ethnography is similar enough to the anthropologically orthodox definition of the term to merit the same label. Malinowksi would not think so, and we should at least address the problems involved in translating this method between Western and non-Western societies, as anthropologists and sociologists have begun to do (Kottak, 1982; Spradley and Rynkiewich, 1975; Rabinow, 1977; Gans, 1962).

Of course, ethnography's original scientific aim has for years been transmuted and challenged, and the activity of ethnography is now seen even by anthropologists as more of an hermeneutic, critical activity, rather than a traditionally scientific one. Yet the intellectual context and specific goals of ethnography in anthropology remain largely different from those in the communication field. Whereas in anthropology ethnographers set out very generally to (in Geertz's terms) "enlarge the universe of human discourse," e.g., to make possible discourse between the inhabitants of diverse cultures, we in communication research operate with more specific questions. Radway, for example, in her study of female romance readers, uses assumptions derived from feminist theory that help her to identify women as an oppressed group and to identify their particular objects of cultural consumption as interesting. These assumptions were critical to identifying women romance readers and their reading practices as interesting and important to study.[6] Researchers influenced by the Centre for Contemporary Cultural Studies in Birmingham (these include Morley, 1986; Willis, 1978; McRobbie, 1978; 1980—each of whom have produced ethnographically based documents documenting popular culture use) frame their questions from a Marxian-derived perspective that emphasizes the importance of actors, located in class-based subcultures, whose cultural actions amount to resisting the dominant culture.

In our country, however, most critical audience researchers have not found it necessary to give us much *sociological* sense of the people who constitute the audience (or of the critic him-or-herself, for that matter). That is, they have slighted the importance of questions such as: Who are they in the context of the society in which they live? What is their history as a group? What is their current social-structural reality? What work do they need to do every day to survive and to reproduce themselves? Assuming an ethnographic approach in a classical sense, these more empirical, scientific questions have been overlooked in this tradition, even though in most cases, critical audience study is motivated by questions that emerge through certain empirically based assumptions about the audience.

Like the current critics of audience research as it now stands, I too believe that audiences do not really exist simply as audiences; they are constructs that we in communication studies have created, abstracting them out of their sociological context to help us study mass media and popular culture reception, usually in a quantifiable, measurable, "scientific" way. But in order to deconstruct this abstraction we need some empirical armor: a sociological sense of the people who have constituted these audiences throughout the decades we have studied them. Indeed, it is this sort of understanding that either implicitly or explicitly gives rise in the communication field to our questions, our studies, and ultimately our sense of who the audience is. Given the complex societies that form the context of our investigations, ethnographic work in our field inevitably moves us away from the classical situation, that somewhat romantic realm depicted in the writings of the classical ethnographers, and into the sociological. We need, in short, both more sense of "us"—who we the researchers are, and what context has given rise to our questions—and of "them"—those groups we construct through our analytical categories. In this enterprise there is nothing to be gained and much explanatory power to be lost by abstracting away this sociological context from our newly broadened notion of who the audience is.

Ethnography as Record

The criticisms I will inevitably draw with this argument include both general critiques of the possibility of objective social observation, and those related to postmodern notions of the individual, the "nomadic" subject that Radway (1984) and others (Grossberg, 1987, 1988; Deleuze and Guattari, 1977) have so elegantly described. Malinowski's claim of anthropology to be science that replaces (or "kills") earlier, impressionistic accounts has clearly been challenged in current writing about ethnography—writing that has benefited from our increasing sophistication in the social sciences concerning the nature of social science inquiry as a whole, the role of writing within it, and the nature of our "objects" (Clifford and Marcus, 1986; Marcus and Fischer, 1986; Nelson and Grossberg, 1988; Geertz, 1973, 1988).

The issue that concerns me here is the relevance of current theoretical debates to the project of constructing a workable qualitative or ethnographic method for audience or related studies. One increasingly feels that part of the impetus for the "turn to ethnography" in communication is the theoreticians' search for an empirical method that is "not too empirical," for example, for a method that avoids the most egregious sins of a theoretical empiricism, such as ignoring the constructed nature of the very concepts of audience and, more importantly, of subjects themselves.[7] These debates are proving to be increasingly irrelevant, in my view. As anyone who has attempted qualitative work of any kind will admit, be it interviews, participant observation, or ethnography itself, there is a contradiction between, for example, current postmodernist questioning of the existence of a unified subject and the experience of conversing with, interacting with, living with, and observing people who at least think of themselves and present themselves consistently as unified subjects, despite the inevitable contradictions that often arise around the edges of their discourse.

While the notion that ethnography is "science" and the impartiality of ethnographic observers may certainly be legitimately questioned,[8] it seems to me that we need to set aside some of the current theoretical controversies particularly concerning the nature of the subject in favor of a working concept of an objective, existing, living, breathing, and creative subject if we are to do ethnographic work or qualitative work of any kind, all of which necessarily involves creating a written record about this subject. There is simply no getting around this basic fact of fieldwork. If we continue our tendency to be more attuned to the influence of postmodernist theory than our anthropological heritage, we will soon have to acknowledge that our investigations are no longer ethnographic at all, but rather of some other less empirical ilk. This may in fact please many critical communication researchers, most of whom eschew an empirical legacy. But rather than mucking about with an empirical method that does not seem too empirical, communication researchers should at least stand firmly behind the tradition they support—either attempting to adapt qualitative social science to communication research or rejecting it altogether.

While I may agree with Foucault that our notions of "experience" have themselves been constructed and that our very notion of the subject itself bespeaks subjection to forces of domination rather than autonomy and freedom,[9] the practice of ethnography (and the academic writing it implies), in contradistinction to postmodernist and poststructuralist challenges, depends—indeed turns—upon both of these concepts and by its inherent configuration no doubt contributes to perpetuating them as concepts basic to qualitative work in the social sciences. If we are to do qualitative work, we are stuck with these concepts, unless the nature of that research is radically redefined—and this has not yet occurred.[10] We need these concepts of experience and the subject not only because they help us to produce finer work by enabling us to conceptualize the role of the observer, the motivations of our work, and the way we discursively and practically interact with those we observe, but also because without them we substitute our own exhaustively formulated critique of subjectivity in modern or postmodern life for the working concept of subjectivity (individuality may be more accurate) held by the people we claim to respect: our informants. We cannot both claim to do ethnography and at the same time discard our informants' working definitions of themselves as individuals.[11] If we would like to place this working concept alongside our critical deconstruction of postmodern subjectivity, so be it. We may continue to redefine the scientific nature of the ethnographic project—but the subject we encounter in our fieldwork must be assured of its place.

Toward an Ethnography of the Audience

In order to forge a workable, qualitative method for audience study, it will help to note the similarities between this kind of work and classical ethnography, as well as the differences between them. I have already mentioned one similarity directly: both rely on a working concept of a unified subject since both rely on the observer's discursive and practical interaction with informants. In spite of the observer's willingness to subject all of his or her assumptions to severe deconstructive analysis, all qualitative research has an unmistakably empirical edge. People and their meaningful discourse are treated, in some respects at least, as objects, the objects of ethnographic record—objects with whom the ethnographer enters into discursive interaction, but objects, in some respects, nonetheless. This objective thrust is mediated epistemologically by the fieldworker's subjective interaction with them and the fact that his or her knowledge of those studied occurs in part through this discursive enterprise. In addition, and perhaps most crucially, their objectivity is ontologically mediated by the fieldworker's respect for the subjectivity of his or her informants, which, I argue here, we must strive to maintain even in the face of the redefinition of this concept in deconstruction theory.[12]

The main differences between classical ethnography and the method we seek lie in the fact that, unlike the classical ethnographers, we often study people living directly within the context of complex, Western societies, and we ourselves share with them this cultural heritage. We need more investigations into the nature of the cultural "whole" within these societies and of the impact of this culture upon ourselves as we attempt to learn more about it through observing others of relatively similar background. Most popular culture research is motivated by personal concerns, often inspired by our own personal experience with aspects of the popular culture we study (and, of course, partly determined by the sociological context of that experience). How could this be otherwise, when we are all products of this same culture? The problems engendered by this origin of our questions, its impact on our framing of problems and our emphases of study, must be more fully addressed than they are at present in our literature. On these points, we can learn from debates in sociology as well, where the problems of observer involvement and observer insider status have received significant attention since the inception of qualitative work in the early years of this century (see note 2).

So where does this leave those of us interested in studying the construction of meaning, and its role in constituting resistance, rebellion, and freedom in relatively repressive societies?[13]—poised unavoidably, as are all researchers, somewhere between societal facts and interpretive analysis. If we are to specify a powerful, qualitative methodology for studying the audience (or for studying what we used to think of as the audience, or whatever new conception may take its place), we must do the following. First, we must learn to supplement our qualitative studies with descriptions of the sociological context within which popular culture is, in the modern world, both created and received. Second, we must make more explicit the empirical frameworks that have given rise to our focus on popular cultural analysis in the first place, in particular the essentially Marxist social critique that leads us to search for freedom and resistance in the realm of leisure. Finally, we must bring these two spheres, our studies of culture and our sociological analysis, both of those we are studying and of our own theoretical basis for framing problems and constructing studies, into more dialogue with one another than we have thus far been able to do.

I agree, in the end, with Lull when he calls for "fewer calls for research and more research" (1987a: 322), that is, less theoretical debate over why such research is impossible and more research itself. In anthropology, it is ironic that those most involved in practicing ethnography are widely recognized to be those least involved in constructing theory (the "cultural theory" that some feel is the desirable developmental result of the accumulation of ethnographic record) and least likely to take part in theoretical debate.[14] Ethnography is often maintained to be antithetical to theory (Geertz, 1973: 24–26). In contrast, those most actively urging the practice of ethnography in the communication field recently (Fiske, 1988, 1990; Seiter, 1990; Grossberg, 1988; Radway, 1988) have been those most embroiled in current theoretical controversies over the nature of the subject, the meaning of "audience," and the nature of culture generally. It is somewhat ironic, too, that those most theoretically astute in our field advocate the use of a method that, in its discipline of origin, has been so staidly antitheoretical—and this situation has contributed to the difficulties in our field that those advocating ethnography have most probably experienced in attempting to practice what they have preached.

This paradox is due in part to the unusually complex origins of the communication field. As a field, we encompass researchers with origins in both the social sciences and the humanities. What is encouraging in recent developments is that members of each camp have become interested in the concerns and methodologies of the other. Different methodological and theoretical traditions and training, however, sometimes cause an uneasiness of fit in this transition. Humanists, trained in the methodology of textual analysis, find that their training gives them insights into qualitative social science, its discursive aspects in particular—especially the "we are all texts" approach that has become a familiar lament in critical communication circles—and many of these have helped social science researchers to become more self-reflective and conscious of their practices. At the same time, however, not all insights translate smoothly between these spheres.

For example, consider Fiske's recently proposed research agenda for communication ethnographers. He proposes that we do a

> semiotic ethnography . . . [in which] there are no texts, no audiences. There is only an instance of the process of making and circulating meanings and pleasures. There are no individuals and no social categories, only those cultural processes that transect the individ ual and the social. . . . What matters is not the audience and not the television text but the generation and circulation of meanings and pleasures throughout our contemporary social formations (1988: 250).

This is a strikingly diffuse model for constructing an actual research agenda. One has difficulty imagining these programs translating into workable research plans. Though I applaud Fiske's desire to escape the theoretical entrapments of "positivism" with his more semiotic version of ethnography (1988: 250), I am not at all sure that a researcher could him-or herself experience a research situation without invoking the concrete concepts of subject and experience that, as I have argued above, have been basic constituents of ethnographic practice. I fear there is an unbridgeable dichotomy between Fiske's theoretical convictions regarding the inherent plurality of subjects, and the requirements of ethnography that subjects be conversed with and observed.

Radway's proposed field project (1988) is a bit more specific, and more oriented toward actual fieldwork, but unfortunately poses similar problems of translation between theory and fieldwork. Acknowledging that ethnographic audience studies (including her own) have thus far been too limited in their focus on subjects as audience rather than as fully articulated individuals embedded within their community, Radway proposes that an ethnographic team investigate

> the production of popular culture within the everyday as a way of trying to understand how social subjects are at once hailed successfully by dominant discourses and therefore dominated by them and yet manage to adapt them to their own other, multiple purposes and even to resist or contest them. By studying the extremely heterogeneous set of practices through which the popular is produced in relation to the legitimate culture or within its interstices, we might render visible the unceasing and heretofore unacknowledged cultural work through which nomadic subjects and dispersed groups confound the unity of domination by articulating together discursive fragments and practices from many different sources and regions. We may be able to see them, then, not simply as audiences, as receivers of the messages of others, but rather as active individuals who productively articulate together bits and pieces of cultural material scavenged from a multitude of sites and who, in doing so, nomadically, perhaps even slyly, take up many different subject positions with respect to the dominant cultural apparatuses (1988: 368).

One is struck by the ambitiousness of the project and the problems this might pose for actually carrying it out. I agree, however, with Radway's attempt to expand the domain of audience research to include a more sociologically situated view of audience members. This is precisely the direction in which audience research must move, away from the "one-aspect" studies of the past, toward a fuller consideration of the context of their media use and practices within practitioners' lives and of how these practices help people to come to know the world and to act within it.

As the reader must anticipate, however, I take some issue with Radway's characterization of her subjects as "nomadic." Her use of this term precludes the main point of the ethnographic inquiry Radway seeks: knowledge about the ability of the subjects we study to actively resist the dominant order. Reducing the cultural activity of these subjects to a collection of unrelated, fragmented "struggles" makes it impossible for us to conceive of them engaging collectively in activity that enables them to truly resist the strictures of the current order. Radway herself well states the disjuncture between the base assumptions of theoretical debates in postmodernism and those of ethnographic method, describing it as "the disjuncture I see and feel between the windless, already lost world described by theories of postmodernism (theories I find disturbingly persuasive) and a world not yet surrendered, still being struggled over by living subjects" (1988: 373). To collapse the latter view and its method into the former, to try to fit the terms of ethnographic inquiry within the confines of postmodern theories of the subject, would ill-serve the purposes, and the potential, of both lines of thought and research. Best, perhaps, to sever the connection.

The debate between Grossberg and Radway in that same 1988 issue of *Cultural Studies* (Radway, 1988; Grossberg, 1988) well illustrates the dangers we face in reconciling these two poles. Grossberg, well known for his theoretical work, lauds Radway in his commentary for attempting to reconcile the distance between the worlds of the ethnographer and postmodern critic. Yet he concludes his discussion by gently chiding Radway (and other recent cultural theorists; see Bennett and Woollacott, 1987, and Corrigan and Willis, 1980) for too strong an attachment to the notion that "it is always the subject who constructs reality," that it is "'historical subjects' who 'articulate their cultural universe' and 'articulately produce subjectivity'" (1988: 387). Here, I fear, Grossberg mistakes the seriousness with which Radway considers ethnographic method and its necessary presuppositions for an expression of her own theoretical convictions. In fact, in Grossberg's research agenda, or rather, in his theoretically programmatic statement, the "nomadic subjects",

> are always empowered and disempowered, shaped and reshaped, by the effectivities of the practices (trajectories, apparatuses, etc.) within which their agency is itself located. Thus a materialist theory of articulation proposes studying not people but practices. For example, even within the plane of ideology, a theory of articulation involves looking neither for the intrinsic meaning of a text nor for people's interpretations/uses of it. Instead, it directs us to look at how a place or contradictory places are made for the text in the wider field of forces (Grossberg, 1988: 387).

This agenda bears an extremely close similarity to Radway's in that both preclude the centrality of an agentic subject, and overlook the fact that ethnographic method itself relies upon this subject by its very nature. Grossberg does not give Radway enough credit for her serious attempt to confront this very difficult disjuncture between our field's theoretical and methodological commitments.

A more recent piece by Fiske (1990) is an extremely useful self-reflection about the process of doing fieldwork. Written in a newly emerging personalistic style (see also Sanjek, 1990; Seiter, 1990; Grossberg, 1989), his account is part of a growing trend that promises to bridge the gap between text-oriented and field-oriented research.[15] Recent personalistic reflections about the process of doing fieldwork, such as those by Fiske (1990) and Seiter (1990), focusing on the experience of the researcher and the way in which the researcher's concerns, context, and experience come into play certainly highlight one critical aspect of the research process, an aspect that has been neglected by both quantitative and qualitative researchers and can only improve the quality of research by both. Rarely have researchers been exposed to this essentially humanistic focus on the individual. And in fact it is precisely on this issue—the approach toward the definition of the individual or subject—that the humanities/social science split makes the most difference. The emphasis in the humanities (until the postmodern present) has traditionally been on the ability of the individual to act in opposition to the group that gives rise to that individual. It is this that makes individuals interesting, motivating the main questions and areas of study in the humanitarian disciplines. In the social sciences, in contrast, most tend to take a certain amount of determinism for granted. We *assume* that we can attribute some part—a large part—of each individual's actions to the impact of society, small groups, the family, or culture, an assumption that often strikes those in the humanities as distinctly uninteresting if not banal. This dichotomy between the humanities and the social sciences makes it difficult to reconcile the aims of the two divisions; perhaps the disjuncture is particularly evident in a field such as communication, which spans both.

I certainly faced aspects of this disjuncture in my own research (Press, 1989; 1990; 1991). I used in-depth interviews to investigate women's television reception in the United States, comparing responses of different social-class and age-differentiated groups.[16] While my research is not explicitly enthographic in the sense I have laid out here, I came up against a number of the basic issues I have discussed. With respect to my informants, I occupied a rather complicated insider-outsider position that current sociological categories helped me to assess. Middle-class in background, I was a "class" insider with respect to my middle-class group; under thirty, I was an "age" insider with respect to my younger group, although many were younger than I. These qualities made me more of a cultural outsider with working-class women, particularly with working-class women in my older category. Yet other planes—race, religion, geographical origin, ethnicity, sexual orientation—made me more or less insider in some cases, outsider in others. All of these factors at least contributed to the different degrees of rapport I achieved with the different women I interviewed.

The categories by which I decided to limit my study–gender, class and age—were in some respects hopelessly inadequate markers of the meaningful group differences within a complex culture like our own. An interest in gender motivated the study and I examined social class and age as the categories most likely to shed light on gender as a primary category of experience, since these are the categories most often used in the current sociological literature about it.[17] While these choices enabled me to fit my study into existing debates within the literature about the interrelationship between gender, social class, and generation, these groupings only superficially describe how people are organized within complex societies. They are a beginning, but remain to be fleshed out by more ethnographic detail about audience groupings in our society, detail which is difficult to gather given that our approach to gathering qualitative data at all in complex societies is necessarily organized by the very categories whose limitations we seek to transcend.

In some respects, I straddle the fence—as qualitative audience researchers, I argue, have no choice but to do—between acknowledging that traditional stratification divisions are important and asserting that these traditions must be radically modified to treat seriously the issue of women and other issues of interest to critical researchers. Basic sociological divisions—the distinction between individual and society, for example, and the very notion of the individual subject him- or herself—are called into question when one takes seriously the issue of women, as postmodernist theorists, feminist theorists, and critical social scientists in all disciplines have long been arguing (Fraser and Nicholson, 1990; Mascia-Lees, Sharpe, and Cohen, 1989; Pateman, 1988).

On another level as well, my categories of proved inadequate. Talking with individual women about their reception of, reactions to, and thoughts about television required that I organize my data around "individual" respondents, at least on the first level of gathering and analyzing it. Yet my decision, based on preliminary interviewing, to organize my data according to social class and generational divisions meant that from the very beginning, as researcher and observer I was grouping, systematizing, categorizing, and summarizing individual responses according to the criteria that divided one group from another. Inevitably I was glossing over the individuality of my informants, which paradoxically I had to respect and appreciate in order to conduct interviews at all. Grouping individual responses, mentally during my interviews and later explicitly as I wrote the book, was a process I found difficult, even ultimately disrespectful to the people who gave me their thoughts and their time. There was a fundamental paradox to my task, one that I can only now, as I look back on and begin to appraise the construction of the study, begin to appreciate.

Current arguments about the interpretive nature of qualitative research are certainly appealing to me at this point. But in a fundamental way, these discussions do not solve the main problem. They do not help me move from a humanitarian appreciation of my subjects' uniqueness to a coherent social-scientific and critical perspective on the way this uniqueness has been hindered, crippled, but also produced in part by the social conditions that both entrap but also facilitate these women's lives. I began my study looking for the ways in which women have developed critical subcultures while existing within a broader and dominant patriarchal culture. But of course in many ways women (like other oppressed groups) have not developed these critical voices at all, or have developed voices whose critical potential is extremely blunted. Sociological categories, however imperfect, helped me to identify the cases—and, yes, some of the groups—within which this blunting was most likely to occur.

Should we lose the critical insight such categories provide us as observers, it is likely that our research, while perhaps becoming more self-reflective, will trade analytic focus for this. Both ends—our analytic properties as well as our power of reflection—must be sharpened simultaneously if we are to avoid the postmodernist abyss and retain and develop the tools necessary for actual critical research. The field of communication, poised precariously between the social sciences and the humanities, is in some respects an ideal home for the emergence of such a critical tradition, one that melds the social-scientific search for regularities and critical purpose with the humanistic celebration of actual critical moments and movements. I hope we take advantage of our interdisciplinary nature rather than allowing it to paralyze our research efforts.

Notes

1. In a recent article, Radway (1988) brings up some of the issues concerning the uses of ethnography in communication research to which this article is addressed. In this discussion I am heavily indebted to her.

2. In the field of sociology there is of course a long tradition of the use of participant observation in the study of both urban and rural communities within complex societies. See, for example, Whyte (1943), Gans (1962), and Becker (1963) for examples of this work. Both Gans (1962: appendix) and Whyte (1943: preface; 1967) explicitly discuss some of the problems and the strengths of using participant observation to study urban communities. See Bulmer (1984) and Kurtz (1984) for overview discussions of the development of empirical sociology at the University of Chicago. "Chicago School" sociology is often associated with the use of ethnographic methods, and participant observation in particular; Kurtz argues, however, that qualitative methodology was never as centrally important in sociology as is often believed (1984: 84–88). Sociological work most commonly proceeded from a concern with social problems, which early on differentiated it from anthropological interest in culture more generally. Our interests in communication transect the two fields, but since the discussion in our field has explicitly centered on the term "ethnography," I deemphasize in this paper related discussions concerning the use of participant observation in urban communities that have occurred within sociology. The reader should be aware, however, that such concerns have been raised and discussed with some relevance by sociologists.

3. It should be noted here that many in anthropology have questioned Levi-Strauss's credentials as a true "ethnogapher" precisely because he presciently interweaved self-conscious philosophical critique throughout his writing, as did other French anthropologists of his time.

4. Benedict felt that in time we would discover the patterns of culture that characterized complex societies, just as we were discovering those characterizing primitive ones. She felt, however, that in complex societies these patterns were more difficult to discern, not

only because of the size of these societies, but also because the analyst was an insider when studying complex societies, and outsider status was crucial to enabling the observer to discern cultural patterns (1934: 54–56).

5. Radway, however, does mention the difference between the ethnographic object in anthropology and in our own more limited domain (1988: 367).

6. It is important to note here that since publishing *Reading the Romance* in 1984, Radway has reflected quite critically and extensively on her own use of ethnography in that work, herself voicing some of the concerns I raise here (see Radway, 1988).

7. Fiske, interestingly enough, criticizes the trend toward ethnographic study of the popular culture audience for its "positivist" aspects, although he wants to retain a more "semiotic" ethnographic method (1988: 250). This distinction is broadly misconceived, as it ignores the necessarily semiotic or interpretive aspects, as well as the empirical or positivist aspects, of all ethnography. See the discussion of Fiske in the concluding section of this chapter.

8. On this point, see especially Kuhn (1962) and Bernstein (1983).

9. Foucault (1973, 1988); see also Dreyfus and Rabinow (1982).

10. Although some have begun this process of redefinition, as I discuss in detail below. See especially Radway (1988) and Fiske (1988).

11. Of course, in the end, the ethnographer can only attempt to incorporate people's own definitions of the subjectivity into his or her interpretation, and must admit to being in control of the final text: "However much multiple authorship is acknowledged, using people's experiences to make statements about matters of anthropological interest in the end subordinates them to the uses of the discipline" (Strathern, 1987: 289).

12. See especially Scott (1988) for a good example of the deconstructionist redefinition of "subjectivity." See Sewell (1989) for a detailed and interesting critique of Scott's position.

13. See Lembo and Tucker (1990) for a fuller discussion of the sociological and historical context that communication researchers ignore when discussing "resistance" in particular.

14. Jackson (1990) writes of how, when she interviewed three hundred ethnographers about their methodology of notetaking in the field, most mentioned that they looked with suspicion at the new cultural theory and self-consciousness about methodology that is coming to characterize anthropology.

15. In the anthropological literature, see also Friedrich (1986), Rosaldo (1980), Shostak (1981), and Rabinow (1977).

16. The data presented are culled from a larger study involving interviews with approximately twenty working-class and twenty middle-class women. See Press (1990; 1991) for a fuller description of the methodology of this study.

17. As is done in most of the sociological literature about gender, I ignored the category of race. This tendency is slowly changing, but the common desire to enter into dialogue with the current literature that influenced many of the choices I made in my work is an extremely conservative force that works toward reproducing the categories of the status quo and militates strongly against radical change in those categories.

References

Acker, Joel (1973). Women and social stratification: A case of intellectural sexism. *American Journal of Sociology* 78: 936–945.

Allor, Martin (1988). Relocating the site of the audience. *Critical Studies in Mass Communication* 5(3): 217–233.

Benedict, Ruth (1934). *Patterns of Culture.* Boston: Houghton Mifflin.

Becker, Howard (1963). *The Outsiders.* New York: Free Press.

Bennett, Tony, and Janet Woollacott (1987). *Bond and Beyond: The Political Career of a Popular Hero.* New York: Methuen.

Bernstein, Richard J. (1983). *Beyond Objectivism and Relativism.* Philadelphia: University of Pennsylvania Press.

Bottomore, Tom, et al. (eds.) (1983). *A Dictionary of Marxist Thought.* Cambridge, Mass.: Harvard University Press.

Bulmer, Martin (1984). *The Chicago School of Sociology.* Chicago: University of Chicago Press.

Clifford, James, and George E. Marcus (1986). *Writing Culture.* Berkeley and Los Angeles: University of California Press.

Corrigan, Philip, and Paul Willis (1980). Cultural forms and class mediations. *Media, Culture, and Society* 2: 297–312.

Dahrendorf, Ralf (1959). *Class and Class conflict in Industrial Society.* Stanford: Stanford University Press.

Deleuze, Gilles, and Felix Guattari (1977). *Anti-Oedipus: Capitalism and Schizophrenia.* Translated by Robert Hurley, Mark Secm, and Helen R. Lane. New York: Viking Press. (Originally published 1972.)

Delli-Carpini, Michael, and Bruce A. Williams (1990). The television audience: implications for political theory. Paper delivered at the June, 1990, Meetings of the International Communication Association, Dublin.

Dreyfus, Hubert L., and Paul Rabinow (1982). *Michel Foucault. Beyond Structuralism and Hermeneutics.* Chicago: University of Chicago Press.

Epstein, Cynthia Fuchs (1970). *Woman's Place.* Berkeley: University of California Press.

Felson, Marcus, and David Knoke (1974). Social status and the married woman. *Journal of Marriage and the Family* 36: 516–521.

Fiske, John (1986). Television and popular culture. *Critical Studies in Mass Communication* 3(3): 200–216.

———.(1987). *Television Culture.* London and New York: Methuen.

———.(1990). Critical response: Meaningful moments. *Critical Studies in Mass Communication* 5(3): 246–251.

Foucault, Michel (1973). *The Order of Things.* New York: Mintage.

———.(1988). *Politics, Philosophy, Culture.* Translated by Alan Sheridan et al. New York: Routledge.

Fraser, Nancy, and Linda J. Nicholson (1990). Social criticism without philosophy: An encounter between feminism and postmodernism. In Linda J. Nicholson (ed.), *Feminism\Postmodernism*, pp. 19–39. New York and London: Routledge.

Friedrich, Paul (1986). *The Princes of Naranja: An Essay in Anthrohistorical Method.* Austin: University of Texas Press.

Gans, Herbett J. (1962). *The Urban Villagers.* New York: Free Press.

Geertz, Clifford (1973). *The Interpretation of Cultures.* New York: Basic Books.

———.(1988). *Works and Lives: The Anthropologist as Author.* Stanford: Stanford University Press.

Giddens, Anthony (1973). *The Class Structure of the Advanced Societies.* New York: Harper & Row.

Grossberg, Lawrence (1987). The In-Difference of television. *Screen* 28(2): 28–45.

———.(1988). Wandering audiences, nomadic critics. *Cultural Studies* 2(3): 377–391.

———.(1989). On the road with three ethnographers. Journal of Communication Inquiry 13(2): 23–26

Hacker, Helen (1951). Women as minority group. *Social Forces* 30: 60–69.

Hall, Stuart (1986a). The problem of ideology: Marxism without guarantees. *Journal of Communication* 10(2): 28–44.

———.(1986b). On postmidernism and articulation: An interview. *Journal of Communication Inquiry* 10(Summer): 45–60.

Hartmann, Heidi, and Amy Bridges (1974) Pedagogy by the oppressed. *Review of Radical Political Economics* 6: 75–79.

———.(1979). The unhappy marriage of marxism and feminism: Towards a more progressive union. *Capital and Class* (Summer).

Jackson, Jean E. (1990). I am a fieldnote': Fieldnotes as a symbol of professional identity. In Roger Sanjek (ed.), *Fieldnotes: The Makings of Anthropology.* Ithaca: Cornell University Press.

Jensen, Klaus Bruhn (1987). Qualitative audience research: Toward and integrative approach to reception. *Critical Studies in Mass Communication* 4(1): 21–36.

Kottak, Conrad Phillip (ed.) (1982). *Researching American Culture.* Ann Arbor, Mich.: University of Michigan Press.

Kuhn, Thomes S. (1962). *The Structure of Scientific Revolutions.* Chicago: University of Chicago Press.

Lombo, Ronald, and Kenneth H. Tucker, Jr. (1990). Culture, television, and opposition: Rethinking cultural studies. *Critical Studies in Mass Communication* 7(2): 97–116.

Lévi-Strauss, Claude (1963). *Structural Anthropology.* Translated by Clair Jacobson and Brook Grundfest Schoepf. New York: Basic Books.

———.(1974). *Tristes Tropiques.* Translated by John and Doreen Weightman. New York: Atheneum.

———.(1975). *The Raw and the Cooked.* Translated by John and Doreen Weightman. New York: Harper & Row.

Lindlof, Thomas R. (ed.) (1987). *Natural Audiences.* Norwood, N.J.: Ablex.

———.(1988). Media audiences as interpretive communities. *Communication Yearbook* 11: 81–107.

Lindlof, Thomas R., and Timothy P. Meyer (1987). Mediated communication as ways of seeing, acting, and constructing culture: The tools and foundations of qualitative research. In Thomas R. Lindlof (ed.), *Natural Audiences,* pp. 1–32. Norwood, N. J.: Ablex.

Lulli James (1987a). Critical response: Audience texts and contexts. *Critical Studies in Mass Communication* 4(3): 318–322.

———.(1987b). Critical response: The audience as nuisance. Critical Studies in Mass Communication 5(3): 239–243.

Malinowski, Bronislaw (1922). *Argonauts of the Western Pacific.* New York: E. P. Dutton.

Marcus, George E., and Michael M. J. Fischer (1986). *Anthropology as Cultural Critique.* Chicago: University of Chicago Press.

Mascia-Lees, Frances E., Patricia Sharpe, and Colleen Ballerino Cohen (1989). The postmodernist turn in Anthropology: Cautions from a feminist perspective. *Signs* 15(11): 7–33.

McRobbie, Angela (1978). Working-class girls and the culture of femininity. In Women's Studies Groups (eds.), *Women Take Issue,* pp. 96–108. London: Hutchinson.

———.(1980). Settling accounts with subcultures: A feminist critique. *Screen Education* 34: 37–49.

Middleton, Chris (1974). Sexual inequality and stratification theory. In Frank Parkin (ed.), *The Social Analysis of Class Structure,* pp. 179–203. London: Tavistock.

Milkman, Ruth (1986). Women's history and the Sears case. *Feminist Studies* 12(2).
———.(1987). *Gender at Work: The Dynamics of Job Segregation by Sex During World War II.* Chicago: University of Illinois Press.
Millett, Kate (1971). *Sexual Politics.* New York: Avon.
Morley, David (1980). *The "Nationwide" Audience: Structure and Decoding.* London: British Film Institute Television Monograph, No. 11.
———.(1986). *Family Television.* London: Comedia.
Nelson, Cary, and Lawrence Grossberg (eds.) (1988). *Marxism and the Interpretation of Culture.* Urbana: University of Illinoise Press.
Oakley, Ann (1974). *Woman's Work: The Housewife, Past and Present.* New York: Vintage.
Parkin, Frank (1971). *Class, Inequality, and Political Order.* New York: Praeger.
———.(1978). Social stratification. In Tom Bottomore and Robert Nisbet (eds.), *A History of Sociological Analysis*, pp. 599–632. New York: Basic Books.
———.(1979). *Marxism and Class Theory: A Bourgeois Critique.* New York: Columbia University Press.
Pateman, Carole (1988). *The Sexual Contract.* Stanford Stanford University Press.
Poulantzas, Nicos (1973). *Political Power and Social classes.* London: New Left Books.
———.(1975). *Classes in Contemporary Capitalism.* London: New Left Books.
Pratt, Mary Louise (1986). Fieldwork in common places. In Clifford and Marcus (eds.), *Writing Culture*, pp. 27–50 Berkeley and Los Angeles: University of California Press.
Press, Andrea (1991). *Women Watching Television.* Philadelphia: University of Pennsylvania Press.
———.(1990). Class, gender, and the female viewer: Women's responses to *Dynasty.* In Mary Ellen Brown (ed.) *Television and Women's Culture.* Newbury Park: Sage.
———.(1989). Class and gender in the hegemonic process: Class differences in women's perceptions of television realism and identification with television characters. *Media, Culture, and Society* 11(2): 229–252.
Rabinow, Paul (1977). *Reflections on Fieldwork in Morocco.* Berkeley and Los Angeles: University of California Press.
———.(1986). Representations are social facts: Modernity and post-modernity in anthropology. In Clifford and Marcus (eds.), *Writing Culture*, pp. 234–261. Berkeley and Los Angeles: University of California Press.
Radway, Janice (1984). *Reading the Romance.* Chapel Hill, N. C.: University of North Carolina Press.
———.(1988). Reception study: Ethnography and the problems of dispersed subjects, *Cultural Studies*, 2(3): 359–376.
Rosoldo, Michele (1980). The use and abuse of anthropology: Reflections on feminism and cross-cultural understanding. Signs 5(3): 389–417.
Rubin, Lillian B. (1976). *Worlds of Pain: Life in the Working-Class Family.* New York: Basic Books.
———.(1979). *Women of a Certain Age: The Midlife Search for Self.* New York: Harper & Row.
———.(1983). *Intimate Strangers: Men and Women Together.* New York: Harper & Row.
Ryan, Mary P. (1975). *Womanhood in America: From Colonial Times to the Present.* New York: New Viewpoints.
Sanjek, Roger (ed.) (1990). *Fieldnotes: The Makings of Anthropology.* Ithaca: Cornell University Press.
Scott, Joan Wallach (1988). *Gender and the Politics of History.* New York: Columbia University Press.

Seiter, Ellen (1990). Making distinctions in TV audience research: Case study of a troubling interview. *Cultural Studies* 4(1): 61–84.

Seiter, Ellen, H. Borchers, G. Kreutzner, and E. Warth (eds.) (1989). *Remote Control: Television, Audiences, and Cultural Power.* London: Rutledge.

Sewell, William (1989). Toward a theory of structure: duality, agency, and transformations. Ann Arbor, Mich.: Center for the Study of Social Transformations, Working Paper No. 29.

Shostak, Marjorie (1981). Nisa: The Life and Words of a !Kung woman. Cambridge, Mass.: Harvard University Press.

Smith, Zena (1972). Maternal aspirations, socialization, and achievement of boys and girls in the white working class. *Journal of Youth and Adolescence* 1(1): 35–57.

Spradley, James P., and Michael A. Rynkiewich (eds.) (1975). *The Nacirema.* Boston: Little, Brown.

Stacey, Judith (1983). The new conservative feminism. *Feminist Studies* 9(3): 559–584.

———.(1987). Sexism by a subtler name: Postindustrial conditions and postfeminist consciousness in the Silicon Valley. *Socialist Review* 17(96): 7–30.

Strathern, Marilyn (1987). An awkward relationship: The Case of Feminism and anthropology. Signs 12(2): 276–292.

Whyte, William Foote (1943). *Street Corner Society.* Chicago: University of Chicago Press.

———.(1967). *On Street Corner Society.* In Ernest W. Burgess and Donald J. Bogue (eds.), *Urban Socioligy,* pp. 156–168. Chicago: University of Chicago Press.

Wright, Eric O. (1976). Class boundaries in advanced capitalist societies. *New Left Review* 93.

———.(1978). *Class, Crisis, and the State.* London: New Left Books.

Notes on Children as a Television Audience

8

Ellen Seiter

Considering children as an audience can help to clarify the ideological issues that surround any discussion of television audiences. Cultural studies has an important intervention to make in the representation of children as a television audience by challenging the adequacy of developmental psychology as a theory and method for studying children and disputing the belief that children's relationship to television is essentially passive. Developmental psychology—the dominant code for representing children in academic research—has pinpointed differences in chronological age but has excluded from the discussion race, class, and gender as meaningful differences within the population of children. Research from Australia and the United Kingdom by Patricia Palmer, Bob Hodge, and David Tripp and by David Buckingham has emphasized the role of television as a form of social communication in children's lives, capturing responses that are more complex and more sensitive to class differences.[1] Hodge and Tripp have suggested that an elitist, denigrating view of television needlessly punishes children who rely on it for recreation. Yet there are problems in translating the work of these researchers to the United States, because of differences both in culture and in television systems. Children's commercial television in the United States provides a good test case for the logical and ethical limits of a cultural studies position that champions popular pleasures. I will explore some of those limits in this chapter.

The middle class belief in the badness of television viewing for children has proven to be exceedingly durable. It circulates constantly in the media targeted to parents: pediatrician's pamphlets, magazines, agony columns, advice literature. "There is a powerful, idealized image of childhood as a time of activity and doing that reinforces some of my misgivings about television," explained Stella Hurd in the anthology *Parents Talking Television*.[2] Complaints about children's television viewing, such as those voiced by Marie Winn and Neal Postman, are backed by a nostalgic mourning for an idealized vision of a "lost" childhood—a childhood that was a time of doing, of direct experience. Blaming television for everything that is wrong with children is a rhetorical strategy that the print media seem to have a special attraction for: *The New York Times* mouths this position numerous times every year. Despite a wealth of sophisticated research by scholars such as Jennings Bryant, Suzanne Pingree, Ellen Wartella, James Anderson, Daniel Anderson, and Elizabeth Lorch,[3] the most quotable and attractive sources for journalists—the story makers—are people who will say that the children's audience are the passive, unwitting victims of the devil television.

In sharp contrast to eighteenth- and nineteenth-century Anglo-American notions about children, where submission, obedience, and docility were prized, passive is today about the worst thing a child can be. This is one reason television viewing is considered so bad by childhood experts, yet is so convenient for parents. Developmental psychology is now the dominant model of childhood in teaching, psychology, social work, and medicine. Passivity is especially problematic, even pathological, according to this model.[4] Children watching television offends the widely held belief in the importance of the child's *actively* achieving developmental tasks. Child experts, television critics, and protectionists are convinced that television deters children from achieving normative agendas of child development: direct interaction with peers and parents, "large motor" skills, socialization, relations with peers and parents, cognitive and physical development. Television is excluded from the list of activities that can "stimulate" growth—and stimulation is something that parents are supposed to provide in endless supply from infancy onward.

Action for Children's Television (ACT), a watchdog organization devoted primarily to the fight to limit the number of commercial minutes on U.S. broadcast television, selects a handful of television programs each year for approval when they meet normative standards of bourgeois taste. ACT has been most active in suggesting standards for "good" television and in calling for choice and quality. For example, ACT has produced two pamphlets advocating children's arts programs and the adaptation of books to children's television. ACT grants its approval to television adapted from more respectable, selective media: "The recipe is simple: start with an imaginative children's book, mix in creative filming and animation techniques and—voila!— a tantalizing addition to television's bill of fare."[5] ACT enthusiastically recommends videotapes over network TV because they do not involve commercial interruption. They often endorse videos that mirror highbrow genres of adult television: programs with an explicit high culture link (Muppets at the Metropolitan Museum of Art), visits to another culture (documentary/travelogue), or anything with an explicit "teaching" goal, covering abstract principles (shapes, numbers, alphabet) or psychological realism (what to do when you're angry, first day at school, death of a pet). Such videos accomplish a dual purpose: introducing children to a better class of culture and shielding them from the onslaught of commercials. Needless to say, what these videos have to offer, in terms of social communication with other children and entertainment values, is rather different from "Slimmer and the Real Ghostbusters" or "My Little Pony." Television is most acceptable to ACT when it is used to recruit children to join other kinds of audiences: museum, orchestra, or opera patrons; tourists; students.

The basic principles of advice literature are that the less television viewing, the better; videotapes are preferable to broadcast television because of the commercials; if you must let your child watch, make it PBS; no viewing is best of all. As Patricia Palmer points out: "The amount of television children watch is still quoted in a way which presupposes adult amazement and disapproval. We speak of 'heavy' or 'light' viewers as if there is indeed a measurable 'amount' of the thing called TV viewing which has entered into the child's system and stays there like a dead weight."[6] Experts advise parents to take on a tough role when urging that children's viewing be monitored: "Establish ground rules, prevent TV from becoming an addiction," say Dorothy and Jerome Singer in *Parents* magazine: "No TV before school, during meals, during daytime hours, or before homework is done. And don't suggest that the child 'go watch tv' whenever you are feeling overwhelmed or need privacy."[7] Obesity, violence, and poor school performance are continually held up as the results of television viewing.

Let me turn now to some of my research with first-time parents and the way they characterize their children's viewing. What I hope to record is a pivotal moment in subjects' media biographies: when parents first watch their own child watch television. I wish to show the trickle-down effect of the academic representation of children's viewing as passive and therefore bad. My goal is to offer not an ethnography of viewing but a study of how parents talk about their child's viewing.

The parents I will discuss here are all members of a self-help parents' group in Eugene, Oregon, to which I myself belong. The group was organized by a child abuse prevention social service agency that contacts all new parents while they are in the obstetrics ward at the hospital. We are from different neighborhoods, different ages, different classes, brought together because our children happen to have been born within the same three-month period. There is, of course, a considerable self-selection factor at work in who joins such a group: these are adults committed to and worried enough about their children to pay a small membership fee, attend monthly meetings, and host some of the meetings at their house or apartment. In the fall of 1990 the children will turn three years old, and the group will have met for thirty-three months. I will quote here from a taped two-hour discussion where the topic was television, toys, and advertising. The discussion took place in the course of a regular meeting, during which we often have had set topics. Television easily fit into an agenda of perceived problems facing parents, taking its place alongside feeding, potty training, and fire safety.

I have arrived at this kind of research design because I believe it best fulfills the goals of ethnography. A long and extensive contact period is the only way to connect attitudes toward television with a meaningful composite of information about the family's social position. As a participant in the group, I have the opportunity to engage in both formal and informal exchanges with other members and to gather solicited and unsolicited comments. My relation to the other parents has changed since they agreed to participate in my research; I have a different kind of self-interest in seeing the group continued. I also have an opportunity to share, debate, and discuss my work with them. Because I am attempting to do a longitudinal study, I have time to monitor and contemplate my commonalities with the other women in the group as a mother, as well as the differences that derive from my status as an intellectual with a relatively well-paid, comfortable job.

Ethnography is a method that has distinct, political limitations: We have to recognize that not everyone is equally likely to welcome an academic into their home for an extended visit. African Americans, for example, are not turning up in many audience studies because nearly all of the academics doing this kind of work are white and because African Americans have a healthy distrust of white researchers and a desire to protect their privacy. Similarly, gay and lesbian parents, whose custody is often challenged, are less likely to invite the invasion of privacy that ethnography entails. *Who* gets represented in ethnography will also depend on *who* is doing the representation: the homogeneity of our samples speaks to the occupational and residential segregation of the United States, the racism in higher education, and the failure of affirmative action to recruit students and faculty, among other things.

We must also recognize that white, English-speaking, middle class is itself a culture, an ethnicity, a class position, and we must try to make that culture strange. For my purposes, the beliefs and values of the white middle class are especially important since the people who usually help to determine what will happen to children—teachers, doctors, psychologists, nurses, social workers—are themselves usually members. One of the peculiarities of my parents' group (perhaps this is typical of the middle class self-help group), is that it seems to exist in order to shore up the identity of the family: we are all very conscious of making public, social presentations of ourselves and our children. Self-help groups are, in a way, devoted to statement-making (and the exchange of information—often about consumer goods and services). The group is not a free speech situation: differences in social power outside the group determine who can speak and what can be said. For example, during our discussion one of the fathers described with great disapproval some friends of theirs who do not turn off the television set when they have visitors and who eat meals in front of the set. Certainly some of us in the group do these very things but no one could confess to such behaviors after that comment. When I acknowledged openly that my own daughter watches lots of television early in the discussion, there was an uncomfortable, embarrassed silence. (The professor as a television fan remains a pretty embarrassing notion outside of a small circle of cultural studies academics.)

Lesley and Wade live in a small two-bedroom apartment downtown above Wade's musical instrument repair business. Lesley works five days a week as a dog groomer. They have a lot of worries over money and cannot afford the same kinds of vacations, housing, or furnishings, and appliances that all of the other families in the group enjoy. Since the beginning, when the babies were only three months old, Lesley has been more consistently open, even confessional about problems she has with her daughter, Kelly (now almost three)—losing her temper, feeling tired, handling tantrums. Lesley often talks about reading books about child-rearing (or seeing child experts on television), which she uses as a source of inspiration and solace, and she offers the tips she has got from them to others. She does not want to hit Kelly, and she is more open than any other mother in the group about the temptation to do so. Lesley is very dedicated to the group; she volunteered to be a coordinator and regularly phones around to everyone to announce the next meeting time and location.

Lesley offered an extremely negative evaluation of television viewing as an addiction, as a mindless, passive activity—*and* was simultaneously very frank about TV viewing's being something she very much likes to do. Wade is usually silent in the group, as though unfamiliar and uncomfortable with the middle class codes that sanction a masculine display of emotional openness and self-revelation. During our two-hour discussion he spoke up only a few times and got the floor only once, to say that since the spread of television, no one plays musical instruments any more, because you cannot play and watch television at the same time. Wade's disapproval of television appears to be one source for Lesley's extremely critical view of her own television watching.

For Christmas, Kelly received a video copy of *Bambi*. Twice during the winter, Lesley took me aside to discuss in a hesitant, confessional way how many times Kelly was able to watch Bambi. I assured her that my daughter did the same thing with many videos, and Lesley said she thought it was good that Kelly verbalized a lot while she was watching it. Lesley felt a responsibility to filter the material Kelly watched on television, but she approved of *Bambi* because it is a story about animals and saving the forest. By June 1990, when I taped the group discussion, her attitude had grown much more negative. She was very concerned that she must curb her own viewing so that she did not teach Kelly by example to be a television zombie:

> It's where I go to lose my mind when I come home from work and I don't want to think and I don't want to do and I don't want to anything. And yet I don't want Kelly. . . [emphatically, but leaving the sentence incomplete]. So I'm really weaning myself off TV lately. I see all these ads for candy and cars and ridiculous-looking dolls and just junky stuff. I want Kelly to want to be entertaining herself.

Lesley was especially troubled by her addiction to the daily sitcoms shown five days a week at 7:00 in the evening. Before Kelly was born she never worried about watching these shows, but now she worries a lot. Lesley, who works a physically demanding job and then comes home to prepare dinner and clean up, was the only mother who offered a picture of her *own* television viewing. The picture was one that is usually associated with masculine modes of viewing: collapsing after work in a chair in front of the television set, beer in hand. Lesley was the only mother who expressed a strong desire to watch television herself—although she was terribly unhappy and troubled by it.

In May Lesley changed day care arrangements for Kelly. Kelly had gone to a home care situation, which Lesley very much liked in every way except one: The television was on all the time, and this worried her a great deal. Kelly's new school was much more expensive and less convenient, but it offered "activities," and there was no television there. Lesley was proud to report that within a month after changing school, Kelly was no longer using her finger as a gun or talking about what she had seen.

Lesley talks about television every day with the women at work. "That's all we talk about," she said. "We all work in one big room and every day it's, 'did you see—?'

But Lesley does not want to resemble these women and she does not want Kelly to grow up to be like them. Every other adult in the group, when asked about it directly, said that television never comes up as a topic for discussion at work, "except maybe a PBS special or something," one woman added. (Whether this statement is accurate is less important than its claim to represent their view of television.)

Lesley differs from the group in other ways too. On a test that a visiting psychologist gave the group to help us sort out our priorities as parents, Lesley marked "spirituality" as most important. I had marked intelligence. After describing how she disapproves of the kinds of toys advertised on television, she went on to proudly recount a story about some of Kelly's play with horses that she had observed:

> I bought a farm set for Kelly and out of all the farm animals she's picked out the horses—thank goodness I love horses—and she makes all kinds of wild scenarios, I mean she's already starting to tell stories. She had a horse going, "Help help help!" and took another horse over to rescue, and made little conversations with the horses and stuff. She's already got this concept of helping each other, [slowly with emphasis] of coming to aid.

For Lesley, the point of the story was that Kelly was acting out a story of helping, not that she was being clever or exhibiting her language skills–the point of most of the stories that the more affluent parents tell.

Carla is the co-coordinator of the group with Lesley. She and her husband, Ron, live in a large home built in the 1920s, and they drive a BMW and a Volvo. Ron started his own business in vitamin and dietary supplements, selling such popular alternative medications as zinc lozenges. Carla and Ron now have two sons, Douglas and Jake, born this winter. He is Jewish, she is Catholic. She is the only full time housewife in the group (they are the only ones who can afford it). Their son Doug was one of the last in the group to attend day care, and he is now enrolled in a preschool about ten hours a week. Both Carla and Ron moved to Eugene from New Jersey in the late 1970s: they were hippies then–something they now laugh about.

Ron and Carla have the most lenient, permissive attitude toward television in the group. Carla laughs when she reports that they have fifteen hours of *Chip-n-Dale Rescue Rangers* on tape—with the commercials zapped out, Ron adds. Carla tells humorous stories about trying to prevent Doug's attention to television from wandering, especially now that she has the new baby to take care of; she talks about the cartoons, trying to get him interested again. Carla is charmed, not frightened by the influence of television on Doug: he regularly has "cheese attacks" like his favorite cartoon characters, and he goes to the refrigerator and serves himself.

When one mother complained about violence on Saturday morning TV, reporting on it as though she assumed that most of us would be unfamiliar with it, *Ghostbusters* came up as an example. Carla broke in to report that it was one of Doug's favorites:

Carla: We watch *Ghostbusters* and we watch *Slimmer.* He hasn't noticed the *Ghostbuster* toys yet—I'm waiting for that shoe to drop [laughs].

What I feel reassured about is he was definitely exhibiting a fear of ghosts and monsters—I mean, really severely afraid of them—but now he's into shooting them with fire.

Ron: Through the media he has found a way to deal with them.

Paula: [laughing] Yeah, now it's shoot Mom with the fire when he's angry at me.

Lesley had said adamantly that she "won't tempt Kelly" by taking her to stores like Toys R Us, and that she is "insulted" by the commercials for toys. In contrast, Carla goes shopping with Doug several times a week for amusement, using the store and its toys on display as a playground. Ron interprets Doug's watching commercials and asking for toys as a positive indication of his developing communication skills—an active process. Because Ron is himself an entrepreneur, the transactions of advertising and consumption are untainted:

> One thing I think is good is that Douglas is understanding that there are choices out there. It opens some form of communication. He knows that it exists and he communicates it to Carla that he wants it.

For Ron and Carla, everything Douglas does is a sign of his intelligence, so they are able extend this to his television viewing as well. Doug's life chances, his economic security, his cultural capital are so secure that his parents can interpret everything he does—including watching television—in relation to the dominant version of childhood as active.[8] (Ron and Carla are using a strategy very similar to the one cultural studies people use to defend the popular audience.)

By contrast, Kelly's future is less assured, and her television viewing is subject to the negative interpretations of manipulation rather than learning, passivity rather than activity, victimization rather than choice. Lesley essentially employs a moral code to interpret Kelly's behavior and the consumer culture that surrounds her. At the toy store, Doug is making choices and communicating; Kelly is being tempted. Doug is using television to conquer negative psychological states (a uses and gratifications position); Kelly is being brainwashed, imitating antisocial values (the bullet model of effects).

In a conversation about styles of play, Ron proudly told the group: "Douglas's new toy is the computer mouse; he's totally jazzed. He was actually manipulating and using it!" If the television is commonly considered the bad screen for children because it causes passivity, the computer is the good screen because it is construed as active and intellectual. Thus public libraries virtually never house television sets that can be used by their patrons; the presence of a television in the children's section of the library would be widely offensive. But computer screens are available in many children's libraries and their use is enthusiastically encouraged by parents and librarians. There is no problem in viewing Doug, a boy in front of the computer screen at age three, as the author of his own intentions.

Different things are at stake for Kelly and for Doug in knowing about television. Kelly's parents are aspiring to be middle class; Doug's parents have very much already arrived. Lesley's fears about Kelly's viewing were confirmed in the discussion when the two mothers in the group who are teachers stated unequivocally that children who watch the most television do the worst in school. (They also reported that teachers never talk about television with their colleagues.) Some evidence suggests that children rapidly learn not to talk about television to teachers. I have observed this firsthand on many occasions at a free preschool for low-income families, where the children, most of whom are much quieter than middle-class children, became very animated discussing the cartoon Ghostbusters among themselves on the playground. But this happened only after I asked them about it, only after I gave them permission to raise the topic. In front of their regular teacher, references to toys or television were politely discouraged. As Hodge and Tripp argue, "Unsurprisingly, those who lose most from the exclusion of television from the school curriculum are those who are channeled by innumerable other features of that curriculum into low-paid, menial positions in the workforce, a position from which come those who habitually watch the most television."[9] Lesley is thus right, in a way, to worry about how much television Kelly watches because this will be taken as a bad sign at school, and Kelly will probably get off to a more precarious start than Doug, who will be extremely verbal (having had adults at home with time to listen to him), already acquainted with the computer, and filled with the expectation that his desires are important ones.

Ed and Laura live in the university neighborhood, in a 1920s home (like Ron and Carla's), which they renovated themselves. Ed is a tax accountant and Laura works half-time in a job-sharing situation as a grade school teacher. Unlike Carla and Lesley, Laura had never spoken to me about the media before, although she has repeatedly asked me about my experiences working at the university. During the taped discussion she chimed in a few times to indicate her total lack of familiarity with Toys 'R' Us, with Saturday morning television, and with commercials. Ed said that they prefer going to the park over watching television. When Laura first got the floor, she told a story about the horrible toys that children in her classroom bring to show-and-tell. Laura finds many toys disgusting and is annoyed at the disruptions caused by their presence in the classroom. As a grade school teacher, Laura is convinced of the terrible effects of television. (The other teacher in the group, a high school English teacher, has moved the television set out of the living room, to prevent the son from watching it at all.) Despite the fact that Victoria only watches PBS, Laura, trained as a teacher and horrified by television, feels deeply that even this is wrong.

Laura: I go to work at 11:00, so between 8:00 and 11:00, I have to get ready and do a lot of things, so I turn on *Mister Rogers* and *Sesame Street* and I have her sit there and watch for an hour and a half and then I feel really guilty—

Ed: [interrupting]—But she usually doesn't watch for that long—

Laura: —Yes she does. You're not there.

Carla: [interrupting] What's wrong with watching *Sesame Street?*

Laura: [crisply] She should be doing other things. [pause] She eats her breakfast while she watches *Sesame Street* and then *Captain Kangaroo* comes on at 10:00, and if I let that go, you know, if I let her *continue* watching, even *then* I feel horrible! So usually at 10:00 I turn it off and I say, "Why don't you find a toy to play with?" That's after *Mister Rogers* and *Sesame Street*—an hour and a half! And she'll sit through the whole thing, . . . depending, she gets up and down. . . .

At this time three other parents interrupted with stories about their child's attention span for television viewing and the different kinds of activities that they engage in while watching television. Several minutes later, Ed gets the floor and reasserts the negative interpretation:

> What bothers me is that it's totally passive, there's nothing for them to do, they just absorb what's coming to them and they don't take part in it."

There is much more discussion and a few minutes pass before Laura, rather emboldened by the fact that it has now emerged that by comparison with many other children in the group, Victoria does not watch that much television, goes back to the earlier question, "What's wrong with watching *Sesame Street?*" Laura is speaking to the group, but her remarks are directed rather pointedly at Ed:

Laura: *Sesame Street,* I think, invites some participation. *Sesame Street* is not a totally passive type of show. One day they were singing in Spanish, and Victoria started singing with them—

Ed: —[interrupting] Well, that's excellent, but I think that when they develop the habit of watching a lot of television as we watch TV, and it's totally passive—you just sit back and it will take your mind wherever.

None of the fathers in the groups felt that monitoring television time was their personal responsibility, although many of them, like Ed and Wade, disapproved strongly of television viewing and increased the guilt that their wives felt about letting the children watch television.

Mothers are both the enforcers of television discipline and the ones who suffer from its exclusion or limitation in the household. When Laura concurs with Ed's idea that Victoria should not watch television because it is passive, it means that Laura must get ready for work in the morning with Victoria at her feet. The price she pays for allowing Victoria to watch television anyway is to feel horrible. Lesley pays perhaps the highest high price for her negative view of television: It led her to have to change day care arrangements, causes more fights with Kelly, puts a strain on Lesley to entertain her or to deal with Kelly's boredom, and removes Lesley's one break during the day ("when she doesn't want to do, she doesn't want to anything"). By interpreting television viewing as passive and bad, Lesley loses the only leisure option available to her during the evening. Carla's solution—to upgrade Doug's viewing to be active—seems to be the happiest one.

Unlike the other parents, Carla and Ron are united in their positive feelings about television. Because Ron works out of the house, they both have an equal stake in the distraction that television offers. Doug's viewing is almost necessitated by the fact that Carla does not work outside the home and Doug is only at preschool a few hours a week. Carla is therefore with Doug and the new baby twenty-four hours a day. The parents have a rule that when Carla goes crazy taking care of the kids, she goes and gets Ron out of his office to take over the child care.

The home is not a haven for adult women; it is a workplace. Almost all ethnographic studies of television audiences point to striking differences based on gender in the opportunities for leisure and the exercise of power over communication technologies. The feminist implications of these findings have not always been emphasized to the extent that they might be, with the notable exception of Janice Radway's *Reading the Romance* and David Morley's *Family Television.* We must put children's television viewing in the context of women's work in the home. Households vary in terms of how much work women do to maintain them, but rather the most important factor in predicting the workload is not income, or education, or employment outside the home, but the presence of children in the home.[10] Of all the types of domestic and reproductive labor assigned by social convention to women, child care is the most strenuous. Women get less help from neighbors, relatives, or older children in doing it, and there are much higher standards for its performance than there used to be.[11]

The use of television as a babysitter must be understood in the context of a gendered division of labor that places enormous strain on adult women living in families. The incessant prohibitions against child television viewing in parental advice literature are aimed almost exclusively at mothers, not at fathers. Very often, television is conceived as though it provides the parent (that is, the mother) with any convenience, any respite. In this context, we can see how the ability to represent the child's viewing as active rather than passive has immediate consequences for the mental health and the workload of mothers. The families I have discussed here are all two-parent families. It can be predicted that the need for television as a babysitter will be more acute for single mothers, whose guilt and sense of anxiety about performing adequately as parents may be even greater. Single mothers and their children are more likely to be scrutinized by teachers, social workers, psychologists—and single mothers know that an adverse expert opinion can even lead to the loss of custody of their children.

Conclusion

Representations of the audience find their way into everyday life in ways that have powerful effects on the domestic economy of child rearing and on the gendered division of household labor. It has recently been argued that ethnography is itself discursive, producing merely another representation of the audience; that is in part true but it is not an argument that can discount ethnographic methods. Discourses, of course, have social affectivity and use value. In the case of the children's audience it becomes apparent that the construction of the child viewer as passive (bad, mindless) has direct punitive consequences for mothers. Every middle-class mother knows that children should not be watching television; they should be doing something else—something else more stimulating, more educational, more creative.[12] Parents who are in a privileged enough position to adopt an interpretation of their child's viewing as active rather than passive win tangible rewards in the form of time. This interpretation of children's viewing as active is not equally available to all parents; it depends, I suspect, on one's fears for the child's future, the degree and nature of social aspiration, and one's moral judgment of popular media and consumer culture. Indeed, the representations of children's television viewing finds its way into the educational system in real ways, essentially punishing children who know too much about television. A studied, conspicuous ignorance about television is a mark of distinction (like all distinctions, it is valued because it is so difficult to maintain).

Cultural studies researchers have participated, along with cognitive psychologists, social psychologists, and mass communications researchers, in upgrading the way we represent the television audience. These are still pretty radical notions. The popular press and schools of education have not caught wind of them yet: they still worry over television as though it were the most important factor in children's lives. When television is represented as all-powerful, all-determining, it directs attention away from more important factors such as schools, housing, transportation, money, and health care. Thus gross discrepancies in money and privilege are covered up by offering the negative image of the child television viewer as the determining factor. When more than 20 percent of children in the United States live in conditions of poverty, it seems criminal to have people believe that television viewing is the primary factor in poor school performance. And it produces in the middle class a sense of righteous self-congratulation and victim-blaming. We must continue to criticize advice literature that insists that if parents merely turn off the set, their children's lives and life chances will magically improve. Cultural studies academics might consider producing some media effects of our own by writing for more popular publications or producing some television programming about these issues. This will entail changes in the way we present our arguments (we might not be allowed to use the words "discourse" and "subjectivity" anymore, for example), but these are compromises that might be worth it if it would earn for cultural studies in the United States a more direct political engagement.

Having introduced a different representation of the audience produced through domestic ethnography, it is important that we move back to the public realms in which these representations circulate. The television industry (which listens more carefully than the moralists to academic research on television) has also come to think of the child audience as active. Some of the consequences are negative. When children failed to show up on the People Meters, they became so "active" that they were in danger of losing all their programming. In a broadcasting system in which quantitative records of audience are used to set the price of commercial time, one network threatened to cancel Saturday morning children's programming altogether, given the slumping ratings (a threat that sent toy manufacturers into despair). Since deregulation, and with increasing competition from video rentals and pay cable (subscriber services like the Disney Channel and Nickelodeon), having any national children's programming at all on network television is at issue.

We need to be explicit about the stakes involved in our representation of children's television viewing. Even though there is some evidence to suggest that children are able to appropriate television commercials for their own uses too, do we really want to line up on the same side as those who allow fifteen minutes of commercials per hour of programming? How can we upgrade the child viewer without abnegating control to others whose politics we do not like in the arena of policy and regulation? If children are already so well defended and assertive, shall we just throw them to the wolves? Most of us want to maintain a critical edge, to say more and different things about the audience than market researchers do; to be effective, however, we must also keep one eye on the political economics of the television industry. In 1990, all of the sponsors of children's television—Hasbro, Toys 'R' Us, Tonka, Mattel—were setting their sights on Europe in 1992, where, as Mattel's 1989 annual report hungrily reports, there are three times as many children as in the United States. This is bound to stir up a great deal of controversy and to change and probably increase the amount of children's commercial television in Europe, with ambiguous and complex results. In the United States, Disney was involved in a block-booking and blind buying scheme in children's syndicated afternoon television. The offer of a package called "The Disney Afternoon" to independent stations only when they promise not to show any other animation before or after the two-hour block, and when they contracted to air the cartoons for three years, is an example of the kinds of industry strategies that threaten to have a homogenizing effect on programming.[13]

Cultural studies has not contributed much on the level of criticism of television programs as texts. Charlotte Brunsdon has pointed out that studying the audience has meant the postponement of saying anything about television itself. As Patricia Palmer argues, "Poor programs are also the consequence of holding low expectations of the television medium for its child audience."[14] The children's schedule consists of more reruns and more commercials, and the same commercials are repeated more frequently than during prime time. Cultural studies has eschewed offering any middlebrow "recipes" for good children's television, such as those put forward by Action for Children's Television. As Brunsdon puts it : "We do not defeat the social power which presents certain critical judgments as natural and inevitable by refusing to make critical judgments." I think Brunsdon is right to argue that it is time that we made explicit critical standards for children's television.[15] Some commercial television has much more going for it in the way of humor and production values (the quality of sound, the complexity of the animation) than "educational" materials; but commercial television is very uneven. If audience studies have proven that children are resilient, innovative, flexible, ingenious, that does not mean that they cannot ask for something more, for something different. We can refute the claim that people (only) want what they already have on television.

Television audience studies should work to change television itself as well as the popular representation of the audience. Audience studies can best be used to politicize the public discussion of children and television in ways that would improve the experience of formal education for children who do watch a lot of television, to remove needless anxieties caused by the overestimation of television's effects, and to make explicit the gains and the losses at stake in promoting different representations of the audience.

Notes

1. David Buckingham, *Public Secrets: "Eastenders" and Its Audience* (London: British Film Institute, 1987); Robert Hodge and David Tripp, *Children and Television: A Semiotic Approach* (Stanford: Stanford University Press, 1986); Patricia Palmer, *The Lively Audience: A Study of Children Around the TV Set*(Sydney: Allen & Unwin, 1986).
2. In Philip Simpson, ed., *Parents Talking Television* (London: Comedia, 1987), 65.
3. Much of this work is summarized in Jennings Bryant and Daniel R. Anderson, eds., *Children's Understanding of Television: Research on Attention and Comprehension* (New York: Academic Press, 1983.)
4. For an extensive discussion of concerns with developmental psychology, see Julian Henriques et al., *Changing the Subject: Psychology, Social Regulation, and Subjectivity* (London: Methuen, 1984).
5. Peggy Charren and Cynthia Alperowicz, *Editor's Choice: A Look at Books for Children's TV,* (Newtonville, Mass.: ACT, 1982).
6. Palmer, *Lively Audience,* 135.
7. Julius and Zelda Segal, "The Two Sides of Television," *Parents* (March 1990): 186.
8. This interpretation was suggested to me by the analysis of a family ethnography presented in Roger Silverstone, Eric Hirsch, and David Morley, "Information and Communication Technologies and the Moral Economy of the Household," International Communication Association, Dublin, June 1990. I am grateful to the authors for making this available to me prior to publication.
9. Hodge and Tripp, *Children and Television,* 138.
10. See Sarah Fenstermaker Berk, *The Gender Factory: The Apportionment of Work in American Households* (New York: Plenum Press, 1985).
11. See Ruth Schwartz Cowan, *More Work for Mother* (New York: Basic Books, 1986); Susan Strasser, *Never Done* (New York: Pantheon, 1982); and Ruby Roy Dholakia, "Feminism and the New Home Economics: What Do They Mean for Marketing?" in A. Fuat Firat, Nikhilesh Dholakia, and Richard P. Bagozzi, eds., *Philosophical and ·Radical Thought in Marketing* (Lexington, Mass.: Lexington Books, 1987), 341–358.
12. Messaris and Lemish have noted that mothers are often fascinated by observing the way their children watch television but do not wish to be seen in a bad light. See Dafna Lemish, "Viewers in Diapers: The Early Development of Television Viewing," Paul Messaris, "Mothers' Comments to Their Children About the Relationship Between Television and Reality," and Ellen Wartella, "Commentary on Qualitative Research and Children's Mediated Communication," in Thomas R. Lindlof (ed.), *Natural Audiences: Qualitative Research of Media Uses and Effects* (Norwood, N. J.: Ablex, 1987).
13. William Mahoney, "The Kid War: Fox-Disney Battle Gets Hotter," *Electronic Media* 12 February 1990, 1, 30.

14. Palmer, *Lively Audience,* 142.

15. Charlotte Brunsdon, "Problems with Quality," *Screen* 31:(1) (Spring 1990): 73. See also Brunsdon, "Text and Audience," in Ellen Seiter, Hans Borchers, Gabriele Kreutzner, and Eva-Maria Warth, eds., *Remote Control: Television, Audiences, and Cultural Power,* (London: Routledge, 1989), 116–129.

References

Berk, Sarah Fenstermaker (1985). *The Gender Factory: The Apportionment of Work in American Households.* New York: Plenum Press.

Brunsdon, Charlotte (1989). Text and audience. In Ellen Seiter, Hans Borchers, Gabriele Kreutzner, and Eva-Maria Warth (eds.), *Remote Control: Television, Audiences, and Cultural Power,* pp. 116–129. London: Routledge.

———. (1990). Problems with quality. *Screen* 31(1) (Spring 1990).

Bryant, Jennings, and Daniel R. Anderson (eds.) (1983). *Children's Understanding of Television: Research on Attention and Comprehension.* New York: Academic Press.

Buckingham, David (1987). *Public Secrets: "Eastenders" and Its Audience.* London: British Film Institute.

Charren, Peggy, and Cynthia Alperowicz (1982). *Editor's Choice: A Look at Books for Children's TV.* Newtonville, Mass.: ACT.

Cowan, Ruth Schwartz (1986). *More Work for Mother.* New York: Basic Books.

Dholakia, Ruby Roy (1987). Feminism and the new home economics: What do they mean for marketing? In A. Fuat Firat, Nikhilesh Dholakia, and Richard P. Bagozzi (eds.), *Philosophical and Radical Thought in Marketing,* pp. 341–358. Lexington, Mass.: Lexington Books.

Henriques, Julian, et al. (1984). *Changing the Subject: Psychology, Social Regulation, and Subjectivity.* London: Methuen.

Hodge, Robert, and David Tripp (1986). *Children and Television: A Semiotic Approach.* Stanford: Stanford University Press.

Lemish, Dafna (1987). Viewers in diapers: The early development of television viewing. In Thomas R. Lindlof (ed.), *Natural Audiences: Qualitative Research of Media Uses and Effects.* Norwood, N.J.: Ablex.

Mahoney, William (1990). The kid war: Fox-Disney battle gets hotter. *Electronic Media* (12 February): 1, 30.

Messaris, Paul (1987). Mothers' comments to their children about the relationship between television and reality. In Thomas R. Lindlof (ed.), *Natural Audiences: Qualitative Research of Media Uses and Effects.* Norwood, N.J.: Ablex.

Palmer, Patricia (1986). *The Lively Audience: A Study of Children Around the TV Set.* Sydney: Allen & Unwin.

Segal, Julius, and Zelda Segal (1990). The two sides of television. *Parents* (March).

Silverstone, Roger, Eric Hirsch, and David Morley (1990). Information and communication technologies and the moral economy of the household. Paper presented to the International Communication Association, Dublin, June.

Strasser, Susan (1982). *Never Done.* New York: Pantheon.

Wartella, Ellen (1987). Commentary on qualitative research and children's mediated communication. In Thomas R. Lindlof (ed.), *Natural Audiences: Qualitative Research of Media Uses and Effects.*Norwood, N.J.: Ablex.

Figuring Audiences and Readers

9

Tony Bennett

In a critical exchange on audience theory, James Lull entered the plea that when speaking about television, we should get rid of the terms "texts" and "readers" (Lull, 1988: 239). It is not difficult to see why. The literary bias of both terms is obvious, as is the inappropriateness of their unqualified application to audiovisual media. Television is not reducible to its texts, and the complexity of our relations to, and forms of involvement in, this cultural technology clearly goes beyond the notion of reading, no matter how generously we might interpret it.

Yet, if this is so, it is not clear why—as Lull implies—the term "audience" should present itself as automatically preferable. There are, after all, other terms that might be used—viewer, for example—just as there have been, historically, other alternatives whose theoretical and political force we should not lose sight of. The early history of sound broadcasting affords ample evidence of significantly contractive views of the relations (actual or ideal) between radio and, as they were initially envisaged, its listeners: the subjects of an action. Lesley Johnson's work on Australian radio, for example, highlights the respects in which listeners were initially conceived as delighting in the act of listening as such, deriving their pleasure from marveling at radio's technical ability to render distant sounds present, rather than from program content (Johnson, 1988). Johnson, like Williams (1974), also stresses the enormous work of social and cultural definition through which listeners subsequently came to be conceptualized as, precisely, audiences—that is, as consumers of centrally produced and broadcast programs—rather than as, say, technical users or as differentiated publics each regulating radio for its own purposes.

The modern concept of the audience (at least before it became "active") as the receivers of messages from a centralized source of transmission, then, was not present at the birth of the modern media but has emerged in tandem with their development and, in part, as a product of their own practices. What *was* present from the outset, however, was the mold in which such a conception of the audience might be cast. As Janice Radway has noted, the term "audience" derives, etymologically, from the contexts of face-to-face communication. It is important to add, however, that those contexts were typically hierarchically organized. To be granted an audience—as in the relations of a subject to a sovereign—was to be granted the right to listen to the enunciations of power, to hear and to take account of a message delivered by and from an authoritative source. If, as Radway suggests, virtually all early mass communication theories "retained the notion of the audience as a unified aggregate of similarly endowed individuals who passively read or hear the words and therefore the message of another" (Radway, 1988: 360), it is equally important to note that such message flows were also conceived as descending the hierarchy of discursive power.[1]

My point, then, is that if it is true that the term "readers" brings with it a particular set of associations, so, too, does the term "audience." Indeed, if the brief analysis hinted at above holds, traditional conceptions of the audience might be regarded as a part of the juridico-discursive conception of power bequeathed to us by the ways of envisaging power associated with absolutist regimes in which power seems to derive from some central and originating source: the sovereign or the state (see Foucault, 1980). However, this is not to suggest that we might either wish or be able to dispense with the term, for the same sort of points could be made in relation to any of the other terms that might be proposed as substitutes. The fact is that the inquiries that are currently conducted under the heading of audience studies—or under such alternative formulations as effects studies, reception theory, or reader response studies—represent, or figure, their objects of study in different ways: as audiences, readers, publics, receptants, interpreters, viewers, spectators, or listeners. Whichever of these conceptions is chosen, moreover, will—if it is accorded any theoretical weight—affect how a particular inquiry is conducted: what it looks for, how it frames its object theoretically, what methods are used, how the results are represented, where and how those results are circulated, and to what effect.

Pierre Bourdieu's enigmatic contention that public opinion does not exist but its effects are real (Bourdieu, 1979) may help to make the point I have in mind here. In maintaining that public opinion does not exist, Bourdieu's purpose is to dispute the supposition, quite common among advocates of opinion polling, especially such pioneers of polling techniques as George Gallup and Saul Rae (Gallup and Rae, 1968), that there exists such a thing as a public whose opinions on any given topic are ready-made, simply lying there waiting to be discovered via the application of an appropriate polling method. The making of any statement regarding public opinion on any particular issue, Bourdieu argues, depends on a specific means of accessing a public and arriving at an assessment of its opinions. Yet the means deployed for these purposes serve more to constitute and to shape a public than simply to discover one, just as they often organize into being the very opinion they seem merely to report.

Thus, whether people are interviewed at home, over the telephone, via methods of random intercept, or by post; whether they are interviewed singly or in groups; who they are interviewed by; what manner of relations of power might exist between interviewer and interviewee; whether they are able to arrive at a means of expressing an opinion via discussion with each other and then voting; whether voting is by means of a public or private ballot; how the options within a questionnaire are structured and how they are phrased: these are among the considerations that can influence the kind of opinion likely to be attributed to a particular population. Whatever the means by which it is arrived at, moreover, such an opinion is subject to a further process of shaping via the discursive forms—from official poll reports through to editorializing comment—that regulate its circulation in the public domain.

Public opinion, then, is always shaped and organized by the very instruments that purport to measure it. No such instrument is neutral, simply revealing a set of opinions that pre-exists its application. Public opinion cannot be spoken of without, in the very process, giving it a determinate form—as a set of statistics, say, or as a set of exemplary views and preferences—which is the effect of the means of its measurement. All of this is to say that public opinion always takes the form of a representation—or, more accurately, a circuit of representations—that has been arrived at in determinative and specific ways. This is not to say that its existence is in the least chimerical or without consequence. While public opinion might not exist in the form that the early pollsters imagined—in the raw, so to speak—it certainly has a very real, if also fractured and contradictory, existence in the many complex and varying representational forms in which it is circulated. Nor is this field of representations in which the public and its opinions are diversely figured without consequence for political processes. Its consequences, however, are not those envisaged by Gallup who, viewing opinion polls as a means of making the will of the people luminously transparent to their political representatives, extolled their virtues as instruments of democracy. Rather, Bourdieu suggests, polling has developed into an instrument for making popular opinion responsive to political direction rather than vice versa. It affords a means whereby opinions, in being organized to conform to the shape in which they are represented, can be mobilized in support of particular political causes or projects.

Broadly similar arguments apply to the researches that are currently conducted under the various headings of audience studies, reception theory, and effects or reader response studies. All such researches require some means of accessing their objects (whether audiences or readers), means that never simply allow "what is going on" in the situations they investigate to be fathomed in ways that are not affected by the means of access selected—participant observation versus postal survey, for example. Indeed, such "means of access" often organize their objects into being, in the sense that, more often than not, audiences are artifacts of the instruments selected for their investigation. There is also little room for doubting the practical consequences of the ways in which the audience is represented, or figured, when the results of such inquiries are written up and circulated. The social and political fields in which such representations can be counted as having consequential effects are many—the relations between commercial broadcasters and advertisers, the activities of regulatory agencies, the investment strategies of publishers, pedagogical practices in the classroom, forms of assessment and examination. They may also occasionally have a bearing on the symbolic currency and strategies of major political movements, as Lawrence Grossberg has argued was the case with the relationship between the "active audience" conceptions of the 1920s and 1930s and American Progressivism (Grossberg, 1989).

Yet, however familiar these perspectives may be, debate in these areas often proceeds as if the point at issue was that of somehow really fathoming out the audience, of finally getting to what audiences actually do with media, how they really interpret media messages—the *real truth* of their media lives. This is especially so where debate crosses the boundary lines between different paradigms. Thus, for James Lull, the value of much of the work that goes on under the heading of "cultural studies" audience research is discounted because, in his view, it is driven more by a priori theoretical considerations than by "descriptions and grounded interpretations of what audiences really think and do" (Lull, 1988: 240). Yet John Fiske, writing in the same colloquium, argues that the academic objectifications of audiences produced by the quantitative methods of the "communications studies" paradigm distort the relations of reciprocal fluidity that, in his view, really characterize text-reader relations. These, Fiske contends, are more accurately revealed by ethnographic approaches, which he sees as allowing audiences/readers more control over the terms in which they describe their own activities (Fiske, 1988).

To suggest that debates conducted in this manner have not proved particularly productive—and, indeed, that they are unlikely ever to do so—is not to suggest that we have no means of deciding between competing approaches to the audience. We certainly do, but never in the abstract. The grounds on which, in practice and in particular regions of debate, such decisions are made are always pragmatic—and provisional and circumspect—rather than abstractly theoretical in the sense of requiring a general validation of a specific method or paradigm. In short, and quite sensibly, they are dependent on the tasks at hand and the instruments that are judged most likely to assist in them. Academic debates concerning the relations between different paradigms of audience research are likely to be more productive if conducted in a similar spirit. This would involve an abandonment of the empiricist dream of some day being finally able to assess which approach had finally got the audience "figured out" correctly and focusing instead on the respects in which different ways of figuring audiences or readers—statistically or ethnographically, for example—are connected to, and calculated to produce effects within, quite different regions of practical activity.

I want, therefore, to develop this argument by considering the different fields of practical activity with which work falling under the heading of the "active audience" approach has been associated. To object, as does Lull, that the empirical protocols for this way of figuring the audience are weak is correct. It would also be correct, in my view, to argue that such constructions of the active audience no more offer an adequate critique of more conventional empirical approaches to audience studies than they provide a usable alternative to many of the practical purposes such approaches serve. Yet these objections are, in a sense, inconsequential. They miss what, in at least some of its variants, this way of representing audiences really aims for: the production of a figure of the audience or reader that, through its pedagogic deployment, can serve as a performative prop for a particular set of exercises through which actual readers—by aligning their reading practices with those associated with the exemplary figure of the active reader—are to be, so to speak, activated. This is said not to defend such approaches but rather to identify more clearly the grounds on which they might more usefully be criticized: If, as I am suggesting, the figure of the "active audience" forms part of a specific technologization of the text-reader relation that aims at the transformation of reading, viewing, or listening practices, then the issue is not its descriptive adequacy but its role as part of an apparatus of textual criticism and pedagogy.

To substantiate this argument, however, will require that the notion of the active audience be, to some degree, deconstructed, for the semantic currency of the term "active" is now an extraordinarily wide and varied one, consequently, it is also an imprecise one. If the audience has recently become "active," its activeness has been differently conceived and constructed in different regions of inquiry. However, the failure to accord these differences the attention they merit means that the image of the "active audience" now uneasily combines often contradictory conceptions arising from quite different disciplinary fields. Some disentangling will be necessary, then, to identify the different kinds of theoretical and political investments that have been made in this figure.

Audiences as Objects and Subjects

Before I pursue these contentions further, let me first summarize my argument so far and explore some of its collateral implications with a view to providing a broader context for the above remarks. My main purpose has been to suggest the incoherence of those empiricist conceptions of audience research according to which an objective knowledge of audiences is eventually to be arrived at via the progressive refinement of research techniques. Ien Ang has recently argued a similar position in disputing the assumption that the audience can be regarded as "a proper object of study whose characteristics can be ever more accurately observed, described, categorized, systematized and explained until the whole picture is 'filled in' (Ang, 1989: 103). Her grounds for doing so, moreover, are broadly similar—that research in the area can only give rise to "historically and culturally specific knowledges that are the result of equally specific discursive encounters between researcher and informants" (p. 105).

Ang is at pains to make it clear that this argument applies with just as much force—neither more nor less—to "cultural studies" ethnographic approaches as its does to empirical survey techniques. Such approaches do not provide a privileged means of access to the real "media lives" of audiences, nor can the statements of viewers' preferences and activities to which they give rise be viewed "as transparent reflections of those viewers," "lived realities" that can speak for themselves" (p. 106). If understanding audience activity is thus "caught up in the discursive representation . . . of realities having to do with audiences" (p. 105), then the politics of audience research have to do with the manner in which those representations are arrived at, the form they take, where they are circulated, and the political projects they can be connected to.

This line of thought might be taken a little further by distinguishing between means of accessing and representing audiences according to whether the political programs they enable and support address audiences as objects or subjects. To clarify what I have in mind: The audience is constituted as an object within governmental or regulatory projects that, typically drawing on statistical representations of audience activities or preferences, aim to influence the nature of the services available to audiences or the conditions within which their activities are situated. In such cases, though audiences constitute the targets of such programs—and, indeed, are often brought into being by them—the programs themselves are conceived, put into effect, and implemented by other agencies (by regulatory bodies like the Australian Broadcasting Tribunal, for example), which therefore also constitute the key reception points at which those representations of audiences effectively circulate. By contrast, where ethnographic studies give rise to representations of audience activities in the form of firsthand descriptions and reported statements whose effective reception points are within the pedagogic apparatus and the various quasi public spheres to which that apparatus is connected, the political programs they support often aim to modify audience behavior directly—via the empowering effect that can arise from making marginalized readings publicly available, for example.

In distinguishing between these programs in this way, I do not mean to imply a preference for the latter on the grounds of some abstract emancipatory potential. Valerie Walkerdine, writing of her experience in researching family uses of video, has noted that, as observer, she became "a 'Surveillant Other' not only watching but producing a knowledge that feeds into the discursive practices regulating families" (Walkerdine, 1986: 190). Though, obviously, from both an ethical and a political point of view, the dangers inherent in this situation need to be kept in mind, it is equally important to avoid the kind of reflex politics in which surveillance and regulation are deemed to be axiomatically oppressive. On the contrary, it is clear that a vast range of progressive political agendas—egalitarian, socialist, and feminist—cannot be even thought, let alone brought into being, without mechanisms of surveillance capable of making visible the fields of activity they address in a form that renders them amenable to political calculation and action.

To draw on an example from a related field of study, it is thus true that the public museum has, from its inception, functioned as a space of surveillance in which conduct has been subjected to regulation by exposing it to a normalizing and controlling gaze—that of security staff or fellow visitors, for example. It is also true (although little noted) that, like the prison and the asylum, the museum has functioned as a laboratory of observation that has served as an incubator of sorts for the development of techniques of observation and behavioral description—intercept studies on exit, the surveying of group behavior patterns, calculations of attention timespan—that have assisted the development of regulatory practices in a wide variety of domains: the design of shopping malls, for example. Yet it is also true that the development of modern political demands for more equitable patterns of access to, and participation in, museums has been made possible only by utilizing this space of observation so as to make visible the sociodemographic profile of its visitors.

Moreover, it is clear that there is now a self-fueling momentum built into the relationship between museum practices and visitor observation in the sense that new fields of museum policy are constantly being generated as the grid of observation to which visitors are subjected becomes more refined. While the notion of the "active visitor" is not yet current within the field of museum studies, it just as well might be. Although no more reducible to their textual components than are film or television, many aspects of museums are now widely recognized to be text-like in their organization: the narrativization of the visitor's route from one gallery to the next, for example. Equally, this recognition is accompanied by an increasing awareness that museums and museum exhibits can be experienced and interpreted in significantly different ways by visitors with different socioeconomic and cultural profiles in accordance with the discursive resources they can draw on to organize their readings of those exhibits or, more generally, to negotiate their relations to the museum environment. And this awareness, finally, is increasingly reflected in attempts to build polysemic possibilities into museum displays so as to enhance the prospects for multiple readings (see, for example, MacDonald and Silverstone, 1990).

Yet the museum visitor is not—or is only rarely—an effective agent within such fields of policy formation. The visitor—and where programs are designed to make museums accessible to broader publics, the nonvisitor too—is always the point of reference for the programs put into action by effective agents (curators, museum directors, boards of trustees, government funding bodies) within the museum sphere. However, owing to the nature of their relations to the institution—infrequent and transitory visits or, for non-goers, none at all—visitors themselves are rarely involved in these policy issues or the politics that accompany them except in the forms in which they are figured via the application of a range of observational techniques. The visitor, moreover, is typically unaware of the ways in which visitor needs, interests, and so on, are represented in policy debates and processes; nor, with few exceptions, are visitors actively enlisted in support of the programs constructed on their behalf. Constantly present figuratively, the museum visitor is yet constantly absent as agent and is seldom addressed as such. Certainly, I know of no studies of museum visitors—whether of their attitudes, behavior, or social and cultural characteristics—that have constituted those visitors as their addressees as well as their objects of study.

By contrast—and to come to the point I want to make by way of this digression—the very point at issue in many versions of the "active audience" approach is precisely how they constitute and, in constituting, address audiences or readers as subjects. Notwithstanding their pretensions to a general theoretical status, such constructions of audiences form part of a rhetorical politics that is concerned less with describing audience behavior than with altering it, less with accounting for reading practices than with changing them. They are, that is to say, parts of larger critical enterprises whose effectivity depends on how they map out and organize a role for the reader and on the mechanisms they deploy to induct real readers into that role and its performative requirements.

Yet this is true only of some versions of the argument. It will therefore now be useful to return to my earlier commitment and, in the light of the preceding considerations, distinguish between a number of positions that are commonly grouped under the "active audience" approach. While the use of a common designation has served usefully to differentiate this group of positions from other and earlier traditions (structuralism, for example) in which the audience was figured as passive, its course mapped out for it as an effect of its textual positioning, the indiscriminate application the designation can prove a barrier to analysis and debate if allowed to occlude equally important differences among the positions it covers. I shall therefore offer a breakdown of the "active audience" approach into three positions. I shall call these the "determined active reader," the "indecipherably active reader," and the "overactive reader" approaches, and I shall argue that they are distinct from one another with regard to both the means they employ in arriving at the figure of the reader they construct and the spheres of practical activity in which those figures of the reader, and of reading, are capable of surfacing and producing effects. This will then allow me to argue that it is mistaken to limit criticisms of the "indecipherably active reader" and "overactive reader" approaches to their inadequacies as research paradigms. It is not that criticisms couched in such terms are wrong. Rather, it is that they fail to see that such approaches, however much they might masquerade as research paradigms, are more accurately viewed as interventions within the politics of reading. The figures of the reader they construct, that is to say, function as a means of rhetorically enlisting readers for certain reading practices while also providing them with a performative prop—a model—that they can use in arriving at readings of a similar kind. This therefore raises questions concerning their implications for the ways in which texts—televisual, filmic, or literary—are to be pedagogically deployed, issues that need to be assessed in other ways and on other grounds.

Activating Audiences

By the "determined active reader" I mean those approaches that, while arguing that a text cannot dictate its readings and thus insisting on the activity of the reader within or vis-à-vis the text, do not construe that activity as being in any way voluntaristic or autochthonous. Rather, if the reader is active vis-à-vis the text, this is by virtue of the other forces or considerations—social position, intertextual relations, discursive mediations, and so on—that are accorded a role in the determination of reading practices. I would cite as examples here David Morley's study of the *Nationwide* audience (Morley, 1980) as well as my own and Janet Woollacott's study of the varying reading formations that have affected the social reception of "the texts of James Bond" (Bennett and Woollacott, 1987).[2] While both attribute a degree of autonomy to the reader vis-à-vis the text, that autonomy is regarded as the result of the *other determinations* that bear upon the social organization of reading practices.

Both, moreover, share a common impetus derived from their opposition to the "hypodermic syringe" model of ideological inoculation implied by the dominant ideology thesis. This is evident in their concern to theorize the social relations and determinations of reading in a manner that would allow for the possibility of resistive or oppositional readings of media texts. That said, however, neither implies that the active reader is, must be, or should be a resistive or oppositional reader. For Morley, the readings of trade unionists, managers, or tertiary students are all equally active in the sense that all are molded by similar configurations of extratextual relations: the patterns of access to the discourse positions that govern their "decoding" activities. Similarly, in suggesting that the texts of Bond might be read in a plurality of ways, the analysis of *Bond and Beyond* allows the "active reader" to be figured as deeply misogynist or conservatively nationalist just as much as benignly restive.

The "determined active reader," then, is one whose activity might best be thought of as being organized by a complex weave of determinations within which the operative influence of the text is always "overdetermined" by the concerted influence of a variety of extratextual forces. The means whereby this figure of the reader may be constructed are various, ranging, for example, from the techniques of focus group discussion to appropriate historiographical methods of inquiry, where it is past reading practices that are at issue (see Parry, 1985). Similarly, the political currency of this version of the active audience is a varied one: whether or not you would like a "determined active reader" as your best friend depends on the nature of the determinations that organize her or his reading practices. Given this, the figure of the reader that is constructed in this way cannot serve as a model for empirical readers. Nor is it the purpose of such approaches to directly politicize reading in this way. On the contrary, their political address is to the determinations that mold and structure the reader's activity; it is only by virtue of transformations in these that the political quality of the audience's textual investments can be altered. Indeed, the political payoff of this version of the active audience approach has consisted in the contribution it has made to the demise of those forms of purely textual politics—the search for radical or progressive texts—in insisting on the need for questions concerning the political effects of cultural practices to be posed in relation to the broader fields of relations (discursive, institutional, intertextual) that organize the cultural terrain of specific text-reader encounters.

If the provenance of the "determined active reader" has been largely sociological, that of the "indecipherably active reader" has been chiefly literary. The product of the uncoupling of the text-reader polarity engendered by deconstruction's insistence on the reciprocal iterability of text and reader (now no more than an intertextual nodal point), the "indecipherable active reader" is one whose activity, while subject to an endless theoretical affirmation, is simultaneously unfathomable since neither the place of the reader that reads nor that of the text that is read is susceptible, even in principle, to a definite determination. Literary in origin, this construction of the reader also constitutes a means whereby the regime of truth that characterizes modern literary studies—a regime that rests on the production of a text whose meaning is forever undecided and undecidable—is extended from the literary text to the domain of its readings.[3] As such its effect, in undermining any attempt to make audiences or readers the objects of a positive knowledge, is to produce an enlarged field in which the protocols of the literary reading might be applied. As a consequence, the type of analysis to which particular reading practices (assuming they could be isolated) might be subject is one that can never be finalized. The reading that is to be read—since, by definition, it exceeds the scope of any particular set of methods—always remains, in some way, indecipherable, out of reach.

Notwithstanding its literary provenance, this figure of the reader has acquired a much broader circulation within the field of audience studies. This has been mainly due to the influence of postmodernist debates, especially as mediated via Ernesto Laclau's and Chantal Mouffe's deconstructive conception of the discursive (and hence undecidable) organization of social relations (see Laclau and Mouffe, 1985). Janice Radway's conception of the dispersal and nomadicity of audiences provides a case in point, and an instructive one in the disjunction between her subscription to such a conception of the audience and her attempts to devise a research method appropriate to it. Noting that most audience studies proceed from the assumption that texts can be categorized as entities of a particular sort, Radway argues:

> Audiences, then, are set into relation to a single set of isolated texts which qualify already as categorically distinct objects. No matter how extensive the effort to dissolve the boundaries of the textual object or the audience, most recent studies of reception, including my own, continue to begin with the "factual" existence of a particular kind of text which is understood to be received by some set of individuals. Such studies perpetuate, then, the notion of a circuit neatly bounded and therefore identifiable, locatable, and open to observation. (Radway, 1988: 363).

Insisting, against such conceptions, that neither texts nor audiences can be treated as simply givens—or as existing in given relations to one another—Radway proposes that the study of audiences should give way to a more diffusely defined object of analysis:

> Instead of segmenting a social formation automatically by construing it precisely as a set of audiences for specific media and/or genres, I have been wondering whether it might not be more fruitful to start with the habits and practices of everyday life as they are actively, discontinuously, even contradictorily pieced together by historical subjects themselves as they move nomadically via disparate associations and relations through day-to-day existence. In effect, I have begun to wonder whether our theories do not impress upon us a new object of analysis, one more difficult to analyze because it can't be so easily pinned down—that is, the endlessly shifting, ever-evolving kaleidoscope of daily life and the way in which the media are integrated and implicated within it. (Radway, 1988: 366).

The challenge, as Radway enunciates it, is thus to design a research strategy that will "provide for a collective mapping of the social terrain equal to the ambitious, majestic scope of our recent theories of subjectivity and intertextuality" (Radway, 1988: 367–368).

The difficulty is that, in seeming to respond positively to this challenge, Radway re-instates assumptions that run counter to the very grain of the challenge they are meant to meet. Radway suggests—although clearly in an exploratory and hesitant fashion—that an appropriate response to such a challenge might take the form of a team-based ethnographic analysis of "the range of practices engaged in by individuals within a single heterogeneous community as they elaborate their own form of popular culture through the realms of leisure and then articulate those practices to others engaged in during their working lives" (Radway, 1988: 368). This does, in truth, sound a little like rediscovering "the real media lives" of small town America after the fashion of the Lynds (1929) or Vidich and Benseman (1960). The more general difficulty, though, is that this solution is bought at the price of suppressing the conditions that generated the problem in the first place. Lawrence Grossberg—on whom I have drawn in much of the foregoing discussion—thus argues that, in spite of her ostensible commitment to the fluidity and transience of audience/media, reader/text relations, the methods Radway appeals to for the analysis of such relations reinscribe them within objectifying frameworks in which their separateness and self-identity are reaffirmed (see Grossberg, 1988).

Grossberg's point in developing this critique is to insist on the need for ways of speaking about audiences or readers that lack the objectifying potential—and especially the division between researchers and audiences, "us" and "them"—inherent in Radway's continuing reliance on ethnography as an acceptable research tool. Mine, per contra, is to suggest that Radway's mistake consists in the supposition that *any* set of research techniques might prove adequate to the task of reconstructing the "real media lives" of "nonaudiences" that would be equal to "the ambitious, majestic scope of our recent theories of subjectivity and intertextuality." The effect of such theories, where they are of literary derivation, is always such as to put the analysis of *any and all* social relations beyond the reach of any determinate research technique and to constitute them, instead, as a resource for a project of reading that—since the possibility of arriving at a correct reading is denied—is simultaneously the project of a rereading that can never be completed.

To ask whether definite and specific research methods might be deployed in a field in which audiences and texts, and the relations between them, have been deconstructed in this manner is, in short, to make a category mistake. The very point of such a deconstruction is, precisely, to place such questions beyond the sphere of positive methods so that the domain of reading and audience practices might be reconstituted for other purposes. In being rescued from the domain of methodology, the "indecipherably active reader" is thereby made available as a resource for extending the sway of literary readings beyond the confines of the literary text. The reader, figured in this fashion, is not a possible object of analysis. Rather, this approach supplies a means for cultivating and exercising a particular reading competence, one that develops the ability to hold to many different meanings at the same time but without being able to (or feeling the need to) decide between them—one that, in literarizing life, extends the spheres in which it is possible to speak without saying anything definite, without any sense of a referential horizon or responsibility. In short, the "indecipherably active reader" forms part of a distinctive technology of reading that—originating within the literary branches of the academy, but now enjoying a broader circulation within specific regions of other humanities and social science disciplines—has to be assessed in terms of the specific intellectual skills and capacities it gives rise to. While this is not a task I shall undertake here, it is surely clear that intellectuals formed by means of such a technology will display a limited ability to contribute in a definite manner to specific endeavors.

The difficulty with the "overactive reader," by contrast, is that of recruiting empirical readers, too easily, and too automatically, for reading practices conducted in the name of resistance. Like the "determined active reader," this figure of the reader was fashioned in critique of the "hypodermic syringe" model of ideological inoculation associated with the dominant ideology thesis in both its capitalist and patriarchal versions. It differs, however, in two key respects. First, the active reader here is always represented as occupying a position of social subordination and as the subject of a reading that is usually valorized in being portrayed as manifesting a resistive, oppositional, or transgressive potential. Second, the methods through which this figure of the reader is produced are hypothetical-deductive rather than empirical. Rather than relying on ethnographic, statistic, or historiographic techniques of inquiry, that is, the more usual procedure here is to deduce what the reading practices of particular subordinate groups might be by positing correspondences between the formal attributes of specific textual regimes and the conditions of life of the groups that are posited as the primary audiences for those textual regimes.

A classic example is Tania Modleski's attempt to explain the nature of women's pleasurable and affirmative investments in soap operas by suggesting a homology between the narrative organization of soap operas and women's daily routines. The fact that soap operas are potentially endless narratives, she suggests, resonates positively with the experience that "women's work is never done." The distracted pleasures women derive from soap opera, Modleski goes on to argue, embody women's resistance to—and prefigure a narrative system beyond—what she posits as the repressive and patriarchal narrative regimes of classic film in its impetus toward speedy narrative resolution and closure (see Modleski, 1982).

The most obvious problem with such hypothetical-deductive approaches is that the same means can be used to arrive at sharply contrasting, but equally convincing (or unconvincing), constructions of the reader and the nature of her or his pleasure. Radway, for example, forges quite a different connection between textual regimes and the conditions of life of her sample of romance readers in attempting "to infer from the women's conscious statements and observable activities other acknowledged significances and functions that make romance reading into a highly desirable and useful action in the context of these women's life" (Radway, 1984: 9). Thus, in this, the hypothetical-deductive moment of her analysis, Radway also construes romance reading as incipiently resistive of patriarchy. It is so, however, for reasons that are diametrically opposed to those that Modleski advances. Reading romances, Radway suggests, is prompted by women's intense interest in and demand for narrative resolution, an interest that she sees as expressing a resistive, protopolitical demand for time on one's own, free from interruptions.

There is, of course, no reason why women readers should not take pleasure in different textual forms for different reasons. What is clearly impossible, however, is that such contradictory pleasures should be accounted for in essentially the same terms and be accorded the same antipatriarchal effects—at least, not unless such readers are to be judged capable of having their resistive cake and eating it too!

Yet, in truth, the reading that is at issue in such analyses is less that of the readers described in the discussion than that of those addressed by it. The former constitute, in effect, a rhetorical device whereby the latter might be inducted into a particular way of reading the textual regime concerned. The "overactive reader" thus serves as a prop whereby real flesh-and-blood readers might be activated into resistive reading practices in embodying a set of exercises and orientations whose emulation will—theoretically, at least—open up the textual regime under discussion to new forms of use and contextualization.

The pertinent questions to put to such constructions of reading thus concern less their empirical reliability than their technological effects; less whether they accurately describe the practices of real flesh-and-blood readers than what they allow, incite, or encourage such readers to do. Viewed in this light, the "overactive reader" occupies the same place and performs the same function as that figure of the reader it is constructed against: the "underactive reader" that informed the practices of textual commentary associated with earlier mass culture critiques or theories of patriarchy. Modleski's housewife thus occupies the same *type* of relation and performs the same *type* of function in relation to daytime soaps as did Richard Haggard's "jukebox boys" in relation to what he dubbed the "newer mass art" of "comics, gangster novelettes, science and crime magazines" (Haggard, 1969: 247). The politics of the positions are, of course, dissimilar. Modleski's housewife serves both to valorize soaps and as a performative prop for, and as a means of legitimating, emulative readings whereby real readers are to derive an enhanced resistive value from such texts. Haggard's "jukebox boys," by contrast, serve as a means of denigrating the texts of mass culture while also organizing, and legitimating, a disdainful practice of reading whereby we disentangle ourselves from and lift ourselves above what such texts do to their unfortunate victims.

I shall not attempt here to evaluate the kind of reading practices—or habits—that this way of figuring the reader serves to promote. It does seem clear, however, that its main legacy has been a series of ideal readers—Modleski's daytime soaps fan, Fiske and Watts's video games player (Fiske and Watts, 1985)—which might best be described as populist inversions of the hapless dupes who formed their erstwhile mass/male culture counterparts. In view of this, reading practices modeled on such figures are likely to tend toward the self-indulgent, politically lazy end of any given set of reading possibilities.

My main purpose, however, has been less to decide between different approaches to audiences and readers than to establish that the field of discourses in which audiences and readers appear is not a unitary one. On the contrary, existing approaches to audiences and readers exhibit vast differences with regard to the means of accessing and representing audiences and readers they deploy. They also differ with regard to the kinds of practical issues that are implicated in their concerns and the ways in which those issues—and the agents involved in them—are addressed. It is only by foregrounding such practical concerns, I want to suggest, that different approaches can be discussed and assessed on grounds that are appropriate to them.

Notes

1. My position is, in this respect, slightly different from Radway's in that she suggests that the modern extension of the term "audience" to refer to readers, viewers, listeners, and so on, rests on the presumed naturalness of face-to-face speech as a model for all forms of communication. This, in turn, leads Radway to suggest that the conception of the audience as passive derives from the assumption of the relations of full presence—and hence the transparency of meaning—which, pre-Derrida, were mistakenly held to characterize the sphere of speech communication. My contention is rather that the deferential position mapped out for the audience derives not from the sphere of speech communication as such but from the historical weight and analogical force of the specific hierarchical sets of socio-verbal relations with which the term was originally associated.

2. I shall not attempt to identify all the sources relevant to the different schools of audience theory I am concerned with. Several excellent general surveys of the field exist with full details of relevant sources. See, especially, Morley (1989) and Moores (1990). On issues specifically to do with the active audience, see Allor (1988), Morley and Silverstone (1990), and Nightingale (1989).

3. For a discussion of the respects in which the literary text's undecidability or unfathomability might be regarded as an artifact of the relations of correction and supervision that characterize modern literary education, rather than as an essential attribute of its literariness (see Hunter, 1988). For a related discussion of the application of these protocols ofliterary reading to the domain of history, see the final chapter of Bennett (1990).

References

Allor, Martin (1988). Relocating the site of the audience. *Critical Studies in Mass Communication*, no. 5.

Ang, Ien (1989). Wanted: Audiences. On the politics of empirical audience studies. in Ellen Seiter et al., (eds) *Remote Control: Television, Audiences, and Cultural Power.* London: Routledge.

Bennett, Tony, and Janet Woollacott (1987). *Bond and Beyond: The Political Career of a Popular Hero.* London: Macmillan.

Bennett, Tony (1990). *Outside Literature.* London: Routledge.

Bourdieu, Pierre (1979). Public opinion does not exist. In A. Mattelart and S. Siegelaub (eds), *Communication and Class Struggle: Capitalism and Imperialism.* New York: International General.

Fiske, John (1988). Critical response: Meaningful moments. *Critical Studies in Mass Communication*, no. 5.

Fiske, John, and Jon Watts (1985). Video games: Inverted pleasures. *Australian Journal of Cultural Studies*, vol. 3, no. 1.

Foucault, Michel (1980). Two lectures. In Colin Gordon (ed.), *Power/Knowledge: Selected Interviews and Other Writings.* New York, Pantheon.

Gallup, George, and Saul Rae (1968). *The Pulse of Democracy.* New York: Greenwood.

Grossberg, Lawrence (1988). Wandering audiences, nomadic critics. *Cultural Studies*, vol. 2, no. 3.

———. (1989). The context of audiences and the politics of difference. *Australian Journal of Communication*, no. 16.

Haggard, Richard (1969). *The Uses of Literacy.* Harmondsworth: Penguin.
Hunter, Ian (1988). *Culture and Government: The Emergence of Literary Education.* London: Macmillan.
Johnson, Leslie (1988). *The Unseen Voice: A Cultural History of Early Australian Radio.* London: Routledge.
Laclau, Ernesto, and Chantal Mouffe (1985). *Hegemony and Socialist Strategy.* London: Verso.
Lull, James (1988). The audience as nuisance. *Critical Studies in Mass Communication,* no. 5.
Lynd, Robert S., and Helen M. Lynd (1929). *Middletown.* London: Constable.
MacDonald, Sharon and Roger Silverstone (1990). "Rewriting the museum's fictions: Taxonomies, stories, and readers." *Cultural Studies,* vol. 4, no. 2.
Modleski, Tania (1982). *Loving with a Vengeance: Mass-produced Fantasies for Women.* New York and London: Methuen.
Moores, Shaun (1990). Texts, readers, and contexts of reading: Developments in the study of media audiences. *Media, Culture, Society,* vol. 12.
Morley, David (1980). *The "Nationwide" Audience: Structure and Decoding.* London: British Film Institute, Monograph 11.
———. (1989). Changing paradigms in audience studies" in Ellen Seiter, et al. (eds), *Remote Control: Television, Audiences, and Cultural Power.* London: Routledge.
Morley, David and Roger Silverstone (1990). Domestic communication: technologies and meaning. *Media, Culture, Society,* vol. 12.
Nightingale, Virginia (1989). "What's "ethnographic" about ethnographic audience research? *Australian Journal of Communication,* no. 16.
Parry, Ann (1985). Reading formations in the Victorian press: The reception of Kipling, 1888–1891. *Literature and History,* vol. 11, no. 2.
Radway, Janice (1984). *Reading the Romance: Women, Patriarchy, and Popular Literature.* Chapel Hill, N.C.: University of North Carolina Press.
———. (1988). Reception study: Ethnology and the problems of dispersed audiences and nomadic subjects. *Cultural Studies,* vol. 2, no. 3.
Vidich, A. J., and J. Benseman (1960). *Small Town in Mass Society.* New York: Anchor Books.
Walkerdine, Valerie (1986). "Video replay: Families, films and fantasy. In Victor Burgin et al. (eds), *Formation of Fantasy.* London: Methuen.
Williams, Raymond (1974). *Television: Technology and Cultural Form.* London: Fontana.

Marginal Texts, Marginal Audiences

10

Larry Gross

Notes on Pornography

The porn "industry" is generally assumed to be both large and rapidly expanding: *Time* magazine in 1987 estimated annual U.S. sales of $8 billion, including 100 million rentals of X-rated videos each year (Leo, 1987: 63). The primary reason for the most dramatic increase in porn profits has been the emergence of the home video sales and rental market. With a VCR in the home men can avoid the sex shop, described by one visitor as a "zone of shame"[1] (Kovel, 1990: 155), and women can become more equal participants in what had been a male enterprise.[2] Video production and distribution has even permitted women to move into the role of producers and distributors of hard-core porn.

The rhetoric of antiporn crusaders describes an iceberg whose visible tip is made up of such mild stuff as *Playboy* centerfolds, but whose lowest depths include such horrors as the infamous *Snuff*, an Argentine film that purported to show a woman being killed and dismembered on camera (it was a hoax; see Birge and Maslin, 1976). Despite the evidence that suggests that millions of Americans are consuming porn, public discourse continues to be overwhelmingly dominated by a presumption of hostility and condemnation. Public opinion polls over the past two decades show a steady or rising level of support for government regulation of pornography and a majority of respondents endorsing the view that pornography leads to "a breakdown in morals" (Smith, 1987).[3]

What follows are notes on the characteristics of pornographic texts and audiences, considered as somewhat special cases within the larger framework of mass media texts and audiences. Rather than to attempt to navigate the treacherous straits of definition, I will consider as pornography those texts (written and, primarily, visual) that most of us recognize as such when we see them. To cite two recent definitions by a lesbian feminist and a heterosexual man, respectively, porn is "work which seeks to arouse, or represent arousal" (Smyth, 1990: 153), "essentially a tool for masturbation, a fantasy enhancer" (Steinberg, 1990: 54).

I will suggest that porn texts are atypical within the larger framework of mass media. Mostly, they are mass media products (examples of tailor-made pornography, such as that written by Anaïs Nin on commission, are relatively rare as far as I know), and they generally fall into the "lowest" category of media: they are low-cost productions, they are distributed outside the mass media mainstream, they are stigmatizing for producers and actors (who rarely can cross over to mainstream media venues, or, if they do, are generally vulnerable to exposure), they are the most liable to be prosecuted (a fate they currently share with rock musicians), and they receive none of the respectful attention of reviewers/critics and infotainment gossip that is so much a part of the contemporary media marketplace. I will suggest that porn audiences are also atypical within the larger framework of mass media audiences. I am not thinking here primarily of the standard concerns of the establishment that focus on the "susceptible" audiences of women, children, and the lower classes who might be subverted and corrupted by porn. Rather I am interested in the prevailing stigma attached to the consumption and the consumers of porn, similar to but probably greater than the stigma attached to romance readers and soap opera fans (if these are still stigmatized passions).

Marginal Texts

As a way to organize the first portion of these notes I have chosen a model synthesized by the late media scholar/anthropologist Eric Michaels (1991). Drawing on recent theories and studies, Michaels referred to television as a "negotiation of texts between producers, technology and audiences" (p. 5). In discussing the social organization of meanings involved in this signifying activity, Michaels cited eight analytically distinguishable "texts," and I will use these as a device for considering porn texts.

The Conceived Text. In the world of mainstream mass media, the conceptual text has an important role and constitutes real and often valuable property. Ideas, concepts, treatments, even titles, can be created, copyrighted, marketed, negotiated and sold (many successful careers have been conducted largely at the level of concepts and treatments, with few if any actual productions). Conceived texts are also the focus of story conferences as well as many other meetings (over lunch or otherwise). In the world of porn it does not seem likely that conceptual texts play quite as central a role, but this remains a hunch to be put to empirical test. Do porn producers get calls from agents selling high-concept ideas? Are there story conferences to work out details of plot and character? It seems probable that the primary feature of porn at the conceptual stage is the choice of performers (or star, as almost any porn performer seems to be called) and some notion of the genre(s) to be featured.[4]

The Production Text. Once embarked upon the long voyage from concept to consumption, a mass media text will undergo numerous, often radical transformations, beginning with the emergence of a script that may bear little resemblance to the concept from which it derives. In the industrial world of the mass media, the author(s) of the script may be very different from the creator(s) of the originating concept. Here, too, porn is likely to be a much simplified venue. Scripts are minimal, plotting focuses mostly on simple and highly conventional (archetypal?) themes, almost all of which lead quickly to the inevitable sexual engagements of the performers (which Linda Williams [1989] analogizes to the "numbers" of musical films, each culminating in the obligatory (for heterosexual and gay male porn) "cum shot."[5]

The Produced Text. As we all know, the final produced text will differ in many ways from the shooting script. "In the best of worlds, actors, directors, cameramen, editors and other participants are expected to contribute inter-pretations, which is to say elaborations on the intended meanings of the text. In the worst of all worlds, these same people may be unable to deliver what is intended for various reasons, including economic and technological constraints and the intrusion of the predictably unpredictable—equipment failure, illness, bad weather, bad temper, etc., etc." (Michaels, 1991: 7).

Porn, beginning as it does with a much less formal production script, is even more likely to evolve in unexpected directions (for one thing, there is an important category of equipment that is notoriously subject to failure). The very marginality of the (nonunionized) production circumstances is also conducive to a more fluid process.[6]

More than most mainstream films, porn depends on editing to provide a narrative of continuous action that is beyond the capabilities of its (male) actors. The steady stream (as it were) of ejaculations, and their repeated appearances from different angles (a kind of cinematic male multiple orgasm), attest to the highly edited form of the produced text (an awareness of which is reassuring to male viewers).

The Transmitted Text. As Michaels points out, "it would be a mistake (often made in experimental media research) to treat the produced text as what the audience actually views" (1991: 8). In most cases the transmitted text will include commercials or other material inserted by broadcasters, the content of which (and consequent "collision" with the surrounding primary text) could not have been foreseen by the producers (even if, as in commercial television production, they construct a program in scenes that end at appropriate moments for commercial breaks). This aspect of television's "flow" of heterogeneous texts is probably less typical of porn, which tends to be consumed in venues that do not feature commercial interruptions (although porn videos now frequently begin with "coming attraction" trailers for other films). However, there are numerous other ways in which the transmitted text might differ from the produced text, the most obvious being censorship of the "hardest" portions of the narrative.[7]

There are a variety of contexts in which porn texts are encountered, chief among them the "loops" of adult bookstore film or video booths, the large screen of the increasingly rare adult movie theater,[8] and currently most frequent, the cassette or cable transmission seen on home video.[9] In these varying contexts the text being transmitted is likely to vary, in the quality of the print and of the projection technology as well as the degree to which the reception is interrupted (by the need to put quarters in a machine, the activities of other audience members, as well as the viewer's own responses).

The Received Text. In Michaels' analytic scheme, the received text "will differ from the transmitted text depending on the viewer's attention, other competing simultaneous information sources and any manipulation of the set itself which affects the text" (1991: 9). Porn films may be the least likely of any media texts to be consumed without interruption, or even granted the respect of being viewed from beginning to end.[10] The home VCR remote control, with its fast-forward, slow-motion, freeze-frame, and sound-mute buttons, allows the viewer to create a received text that may bear little relation to the transmitted text or the text received by other consumers of the same film.

Michaels notes that this model "contradicts the usual academic and popular idea that TV is a mass medium, which allows the same thing to be in many places at once. . . . The model suggests instead that it is many things which are in many places at once." Yet, Michaels continues, speaking of mainstream television, "we know that this must be an overstatement; generally there is some agreement that when people discuss watching TV, they presume they have had a common experience, that they have 'seen' the same thing. It must be, then, that the experience of TV includes some further processes by which the diversity of received texts is reduced" (1991: 9). But here, in particular, we may find that the marginal status of porn texts and porn audiences is most marked, as we cannot assume that any such processes occur in these instances. Not only do viewers consume different portions of the transmitted text(s), but, as we'll discuss below, they are also not likely to engage in interpretive exchanges with other audience members.[11]

Michaels distinguishes three further categories of text (i.e., interpretations) beyond the received text: perceived, social, and public; as we shall see, porn texts will not fit neatly alongside mainstream texts in any of these categories (in fact, I will collapse the latter two categories for present purposes).

The Perceived Text. It has often been noted in studies of cross-cultural reception or viewing by children that viewers may arrive at interpretations of a text that differ greatly from those intended by the creator(s) or derived by other audiences, depending, among other things, on whether these viewers share the conventions assumed by the creator(s) and relied upon by other audiences (see Worth and Gross, 1974). It is a commonplace of current theory that there is no essential, privileged interpretation that constitutes the true meaning of the text; yet in the case of porn we may have the most extreme imbalance of power in favor of the beholder's share in determining that meaning.

Porn texts are received, as I have noted, under conditions that maximize the engagement of the viewer in an "independent" interpretive response (to coin a euphemism), and this may be both partial cause and effect of porn's lowly status on the aesthetic hierarchy: The work is not granted the sort of respectful attention that art supposedly demands and receives. The viewer may well be engaged and aroused, but also and simultaneously distracted and self-absorbed. And, of course, porn texts may well be the best demonstration of the relativity of evaluation. Carole Vance (1986) is fond of citing what she calls her "Rule of Three": whenever she describes a sex practice one portion of her audience finds it hot, one portion finds it disgusting, and the third portion think it's funny. *De gustibus,* indeed.[12]

The Social/Public Text. Despite the variability of individual attention to mainstream media texts, "when people speak about a TV program or event, they generally assume, and rarely discover otherwise, that they have 'seen' and 'mean' the same thing" (Michaels, 1991: 12). There are several reasons why this agreement can be assumed to occur, beyond the basic fact that most members of an "interpretive community" share fundamental conventions of discourse. One of these that Michaels mentions is "that people do talk about TV programs. In so doing, they presumably negotiate and revise, within some limits, the meanings they have ascribed to what they have seen. In fact, not only do people talk about TV, they read about it (in newspapers and magazines); they imitate and repeat certain elements; children add its characters and actions to their free play, the public may even hold meetings and demand community response, censorship, etc." (p. 12). In the case of porn we are not likely to find counterparts of all of these forms of interchange and negotiation, although some parallels probably do exist. Given the highly stigmatized status of porn (most notably of hard-core porn), few consumers are likely to engage in extended discussion with others about the films they have viewed.[13]

Contemporary popular media programs and performers are surrounded by a penumbra of discussion, evaluation, and gossip. In this landscape one has to look hard to find any representation of hard core pornography. Porn films are not reviewed by Siskel and Ebert,[14] porn performers are not profiled in *People Magazine* or *Entertainment Tonight,*[15] or interviewed by Jay Leno or even David Letterman (though they are likely to appear on the Donahue-Oprah-Geraldo circuit once characterized to me by a television station general manager as "nuts and sluts").

In recent years the fans of various low-status genres (i.e., genres assumed to have predominantly female, young, or lower-class audiences) have emerged from their respective closets and demanded a share in the interpretation and evaluation of their favored texts, and their voices have influenced our views of the contents and likely effects of daytime and prime time soaps, romance novels, music videos, or even rap songs.[16] But porn audiences are both silent and silenced. The stigma attached to porn (in this case a mostly male-oriented genre) prevents its adherents from coming forward to join the discussion, and the public debates that raged through much of the 1980s featured exchanges between antiporn and anti-censorship voices, with proporn voices largely absent.[17] As the battlefield of the sexual counterrevolution shifted from the adult movie theater and the video rental store to the art gallery the same imbalance could be seen.[18]

Pornography is often discussed by people who make no pretense of having viewed much or any of it (quite the contrary in many instances, they know it when they don't see it), and those who do consume it are scorned, pitied, and subjected to experimental and/or clinical research. While something similar could be said about many media genres, porn consumers seem the most silent and silenced at present.

Marginal Audiences

In examining pornography we are confronted with the seeming paradox of a mass market without any (well, hardly any) visible, self-identified consumers. This is not an unknown problem in media studies.

The first lesson I learned when I began studying television in the early 1970s was that no college-educated, middle-class person ever watched television, but everyone "just happened to see a program last night" and was happy to talk about it at length. While the need to make this sort of disclaimer seems to have diminished in the past twenty years, and may never have afflicted those born well into the age of television, I suspect that "avowed" porn consumers are as rare as television viewers were in 1970, while many people "just happen" to pick up a magazine or rent a video out of curiosity.

In 1977, even before the porn wars had broken out in full force, Gloria Steinem excommunicated pornography consumers (as contrasted with consumers of acceptable "erotica")[19] and those who tolerated them: "We now have the courage to demonstrate publicly against pornography, to keep its magazines and films out of our houses, to boycott its purveyors, to treat even friends and family members who support it seriously as we would treat someone who supported and enjoyed Nazi literature or the teachings of the Klan" (quoted in Nobile and Nadler, 1986: 15).

Over a decade and a half later, after the timely demise of the Dworkin/MacKinnon legal strategy (ruled unconstitutional by the federal courts) and the anticlimax of the Meese Commission, the sexual counterrevolutionaries have not lost their energy or determination; they have merely shifted their primary focus from the gutter of porn to the precincts of elite art (see Vance, 1989). Today, porn consumers are more likely to be treated with clinical condescension. "Dear Abby" recently headlined a letter from a worried wife, "My Husband's a Porn Fan," and counseled the wife—whose husband "started out with girlie magazines, but now (is) seeing porno VCR tapes, and (is) spending more and more time on this stuff"— that, "yes, there is such a thing as a 'sexual addiction,' and your husband has it" (Van Buren, 1989). She concluded by providing the worried wife and any other readers in a similar predicament with the address of Sexaholics Anonymous.[20]

The real story is probably a lot less extreme or dramatic, though possibly even more worrisome to the opponents of pornography. Recent articles reporting the rise in consumption of porn available for home video viewing have emphasized the cross-sectional heterogeneity of consumers: "Millions of Americans are routinely bringing naked strangers into their homes to perform sex acts on the small screen. . . . Steven Apple, the executive editor of Video Insider . . . said it was impossible to describe a typical adult-video customer. 'I suspect it goes across the board. . . . I worked in sales once, and every walk of life walked in and rented adult video. I don't think there's a class distinction'" (Ravo, 1990).

What is more, this profile of porn consumers might not be as recent as these articles suggest. Although it is certainly the case that few women were regular consumers of porn in earlier decades, the popular image of the "raincoat brigade" of sad men in adult theaters has little factual support. One of the rare empirical studies of porn users was conducted by Harold Nawy in San Francisco for the 1970 Commission on Obscenity and Pornography. Nawy and his associates made contact with several thousand individuals who "were involved with the production, distribution, sale, and consumption of sexually oriented material" (Nawy, 1973: 147). Data from over 4000 observations and 251 questionnaires were used to draw a profile of porn users.

> The average consumer of erotica in San Francisco appears as a white, middle-aged, married male, neatly attired and shopping alone. Questionnaire data reveal that these patrons are also well-educated, highly paid, and employed in white-collar or above occupations. Most consumers lead an active and varied sex life and report that involvement in the erotic marketplace increases their social and sexual interaction (Nawy, 1973: 147).

Yet, of course, the public position has been, and may still be, that normal, sexually well-adjusted people do not need or use pornography.[21] However, if we assume for the sake of argument that porn consumers are normal, sexually well-adjusted people just like us, how might we characterize the porn audience and its consumption practices?

It seems fair to say that the porn consumer actively seeks out material, and is probably more selective in this search than the average heavy viewer of television, who watches by the clock rather than by the program. The porn consumer, as I have noted, will have braved the potential embarrassment of being caught in the act—although I have also noted the safety increasingly provided by the numbers of coconsumers.[22]

Having secured the tape, or book, or magazine of his or her choice, the porn user is likely to engage in an active construction of meanings and, of course, pleasures, while consuming (this is sometimes referred to as "one-handed" reading or viewing). Porn consumers may well be the best example of active audiences to contrast with the supposedly passive couch-potato audiences of mainstream media fare.

If these characterizations are at all accurate, as I think they probably are, why is the porn consumer so readily despised and rejected, or at best treated as a bit sad and ridiculous? Why is their voice the least often raised, or listened to, in the endless discussions and debates about pornography?[23] Obviously, the answer lies in the nature of pornography's content: sexuality. Simply put, pornography is a special case because it consists of sexually explicit images, and as Susan Sontag once noted, since Christianity focused on "sexual behavior as the root of virtue, everything pertaining to sex has been a 'special case' in our culture" (1969: 46). As Gayle Rubin put it, "Sexual acts are burdened with an excess of significance" (1984: 279).

The Bill of Indictment

As I have already suggested, indulgence in mass media consumption—at least for educated, middle-class people who are not officially authorized by a cultural studies research license—requires some sort of justification beyond the pleasures thus generated. The underlying implication would seem to be that media consumption is presumed guilty until proven innocent. This, of course, is reminiscent of the traditional view of sexuality in Western culture, a view that I believe is still dominant despite the important achievements of the sexual revolution, feminism, and the lesbian/gay movements.

The traditional grounds for the exculpation of sexual activity—married heterosexuals with procreative intent (in the missionary position, at night, in the dark, in bed)—have gradually been expanded (or replaced) to permit contraception, unmarried partners (preferably in a long-term, committed, monogamous relationship), Kama Sutra—esque positions, and nonprocreative acts. But while marriage is no longer a requirement, love and emotional commitment are still deemed essential if sexuality is to be considered moral (see Rubin, 1984, for a full exposition of the sex hierarchy). The sin that cannot be redeemed (and AIDS has of course added to its infamy) is "promiscuity," although it might be impossible to find a generally acceptable definition of it (more than one partner in a lifetime? during the same period? at the same time? etc.).

In all of these shifting definitional sands the status of masturbation remains notably ambiguous; and this ambiguity, of course, lies at the heart of our attitudes toward pornography. While masturbation is no longer scorned as self-abuse—that obsession of nineteenth-century doctors and moralists (and twentieth-century priests and nuns)—we are still ambivalent about its moral status. It is undeniably nonprocreative (aka "the sin of Onan"), and it is difficult to argue for it on the basis of emotional commitment to one's sex partner (although a recent manual on autoeroticism is entitled *Self Love*); thus it leaves sexual pleasure alone and unadorned with any of the various fig leaves currently in fashion.

Pornography thus starts out on shaky moral grounds, given its prime function as a masturbation tool. It has also been loudly (and often justly) condemned for reflecting and playing some role in cultivating sexism.[24] What else is it accused of? Some of the specifics of the indictment may help illuminate the murky territory of our sexual attitudes.

The recent Meese Commission represented the voices of the sexual counter-revolutionaries and quite possibly of many bystanders who are troubled by the changing sexual landscape of recent decades. Despite the somewhat varied views expressed, among the commissioners were some who fearlessly proclaimed a total condemnation of pornography. Father Bruce Ritter[25] was disturbed by the willingness of (some of) his fellow commissioners to postulate "the existence of a . . . category of sexual materials designed to arouse that was neither violent nor degrading, and that was in some vague and unspecified sense, permissible to some extent" (Ritter, 1986: 95). In contrast, Father Ritter saw that

> A much larger issue is at stake here than the harm or degradation of a particular man or woman, or even of society itself caused by [pornographic] materials. . . . The question may be posed: does pornography, of any category, so degrade the very nature of human sexuality itself, its purposes, its beauty, and so distort its meaning that society itself suffers a great harm?
>
> The message of pornography is unmistakably and undeniably clear: sex bears no relationship to love and commitment, to fidelity in marriage, that sex has nothing to do with privacy and modesty and any necessary and essential ordering toward procreation. The powerful and provocative images proclaim universally—and most of all to the youth of our country—that pleasure—not love and commitment—is what sex is all about. What is more, that message is proclaimed by powerful self-validating images, that carry within themselves their own pragmatic self-justification (1986: 95).

Father Ritter's views rest on more than his credentials as a Catholic priest with a doctorate in medieval dogma; they are echoed in the conclusions drawn by Dolf Zillmann from his experimental studies of the effects of "prolonged consumption of pornography" (1989). A series of studies by Zillmann and Jennings Bryant[26] that were presented to the Meese Commission (and subsequently to the academic world) report on the consequences of exposing groups of college students to "massive" doses (about an hour a week for six weeks) of pornographic films that "focused on the exhibition of heterosexual activities in all conceivable manifestations"[27] and then assessing their views about various aspects of sexuality and marriage.[28] The results are clear, to Zillmann and Bryant, anyway, and they are quick to draw the conclusions Father Ritter already foreshadowed.

Among the central "effects" that they believe have been demonstrated are:

- Prolonged consumption of common pornography alters perceptions of sexuality. Specifically, it fosters presumptions of popularity for less common sexual practices.[29]
- Prolonged consumption of pornography promotes acceptance of pre- and extramarital sexuality. Although it decreases trust among sexual intimates, it increases tolerance for violations of sexual exclusivity. Moral condemnation of sexual improprieties diminishes sharply.
- Prolonged consumption of pornography prompts beliefs of health risks from sexual hypoactivity.[30]
- Prolonged consumption of common pornography spawns doubts about the value of marriage as an essential societal institution and about its future viability (1989: 154).

The true danger of pornography is thus unmasked by value-free social science, and reveals the very face of moral decay so familiar from the warnings of preachers and politicians:

> The nuclear family is generally considered vital for societal welfare. In terms of educational efforts, the family concept seems universally endorsed. Its values are rarely allowed to be challenged. Yet the values expressed in pornography obviously clash with the family concept, and they potentially undermine the traditional values that favor marriage, family, and children (1989: 140).

Pornography, then, is charged with much more than promoting the sin of Onan; it is accused of undermining the nuclear "family concept" by encouraging variety in sexual acts and partners.[31]

The Spice of Life?

Writing in *The Journal of Gastronomy* Mireille Johnston describes her conscious efforts to develop her daughters' appreciation and love of food:

> Delight with food began early. . . . Suddenly feeding, already a source of shared joy throughout the day, took on another dimension, improvements were added, and salty, bitter, sweet and even sour taste brought excitement. "Astonish me, but gently" was the message of my daughters. I was to remain vigilant to please the older one and imaginative to interest the young one.
>
> Children, like cats, hate to be startled; they like lukewarm food; anything hot or spicy awakens them too much, too fast. Diversity, gentle surprise is welcome; anything bumpy, large, hot is not. Pleasure blossoms in diversity and lightness. . . . I soon realized that in educating a palate, just as in educating a soul, nothing is more dangerous than to follow one single idea too stubbornly, even a good one. My impeccably balanced diet led to monotony. . . . I decided to let them discover and enjoy food their way. . . . I enjoyed the enthusiasm, the gusto of their appetites, the violent likes and dislikes, the moment of complete satisfaction. . . . Slowly an art of happiness, as art of living emerged. . . . Diversity, quality in the food presented, in the combinations suggested and tools to sample it made for an exciting time. . . . We concentrated on what we tasted, we commented, we praised, we compared previous treats. . . . But we all know that a man is not a gourmet just by wishing it; to awaken a palate to new things takes time and patience. . . . Children had to try a tiny bit of each new dish, then I bet on their curiosity and the emulation around them to do the rest (1985: 5–8).

Compare that to the recollections of another writer in the same journal, who overcame a much more limited early culinary environment:

> Although my birthplace, New York, offers as much diversity of food as any place one can think of, those riches were not part of my early memories. My parents whisked me off to . . . Trenton, New Jersey, hardly a gastronomic pleasure dome. The high point of my mandatory seventh-grade cooking class there was creamed chipped beef on toast. . . . My mother was a businesswoman who entered the kitchen only to give instructions to the current housekeeper-cook or to prepare

> one or another of a very limited repertoire of dishes. . . . Every night of the week had its unfailingly same menu like the table d'hôte dinners at small ethnic restaurants. . . . (Muscatine, 1984: 78).

Need we ask which of these homes would receive the Ritter-Zillmann seal of approval?

It is instructive to contrast the rather different ways in which our society views an interest in sexual variety, exploration, and experimentation in contrast with the same openness to novelty and creativity in cuisine.[32] On the face of it, we might take conclusions such as the following as endorsements of pornography: "Prolonged consumption of common fare fostered more favorable evaluations of portrayals of uncommon sexual practices . . . (and) can be expected to promote tolerance towards behavior deemed 'deviant' by others" (Zillmann, 1989: 132, 135). But we know they were not intended as such.

If I open the San Francisco *Yellow Pages* to the "restaurant" listings (and I do, frequently), I find forty pages of names and addresses, as well as a "neighborhood guide" and a "convenient guide to restaurants by type of cuisine" that fills six pages, from American to Vietnamese by way of Arabic, Australian, Basque, Chinese (about a page of these), through Lebanese, Mexican, Sushi, Thai, and Vegetarian. There are no comparable phone book listings for sexual services, though the classified ads of many sophisticated and/or minority papers provide somewhat analogous information. A recent issue of the *Village Voice* contains listings for "adult entertainment" ("Waiting to satisfy you!!"), "escort services" ("A beautiful experience: Oriental delight"), "Massage" ("Samson—Man wonder of massage"), and the increasingly popular phone lines for recorded or live conversation. The phone line listings now take up pages in the *Voice* and in every gay paper in the country,[33] and they have much of the appeal to diverse tastes of the restaurant listings.[34] On the other hand, letters to the editors of these gay papers often complain that the presence of these ad pages prevents them from leaving such papers around where (straight) friends or family members might see them.

I will give the last word to a sex worker who wrote to the *Village Voice* in response to an article about phone sex workers: "I am not sure why Simakis says each phone sex fantasy girl has her own 'rationalization' for her job. Does a chef need a rationalization for providing customers with food they won't find at home? Sex professionals are the chefs, short-order cooks, sidewalk vendors, and caterers of erotic pleasure"(Quan, 1990).

Notes

1. "One can see this in the furtive looks of the customers, the averted eyes, the monadic relations between the men. I find it hard to believe that a man would feel better about himself after a visit to the porn parlor" (Kovel, 1990: 154).

2. "One lesson learned by the porn industry is that traditional female repugnance to porn can melt when the product is cleaned up a bit and presented at home, where the woman can feel safe and treat the movie as a prelude to lovemaking" (Leo, 1987: 63). "A

1986 survey of one thousand video stores . . . revealed that women and couples rent 63 percent of all porn tapes" (Fraser, 1990: 33).

3. Interestingly, as Smith (1987) shows, the Meese Commission systematically misrepresented these poll results and claimed decreasing public hostility to pornography.

4. Bill Clayton, a prominent gay porn producer/director gives an interviewer the following account of the genesis of his favorite film: "*Aspen I* was shot at a ski resort, because I wanted to go skiing, and I figured it was a way to get some work done at the same time. So we took everyone up to this resort, and most of the shoots were in the evening. In between, I skied. And we got one of the hottest films that's ever been made. It is the largest selling video in the history of this business. Sold over 40,000 pieces in retail, wholesale, and mail order" (Douglas, 1989: 50).

5. Clayton is asked, "When you make a feature-length film, how much of a scenario do you start with?" He answers, "Oh, we've got a pretty clear scenario. We know where we're going. We know who's going to be together, although that can change. If we have a problem, the top may become the bottom, the bottom may become the top. You have to be flexible, because you can script and plan till the cows come home, and it will never come out that way" (Douglas, 1989: 23).

6. Clayton explains why there are no credits listed on his films: "Because everyone here contributes their maximum effort, and that's why the videos are what they are. If the art director, the still photographer, the production manager, or even the gaffer who's lugging the lights wants to offer a suggestion, if it sounds reasonable to me, I'll call time out and we'll have a production huddle" (Douglas, 1989: 20).

7. I have seen porn films shown on cable TV in Belgium (a country in which porn is illegal), in which there are cuts whenever the camera begins straying toward the performers' genitals—a practice that results in a jumpy and nearly incoherent film. I have also seen films produced for Belgian TV that were shot at the same time as a "normal" hard-core porn film, but in these the camera is always placed so that the genital areas are hidden–the opposite of the standard porn film shooting convention.

8. According to *Time*, "There are now about 350 porn theaters in the U.S., half the number of a decade ago. The remaining theaters have trouble getting new X-rated fare, since many, perhaps most, 'dirty' films are now shot on videotape and cannot be projected clearly on theater screens" (Leo, 1987: 63).

9. As reported in *Time* magazine, "'The VCR put porno where it belongs, in people's bedrooms,' says an executive vice president for one porn house" (Leo, 1987: 63).

10. In one of the rare published accounts by a porn consumer, written before the takeover by video and cable, Scott MacDonald describes his viewing of "theatrical" porn films: "Once I've decided to go to a porn theater, I go immediately; without checking to see when the movies begin or end; as often as not, I arrive in the middle of a film. . . . With very rare exceptions, I've always left before a show is over; after one film has led up to and past its most stimulating motifs, I've waited only long enough to calm down and not leave the theater with a visible erection. I've never sat all the way through a double feature of porn films" (1983: 16).

11. A possible exception here might be couples viewing porn as a stimulus/guide to sexual activity that accompanies or follows the viewing; although even here each of them might be creating a different fantasy-interpretation that is not shared with the other.

12. This very relativism creates interesting juridical problems, especially under the current Supreme Court criterion (from *Miller v. California*), which asks whether "the average

person, applying contemporary community standards, would find that the work . . . appeals to the prurient interest. . . . " For example, in a 1987 federal case in which a porn distributor was convicted for sending across state lines such films as *Snake Fuckers, Horny Boar,* and *Horsepower,* there was considerable discussion about whether there was such a thing as an "average zoophiliac" to whose prurient interests these films would appeal. The Court of Appeals ruled, however, that "The average person comes into the test not as the object of the appeal but as its judge. . . . There is no explicit requirement that the average person determine that the material appeals to the prurient interest of the average person. If that were the case, the average juror would answer the question by assessing his own reaction. . . . (T)he reaction of most people to these films would be one of rejection and disgust, not one of sexual arousal, but that cannot lead to the conclusion that the most offensive material has constitutional protection which less offensive material does not" (*U.S. v. Guglielmi,* 819 F. 2d 451 [4th Cir 1987]).

13. Although there are likely to be subcultural differences in the degree to which viewers are intimidated by this concern. I have been told by the desk clerk at a major video rental store in Philadelphia that most heterosexual men avoid eye contact when renting porn, while many gay men freely discuss the quality of films they have seen and ask for advice on new rentals.

14. Reviews of porn can occasionally be found in some marginal media, such as in some gay newspapers; but these reviewers seem to feel a need to adopt an ironic, role-distancing stance toward their subject. For example: "Matt Guinther('s) hypnotic performance has clearly been informed by scientific studies on post-trauma survivors. Less experienced actors might too-quickly show passion. . . . Matt remains aloof and represses any signs of excitement. He brings the ennui fuck to state-of-the-art. . . . This is good bad porn. Leonetti, however . . . is simply bad; he gropes aimlessly, fails to get hard, and worst of all, has a boring hair color–this is important when you have to watch his head bob up and down for ten mantraesque minutes" (Finch, 1990: 32).

15. The one exception I can recall was Traci Lord, but she was featured in the media gossip system when she was attempting to cross over to the mainstream, and had been exposed as someone who performed in porn at the age of fifteen.

16. It might more accurately be said, in most instances, that their territories have been invaded by researchers flying the flag of cultural studies.

17. The disparity of attention to proponents of opposing positions on porn was dramatized in the Meese Commission and its public hearings. "Prior to convening, seven of the eleven commissioners had taken public stands opposing pornography and supporting obscenity law as a means to control it. . . . The list of witnesses invited to testify was no more open than the commissioners' minds: 77 percent supported greater control, if not elimination, of sexually explicit material. . . . Commissioners accepted virtually any claim made by antipornography witnesses as true, asking few probing questions and making only the most cursory requests for evidence or attempts to determine witness credibility. Those who did not support more restriction of sexually explicit material were often met with rudeness and hostility and their motives for testifying were impugned" (Vance, 1986: 76).

18. In the extensive Senate debates over Helms' various attempts to curtail arts funding not a single senator spoke in defense of the art works singled out for attack; Helms' opponents were careful to express their disgust with the work of Mapplethorpe, Serrano, et al., while wrapping themselves tightly in the First Amendment.

19. The (in)famous distinction between porn and erotica was neatly dissected by Ellen Willis: "The distinction sounds promising but it doesn't hold up. . . . Pornography expressed in literary language or expensive photography and consumed by the upper middle-class is 'erotica'; the cheap stuff, which can't pretend to any purpose but getting people off, is smut. The erotica-versus-porn approach evades the (embarrassing?) question of how porn is *used*. . . . In practice, attempts to sort out good erotica from bad porn inevitably comes down to "What turns me on is erotic; what turns you on is pornographic" (1983: 463).

20. For those interested, the address is P.O.B. 300, Simi Valley, CA 93062.

21. In Scott MacDonald's "confessions," he notes that the fear and embarrassment he felt when visiting the adult theater had "less to do with guilt than with a fear of being misunderstood. Even though the frequency of my experiences with pornography has nothing at all to do with the success of my sex life–I'm at least as likely to visit a porn arcade when I'm sexually active as when I'm lonely and horny–I always feel the power of the social stigma against such experiences. Unless the people who see me have been in my situation, I'm sure they'll deduce that my visit to the arcade reflects my inadequacy or some inadequacy in the person I'm living with, that either I 'can't get any' or I'm not satisfied with what I can get" (1983: 11).

22. "Mr. Schnaubelt, a 24-year-old computer programmer who lives in Stamford, wasn't self conscious about the transaction. . . . He said he rents such tapes about once a week. 'The fact that more and more people are buying them alleviates any feelings of guilt or embarrassment,' he said" (Ravo, 1990).

23. Dr. Judith Becker, one of the two (women) members of the Meese Commission who entered a (mildly) dissenting statement, later expressed her regret that the Commission had not paid any attention to satisfied porn users (as opposed to porn victims/survivors): "The fact is that millions of Americans use this material, . . . but I wonder what government agency would ever make monies available to a researcher to discover what customers get from it. On a number of occasions I requested that we hear from people who consider pornography sexually beneficial. But the issue was too charged. I mean, can you imagine people coming forth, even behind a screen, and relating their positive experiences?" (Nobile and Nadler, 1986: 296).

24. This is not the place for an extended discussion of pornography's role in sexism and/or violence against women. Let me merely note that in a society drenched in misogyny most media products will reflect and amplify sexist attitudes; I do not, however, consider pornography to be the primary source of sexism or violence against women, and it can also function to subvert and reduce sexism.

25. I can not resist noting his subsequent fall from grace into the select company of Jim Bakker and Jimmy Swaggart.

26. It might be more accurate to say one or two studies that have been carved up and presented in a series of articles (this is not a trivial methodological point).

27. One must question Zillmann and Bryant's conceptual range, as this statement is immediately followed by the assurance that "None of the activities involved sadomasochistic acts or anything, such as bondage, that could be construed as nonvoluntary or coercive behavior on someone's part." These would be inconceivable, we must then presume, like homosexuality.

28. They were also presumably asked to rate these films on aesthetic grounds, which is the usual "cover story" used in Zillmann and Bryant's studies. What their subjects think of this cover story is not revealed to us.

29. Specifically, this means that the group exposed to "massive doses" of porn were more likely to give what the researchers considered high estimates of the number of people engaging in (in descending order of estimated frequency) oral sex, anal sex, group sex, S&M, bestiality. Zillmann noted that "This shift in the perceived normalcy of sexual behaviors can be expected to promote tolerance towards behaviors deemed 'deviant' by others" (1989: 135). It might be worth knowing that the "high" estimates given by his experimental subjects ranged from around 65% for oral sex (versus around 35% for controls), to 30% for anal sex (10% for controls), and 18% for S&M (9% for controls).

30. Translation: pornography makes you think that it is unhealthy not to have sex. This puts Zillmann in the august company of many nineteenth and early twentieth century authorities, such as F. H. Gerrish, Professor of Surgery in Bowdoin College, whose *Sex-Hygiene: A Talk to College Boys* includes the admonition that, "First and foremost, there is no greater error than that sexual intercourse is an important health measure. I know of no competent authority who justifies the idea. On the contrary, it is the uniform opinion of those who have most carefully and judicially examined the subject that continence is beneficial" (1917: 35).

31. The full import of Zillmann's jeremiad is worth noting if we are to understand why limitations on our sexual imaginations are a necessary price to pay for civilization: "Sexual gratification in pornography is not a function of emotional attachment, of kindness, of caring, and especially not of continuance of the relationship, as such continuance would translate into responsibilities and curtailments, and costs. Irrespective of the merits or demerits of the projection that much gratification is accessible from sexual activities involving unattached others, the projection is diametrically opposed to the values that promote enduring social aggregations, especially those that are to serve reproduction. Enduring intimate relationships curtail personal freedoms to some degree. Relationships that provide economic and emotional security are based on responsibility, if not on sacrifice. And where in such a relationship sexuality is vital and valued, partners tend to lay claim to exclusive sexual access. Finally, the decision to have a child or children, whether by a married couple or by persons otherwise aggregated, is probably the greatest responsibility that human beings accept. It amounts to restricted freedom, servitude, and to enormous expenditures for a good portion of adult life. If sexuality is considered part and parcel of such enduring relationships, there can be no question that it comes at a forbidding price. In terms of sheer recreational sexual joy, then, these relationships compare poorly with the short-lived ones that are continually exhibited in pornography–those that invariably show that great pleasures can be had at next to no cost. Prolonged consumption of entertainment with clear messages of this kind must be expected to impact profoundly the perception and evaluation of sexuality and its social institutions and arrangements" (1989: 140).

32. Although the traditional Jewish culinary restrictions and requirements can match any puritan's desire for control of sexual activities.

33. The phone sex companies' ads have become the financial mainstay of much of the gay press; consequently the Helms-sponsored legislation to restrict these services would have threatened the survival of these papers had it not been thrown out by a federal judge.

34. A 1990 issue of the gay paper *Outweek* included ads ranging from the romantic ("If you really want to meet someone special . . . there's only one number to call! 540-M-E-E-T") to the less coy ("550-BODY, Hot, Hard Muscle . . . We've got the beef!") and the blatant ("550-HARD: N.Y.'s Premiere Raunch and Sleaze Line–hardcore group scenes" or "1–900–999-OK-SM").

References

Birge, Peter, and Janet Maslin (1976). Getting snuffed in Boston. *Film Comment* 12(3): 35, 63.

Douglas, Jerry (1989). Inside Falcon Studios. *Manshots* (January): 19–23, 50.

Finch, Mark (1990). Mildred, Pierce! (Video Column). *Bay Area Reporter* (August 23): 32.

Fraser, Laura (1990). Nasty girls. *Mother Jones* (Feb/March): 32–35, 48–50.

Gerrish, Frederic Henry (1917). *Sex-Hygiene: A Talk to College Boys.* Boston: Gorham Press.

Johnston, Mireille (1985). Educating a palate. *Journal of Gastronomy* 1(3): 5–16.

Kovel, Joel (1990). The antidialectic of porn. In Michael Kimmel (ed.), *Men Confront Pornography,* pp. 153–167. New York: Crown.

Leo, John (1987). Romantic porn in the boudoir. *Time,* 30 March, p. 63.

MacDonald, Scott (1983). Confessions of a feminist porn watcher. *Film Quarterly.* 36(3): 10–17.

Michaels, Eric (1991). A model of teleported texts (with reference to Aboriginal television). *Visual Anthropology* 4: 301–323.

Muscastine, Doris (1984). On the road from Mushkaboola. *Journal of Gastronomy* 1: 78–85.

Nawy, Harold (1973). In the pursuit of happiness? Consumers of erotica in San Francisco. *Journal of Social Issues* 29(3): 147–162.

Nobile, Philip, and Eric Nadler (1986). *United States of America vs. Sex: How the Meese Commission Lied About Pornography.* New York: Minotaur Press.

Quan, Tracy (1990). Letter to the Editor. *The Village Voice,* 28 August, p. 4.

Ravo, Nick (1990). A fact of life: Sex-video rentals. *The New York Times,* 16 May, p. C1/8.

Ritter, Bruce (1986). Statement. In *Attorney General's Commission on Pornography: Final Report,* pp. 89–114. Washington, D.C.: U.S. Department of Justice.

Rubin, Gayle (1984). Thinking sex. In Carole Vance (ed.), *Pleasure and Danger: Exploring female sexuality.* Boston: Routledge & Kegan Paul.

Smith, Tom (1987). The use of public opinion data by the Attorney General's Commission on Pornography. *Public Opinion Quarterly* 51: 249–267.

Smyth, Cherry (1990). The pleasure threshold: Looking at lesbian pornography on film. *Feminist Review* 34 (Spring): 152–159.

Sontag, Susan (1969). *Styles of Radical Will.* New York: Farrar, Strauss and Giroux.

Steinberg, David (1990). The roots of pornography. In Michael Kimmel (ed.), *Men Confront Pornography,* pp. 54–59. New York: Crown.

Van Buren, Abigail (1989). Dear Abby: My Husband's a Porn Fan. *San Francisco Chronicle,* 29 May.

Vance, Carole (1986). The Meese Commission on the road. *The Nation,* 2/9 August, pp. 1, 76–82.

———. (1989). The war on culture. *Art in America* (September): 39–45.

Williams, Linda (1989). *Hard Core: Power, Pleasure, and the Frenzy of the Visible.* Berkeley: University of California Press.

Willis, Ellen (1983). Feminism, moralism, and pornography. In Ann Snitow, Christine Stansell, and Sharon Thompson (eds.), *The Powers of Desire.* New York: Monthly Review Press. (Originally published *Village Voice,* 1979.)

Worth, Sol, and Larry Gross (1974). Symbolic strategies. *Journal of Communication* 24: 27–39.

Zillmann, Dolf (1989). Effects of prolonged consumption of pornography. in Dolf Zillmann and Jennings Bryant (eds.), *Pornography: Research Advances and Policy Considerations,* pp. 127–157. Hillsdale,: N.J. Erlbaum.

Notes on the Struggle to Define Involvement in Television Viewing

11

Tamar Liebes

What we take for granted about so-called cultural imperialism—*Dallas,* for example—is the assumption that an alien text can make its way across cultural frontiers. Even before one raises the issue of effects, one must answer the question of how it is that so many different kinds of people and peoples are giving their attention to such texts? Semiotic theory suggests that we should look to the concept of "openness" to find an answer.

All texts invite different modes of relations with readers, but some texts are more open than others. In this sense, Eco (1979) goes against the grain of most semioticians, arguing ironically that many literary texts are open only to initiated readers—in that the author invites the reader to commute between the several levels intended by the author—and that popular texts are closed in that they force a standard reading, in their arbitrariness, on all kinds of readers.

Almost everybody else in this business, however, seems to think that it is in the nature of popular texts to allow for multiple readings. One way to understand this is that certain genres, mostly on television, are organized as loosely structured collages open to different constructional logics. MTV, for example, is made of a series of disconnected images that break away from the traditional constraints of narrative, allowing viewers to project a variety of meanings (Kaplan, 1987). If this sounds like a far-out example, consider that the very idea of the television "strip" is a loose string of disconnected stimuli, and that the news, spot advertising, and children's programs are all segmented in form (Williams, 1974; Newcomb and Hirsh, 1983).

Soap opera is the example that interests me. As Sonia Livingstone (1990) puts it, soap operas may be considered closed in the sense of their narrative form, their realism (which excludes different points of view, such as those of author and character), the continuity of characters and plot, their steadfast individualism, and when it comes down to it, their reinforcement of the traditional values of community, patriarchy, stability, and so on. Even without invoking *Dallas* and *Dynasty,* however, Livingstone also makes a strong case for openness. The soaps are not striving linearly toward resolution; the opposite is necessarily true. They are peopled by multiple characters, and by characters whose personalities are necessarily ambiguous and variable. They entertain multiple perspectives on social problems, and they are a forum for multiple moralities.[1] As in the open lives of their viewers, who are always in the midst of multiple ongoing stories that do not have clear solutions, so the soap operas lead viewers through a maze in which anybody's momentary solution is as good—narratively and morally—as anybody else's.

The argument about openness and closedness, even when it is based on textual or genre analysis, is obviously made from the point of view of the reader, whether one ideal reading is proposed or more than one. Some analysts leave it at that, and others go into the field to see how it actually works. Studies deriving from different traditions of communications research are full of dichotomies. All of them, however, emerge with dichotomous schemes that seem to be inspired—at least in the researchers' minds—by some notion of the dichotomy of open and closed. Closer comparison of these pairs, however, suggests that the field requires a more complex scheme, which I will try to propose here.

First, let us examine these dichotomies, beginning with Eco's. Regardless of whether he would grant any amount of openness to the soap-opera text, he would allow for two kinds of readers, which he calls "naive" and "smart" (1985). The first decodes the "mythic," "primordial" dimension of the text; the other, its "strategic" or "serial" dimension. Obviously, the naive reader, who takes reality as given, is a closed reader, and the smart one, who discerns the constructedness of the text, is capable of reopening the narrative and recombining its elements. For Worth and Gross (1974), the corresponding types of readings are labeled "inferential" and "attributional." The first does not acknowledge an external author and infers the meaning of the text from reference to real life, regarding characters and events as natural phenomena.

The attributional code is a fictive one, attributing the meaning of the text to an author. An empirical study based on Worth and Gross's formulation was conducted by Sari Thomas and reported upon by Allen (1983), who relabels "inferential/attributional" as "realistic" and "fictive," adding that realistic decoding is "ideological," not because the viewers regard the text as real but rather because they make sense of the narrative through integrating the characters into their own world of knowledge, values, and experience. This relabeling reflects the researcher's—not the readers'—judgment that the taking-for-grantedness of such decoding and the suspension of critical judgment is "ideological." The attributional thereby becomes nonideological, in that it deals in textual and aesthetic characteristics such as acting conventions, narrative expectations, and intertextual codes that reflect the reader's freedom from the dictates of the text.

The labels "inferential" and "attributional" were chosen by Worth and Gross to reflect their concern with the role of the reader from an aesthetic perspective. Thomas and Allen's relabelings show that they are more concerned with the political influence of the text. This is also Stuart Hall's (1980) concern in making his distinction between "hegemonic" and "oppositional" codes.[2] Coming from a critical position that accorded the text hegemony over viewers, Hall is suggesting in this distinction a possible shift in the balance of power between text and reader. In an empirical application of Hall's approach to the news genre, Morley (1980) is able to confirm that oppositional readings indeed exist, and that they take both aesthetic and ideological forms.

In a more positivistic tradition, similar distinctions are made between different types of readings. Neuman (1982) distinguishes between "interpretive" and "analytic" decodings, that is, between a reading that relates the program to viewers' lives or to broader issues of society and culture and a reading that focuses on the syntactic elements of that construction and the quality of scriptwriting and acting. Clearly oppositional readings for Neuman are aesthetic and express a concern with the viewers' cognitive ability, not ideological opposition.

This brief review of these dichotomies—naive/smart, inferential/attributional, realistic/fictive, ideological/nonideological, hegemonic/oppositional, interpretive/analytic—suggests that they do not cover exactly the same ground. I have already noted that there are two dichotomies rather than one: that between the real and the aesthetic readings, and that between the real and the ideological (not in Allen's sense).

While it is correct to say that both the aesthetic and the ideological address the constructional aspect of the text, concentrating on an awareness of either the syntax or the values from which the text is constructed, it seems incorrect to toss out the difference between them in the scramble to label both as metalinguistic or critical or constructional. Doing so will show that Eco's concern is with the aesthetic—that is, with the opposition between naive semantics and sophisticated syntactics—while Morley's concern is with the ideological—that is, with the acceptance or rejection of the producer's values.

In this perspective, Livingstone's "aberrant" readings are not at all on the metalinguistic side. Unlike Hall and Morley, who see hegemony at work when the viewer incorporates media representations into everyday knowledge, Livingstone sees selectivity and creativity—and thus viewer-power—at work when viewers bend media representations to their own social understandings. Asking viewers to retell a strand of a British soap she finds viewers offering various causes and motivations to explain characters' actions and finds it necessary to distinguish between the text's "preferred" meaning and the multiple aberrant readings of viewers.

The two dimensions combined yield four types. Referential and Closed defines the *Real*, in which characters and situations are incorporated uncritically into viewers' lives and vice versa. Still within the Referential, there is a more Open reading, in which viewers "try on" different options for the rewriting of the program and their lives by relating to characters subjunctively and playfully. I call this type of involvement *Ludic*. Likewise, there are Closed and Open types on the Constructional side. Constructional/Closed has to do with the discerning of the *Ideology* that pervades the producers' message (thus assuming one hidden meaning of a metatext). Constructional/Open is again the more playful of the two, whereby viewers discern *Aesthetic* building blocks of which the narrative is composed, and which show the readers' ability to construct alternatives. (See Table 11.1.)

In these terms, Eco's "smart" and Worth and Gross's "attributional" are clearly Aesthetic readers; Hall and Morley's "oppositional" are Ideological. Although Morley is aware that there is an Aesthetic option that is discreet, insofar as it fails to lead to the Ideological, he discards it. The subcategories of Neuman's

Table 11.1 Types of Involvement in Television Texts

	Referential	Constructional
	Real	Ideological
Closed	hegemonic/ confrontational	diagnostic/ oppositional
	Ludic	Aesthetic
Open	subjunctive	analytical/ critical

"interpretive," however, do not betray an awareness of this distinction. Livingstone, as noted earlier, is the only one who notes that "aberrant" viewers take the story into their own hands, and thus fall into our Ludic category. Everybody's opposite pole is obviously our Real. Aberrant readings stay within the referential: Viewers may invent their own explanations but these are explanations of the characters' actions, not of the producer's choices.

In studying viewers' readings of *Dallas* in different cultures Elihu Katz and I (Liebes and Katz, 1990) encountered the full gamut of these readings and have been struggling with an effort to find a mapping that will give them systematic expression. Thus the two dimensions we have been working with are, first, the now-familiar referential/constructional dimension (or if you prefer, referential/critical, or referential/metalinguistic),[3] and second, what we have variously called emotional/cognitive, hot/cool, indicative/subjunctive, semantic/syntactic, and, in the present context, closed/open. The second dimension is intended to express a difference between McLuhan's hot and cool from the viewers' point of view—that is, between the acceptance of intense/unambiguous directives and the acceptance of an invitation to ambiguity.

For the *Dallas* viewers, whose small-group conversations about the program we recorded, the Real[4] predominates. Following an episode of the serial, we found that most program-related statements took the form of using characters as pegs to discuss viewers' lives. Note, however, that we did not classify viewers but viewers' statements, many of which fell into the other categories; that is, discussion of the program moved back and forth among the four types. Ideological statements were also quite frequent, and they were, in fact, a speciality of our Russian respondents, who insisted that they were being manipulated by the message that the rich are unhappy, even while American viewers denied that there was any message at all. Americans, for their part, as well as kibbutzniks, specialized in the Ludic, enjoying the game of re-creation. We speculate that these are the two most secure of the groups in the study in that they have the least need to prove themselves to the interviewers either as experts on, or as superior to, the program (Liebes, 1986). Japanese, who rejected the program on artistic grounds, are concentrate in the Aesthetic region.

In spite of these specializations, it is perhaps more useful to think of viewers as starting from a Real reading, and having acquired knowledge of the characters or expertise with the genre or with the overall message of the text, they may proceed to more sophisticated types—specializing in what is more compatible with their social or cultural position.

Thus, the observation, made by all, that the rich are unhappy is assigned different meanings. Real readings perceive this as a fact, a natural phenomenon that, some say, should be treated as a morality tale. Ideological readings regard it as a message, often false, planted intentionally by producers in order to undermine the viewers' interests. For aesthetic readers who regard any content as a more or less efficient dramatic device, it is a good conflict, interchangeable with any other. Ludic readers experiment with the statement that the rich are unhappy by playing subjunctive games, mostly in interaction with other, which take the form of: What would I do if I were so rich? or if I were a rich man's wife, or mother? In such interactions—often between partners—viewers may enter into the scene of the story or abduct a character into hypothetical real-life situations.

I wish to point out that each of these types of involvement has a positive and a negative aspect. Thus, one can be involved in the Real by arguing with it and resisting the implied analogy between "them" and "us," as did the Moroccan and Arab groups in our study, protesting that they are better off in spite of being poorer. It might be argued that this confrontational form, which is quite frequent, is in fact oppositional. But we reject this proposal inasmuch as this form constitutes a confrontation with characters whose reality is not only unchallenged, but also considered an actual threat to lifestyle and culture. Moreover, as critical theorists would argue, it would be precisely the producers' intention to create the illusion that the viewers are better off than they are, the discerning of which would be indeed ideological. Failing to discern it, however, makes the confrontational reading the most hegemonic of all.

As for the other three types, each expresses—compared to the Real—some kind of critical distance or opposition, but not necessarily in a negative sense. Thus, detecting the message of the producers, ideological readers sometimes condone rather than condemn. An example of this, although untypical, comes from an American viewer who praises the producers for demonstrating that motherhood does not contradict egoist self-interest. In the same way that Real readings may criticize or accept characters morally, and Ideological readings may criticize or accept the message of producers, Aesthetic readings may accept or reject the producers' artistic product. Aesthetic interpretation of the unhappiness of the rich in *Dallas* argues that if the rich were happy, the narrative would lack dramatic tension ("then there would be no *Dallas*," as Cecile, the *Dallas* expert in one Morrocan group, observes). Ludic viewers may also reject the characters they play or play against, but this is of little interest in games of fantasy. Altogether, one can say that the negative predominantes over the positive in the Real, the Ideological, and the Aesthetic, and that the Ludic on the whole is neutral in evaluative terms.

Considering that involvement is often understood as some kind of identification with one or more characters, the analysis of these types of engagement with a television program shows that it is possible, even probable, to be negatively involved. Not only are the confrontational, oppositional, and critical modes typical of, respectively, referential, ideological, and aesthetic involvement, they are also more emotionally expressed. In other words, defining the relationship between text and viewers in terms of how the text positions viewers, one can argue that the viewer formulates and elaborates different aspects of his or her identity vis-à-vis the text. The I or We of the viewer is experienced as different from, or contradictory to, the Him, Her, or Them on the screen or behind it.[5] This might sound like interactivity between viewers and television set (Horton and Wohl, 1956), but in fact these negative relationships emerge typically not from talking back to the characters, but from talking about the characters with intimate others. Rather than establishing closeness with the characters, it is used to create an otherness than contributes to we-group solidarity.

The active role assigned to the reader in this kind of engagement is thus conducting exposés of the evils of characters and producers. Confronting characters through gossiping about their immorality as if they were real often affirms the viewer in his or her own moral values. Likewise, oppositional viewers affirm their own ideology by exposing the producers' manipulative intentions, just as aesthetic critics confirm their superior artistic judgment by exposing the flaws in the production. By trying on different psychological and moral options, subjunctive readers may expose and examine their own moral norms, reflexively. For Arlen (1980) at least, this last type of viewer is the one addressed by postmodern series such as *Dallas* and *Dynasty.*

In attributing these patterns of involvement to viewers, I argue that the type of pleasure that Csikszentmihalyi (1975; cited in Clarke and Evans, 1989) attributes to active pursuits can be equally applied to television viewing, which, for him, is the most passive activity (Kubey and Csikszentmihalyi, 1990). In this sense, television is like playing chess, for example, where viewers exercise skills, experience uncertainty about outcome, and, once they have entered the "flow," lose the distinction between I and the environment. We are aware, of course, that not all television viewing is so active. But particular texts, such as soaps, news, and live broadcasts all may involve the viewer in active ways (Dayan and Katz, 1992).

My analysis of the roles viewers take vis-à-vis the text raises the question of why viewers go on watching after they have defined their We as unlike Them—characters and producers—either morally or aesthetically. If these statements were unequivocal, why not turn off the set? That they do not suggests that these patterns are more ambiguous than they appear.

Even without analyzing viewers' discourse, I argue that the text of television serials has to be multidimensional as far as characters and their relationships are concerned. The plot itself is not enough to attract and sustain attention because its central characters have to remain in their places, more or less, and the subplots tend to the formulaic and repetitious. Therefore, as television theorists (Thorburn, 1982; Fiske, 1988; Allen, 1985) have observed, the locus of dramatic tension moves to the personalities of characters. If these were entirely good or bad, brutal or gentle, aggressor or victim, there would be no way left to generate interest or curiosity in the narrative. Thus, more than in linear, goal-oriented stories, the characters of television serials shift among moral and psychological positions, just as in real life. Both over time (diachronically) and synchronically, they display different aspects of their personalities in their relationships to the other characters.

It may be useful to think of viewers' involvement not as polarized—negatively or positively—but, in parallel with the text and with everyday social relations, as ambivalent and ambiguous, as well as changing over time (diachronically) and synchronically. Indeed, one can argue that attention to this kind of story can be sustained only when there is a tension, within viewers, between acceptance and rejection of characters (Herzog [1941] says, do it with different characters).

There are several forms that this ambivalence may take. When rejecting a character, viewers may claim that

1. we are not like them
2. we don't like them
3. we don't want to be like them

However, if all three of these forms of relating were negative, viewers would have no incentive to go on viewing. Rather, viewing itself means willingness to entertain the characters positively in at least one of these ways. The same thing can be said for the case in which all three forms of liking are positive.

Thus, involvement in the text means that some type of psychological conflict should exist, for example, a conflict between elements (I like him/I don't want to be like him), or an ambivalence within one element (a love/hate relationship). When it happens that all three are either negative or positive, the conflict may occur in a love-to-hate or hate-to-love relationship, based on ego's evaluation of itself experiencing these emotions (Zillman and Bryant, 1986). (See Table 11.2.)

Table 11.2 Types of Identification with Television Characters

	I am like him/her		**I am not like him/her**	
	I want to be like him/her	I don't want to be like him/her	I want to be like him/her	I don't want to be like him/her
I like him /her	+ + +	+ − +	− + +	− − +
I dislike him/her	+ + −	+ − −	− + −	− − −

As an example of these possibilities, consider viewers' relationships with JR, the charming manipulator of *Dallas*. A viewer's statement that she hates JR may mean I hate him/I am like him/I don't want to be like him, or I hate him/I'm not like him/I want to be like him. Both of these types exist in our data and express a kind of self-hate—hating oneself for thinking of oneself as being like him, or hating oneself for wanting to be like him (perhaps because one perceives JR's model as the only road to success). The reverse consists of admiration of two kinds: Liking him/wanting to be like him but feeling that there is no chance of achieving this because one is unlike him (possibly because the viewer perceives himself or herself as not tough enough), or liking him/thinking of oneself as like him/but not wanting to be, perhaps because this is perceived as morally wrong.

Consider this last example, for a moment, as expressing a relationship between a viewer and Sue Ellen at the time that she was being portrayed as her husband's victim. Relating to her, a kibbutz member explains: I like her (as a woman), I am like her (at least inasumuch as we are both married to domineering husbands), but I certainly don't want to be like her ("in her place I would never agree to be his slave.") This involvement with the character can be described as identification (I am like her), sympathy (I like her), and downward comparison, perhaps even negative judgment (I don't want to be like her).

The more familiar form of ambivalence, of course, is within any one of the three elements. People do, indeed, love and hate, as we know. In fact, Freud (1921) himself says that "love is so often ambivalent that it appears in conjunction with feelings of hatred towards the very same object."

There remains the third possibility where the viewer's professed relationship to the character arouses his own antagonism. Thus, I can reject JR on all three account and have a favorable image of myself precisely for doing so. And the same thing would hold for admirers, who may hate themselves for their admiration.

All of these cases of dissonance in the appraisal of characters best fit the Ludic form of involvement, which, by definition, involves experimentation with different forms of identification and rejection. It also applies, in part, to Real and Aesthetic forms. But the larger point I am making is that ambivalence contradiction, superego judgments, and other complex forms of relating to characters may be better descriptions of viewers' involvement in television fiction than single-minded identification or rejection. At least this approach has the advantage of making sense of seemingly strange viewers' discourse such as love/hate, love to hate, hate to love, and their rapid turnabout from one to the other.

To summarize, I propose:

1. that popular television texts—such as soaps, perhaps even news—can be explained in terms of their openness (not in Eco's sense);
2. that such openness invites multiple readings that imply several different types of involvement of viewers with the text;

3. that the literature on decoding, which acknowledges only two types of reading, in fact points to the types presented here, on the basis of the referential versus the constructional and the open versus the closed;
4. that these types may engage the same people, rather than different ones;
5. that each of these types may be further subdivided into positive and negative, thus calling into the question the commonplace assumption of positive identification (as the motivating form of involvement) while proposing that negative relating may be more frequent, and more effective, as a type of involvement;
6. that such negative relating strengthens the viewer's moral and aesthetic superiority of self- and group-identity;
7. that expression of rejection and identification are themselves complex concepts that may be combined in various ways in the multiple elements that constitute the concepts themselves and the interaction of these elements; and, finally,
8. that the seemingly dissonant talk of viewers about programs and characters is all expressed in these more complex terms.

Notes

1. British soaps are shown by Liebes and Livingstone (1992) to qualify as forum more than their American counterparts.

2. Hall, in fact, mentions a third—"negotiated"—code, situated somewhere between the two, which accepts the "preferred" meaning as a rule but not when it comes to issues in which one is personally involved.

3. By adding a second dimension we can limit Jakobson's (1972) "referential" to characters seen as real and separate it from the "message" drawn from the text, which becomes Constructional/Closed.

4. Compare also Krugman's (in Clarke and Evans, 1989) definition of involvement in television advertising. In trying to explain the apparent contradiction between the lack of evidence for cognitive effect on individuals by advertising and the changes in consumption, Krugman (1967) suggests that the intervening variable is involvement (experiences), which can be measured by the number of conscious bridging experiences the subject makes between the text and his or her own life.

5. We find this to be even more central in the involvement with television news. In a study of family conversations following the television evening news, positioning oneself and definine one's social and political identity vis-à-vis actors in the news is much more unequivocal and consistent than in relating to TV fiction (Liebes and Ribak, 1991).

References

Allen, Robert C. (1983). On reading soaps: A semiotic primer. In E. Ann Kaplan (ed.), *Regarding Television.* Frederick, Md.: University Publications of America.

———. (1985). *Speaking of Soap Opera.* Chapel Hill, N.C.: University of North Carolina Press.

Arlen, Michael (1980). *Camera Age: Essays on Television.* New York: Farrar, Strauss and Giroux.

Clarke, Peter, and Susan H. Evans (1989). Striking deep chords: Increasing audience involvement with communication. Unpublished manuscript.
Czikszentmihalyi, Mihaly (1975). *Beyond Boredom and Anxiety.* San Francisco: Jossey Bass.
Dayan, Daniel, and Elihu Katz (1992). *Media Events.* Cambridge, Mass.: Harvard University Press.
Eco, Umberto (1979). *The Role of the Reader.* Bloomington, Ind.: Indiana University Press.
———. (1985). Innovation and repetition: Between modern and postmodern aesthetics. *Daedalus* 774: 161–184.
Fiske, John (1988). *Television Culture.* London and New York: Methuen.
Freud, Sigmund (1921). Massenpsychologie und Ich-Analyse.
Hall, Stuart (1980). Encoding and decoding. In Stuart Hall et al. (eds.), *Culture, Media, Language,* pp. 128–138. London: Hutchinson.
Herzog, Herta (1941). On borrowed experience: An analysis of listening to daytime sketches. *Studies in Philosophy and Social Science.* The Institute of Social Research.
Horton, Donald, and Richard Wohl (1956). Mass communication and parasocial interaction. *Psychiatry* 19: 215–229.
Jakobson, Roman (1972). Linguistics and poetics. In Richard T. De George and Fernande M. De George (eds.), *The Structuralists: From Marx to Levi-Strauss.* New York: Anchor Books.
Kaplan, E. Ann (1987). *Rocking Around the Clock: Music Television, Postmodernism, and Consumer Culture.* London: Methuen.
Krugman, H. (1967). The measuring of advertising involvement. *Public Opinion Quarterly.* 30: 583–596.
Kubey, Robert, and Mihaly Csikszentmihalyi (1990). *Television and the Quality of Life: How Viewing Shapes Everyday Experience.* Hillsdale, N.J.: Earlbaum.
Liebes, Tamar (1986). Importing culture: "Readings" of an American program in various social settings. Ph.D. dissertation, Hebrew University, Jerusalem.
Liebes, Tamar, and Elihu Katz (1990). The Export of Meaning: Cross-Cultural Readings of "Dallas." New York: Oxford University Press.
Liebes, Tamar, and Sonia Livingstone (1992). Mothers and lovers: How British and American soap operas cope with women's dilemma. In J.G. Blumler et al. (eds.), *Comparatively Speaking.* Newbury Park, Calif.: Sage.
Liebes, Tamar, and Rivka Ribak (1991). A mother's battle against TV news: A case study of political socialization. *Discourse and Society* 2: 203–222.
Livingstone, Sonia M. (1990). *Making Sense of Television: The Psychology of Audience Interpretation.* Oxford: Pergamon Press.
Morley, Dave (1980). *The "Nationwide" Audience.* London: British Film Institute.
Neuman, W. Russell (1982). Television and American culture: The mass medium and the pluralist audience. *Public Opinion Quarterly* 46: 471–487.
Newcomb, Horace, and Paul M. Hirsch (1983). Television as a cultural forum: Implications for research *Quarterly Review of Film Studies* 8: 45–56.
Thorburn, David (1982). Television melodrama. In Horace Newcomb (ed.), *Television: The Critical View.* New York: Oxford University Press.
Williams, Raymond (1974). Television, Technology, and Cultural Form. London: Fontana.
Worth, Sol, and Larry Gross (1974). Symbolic strategies. Journal of Communication. 24: 27–39.
Zillman, D., and J. Bryant (1986). Exploring the entertainment experience. In J. Bryant and D. Zillman (eds.), *Perspectives on Media Effects.* Hillsdale, N.J.: Erlbaum.

On Not Finding Media Effects: Conceptual Problems in the Notion of an "Active" Audience (with a Reply to Elihu Katz)

12

Robert Kubey

Since the mid-1970s, I have been conducting research on how people use and experience the mass media. The primary focus has been television, but I have also studied popular music, music videos, video games, and reading.

Many of these studies have employed the Experience Sampling Method (ESM). In this method, respondents are given a small booklet of self-report forms and a paging device or beeper. Each research subject carries both for a week and fills out a self-report form—it takes about one to two minutes—each time they are randomly signalled, usually six to eight times a day between morning and late evening.

With the help of colleagues in Canada, Italy, and West Germany, by the time *Television and the Quality of Life* (Kubey and Csikszentmihalyi, 1990) was written we had been able to assemble nearly 50,000 ESM records from approximately 1,200 people participating in nine different studies.

In the ESM we use open-ended items that ask subjects to tell us where they were when signalled, whom they were with, what time of day it was, what they were doing, secondary activities that might have engaged them, and with standard semantic differential items we ask how they were thinking and feeling. In none of our studies have subjects had any idea that media experiences would be the object of study. In fact, very little of the data was originally collected with the study of media in mind. This was because one of our interests was how different daily activities would be subjectively rated by our respondents. Indeed, comparing subjective reports of different activities in this way had not been done before. As a result of this approach, media were in no way singled out in the respondent's mind. The activities and situations that persons found themselves in occurred naturally and normally. In the case of television, because we sampled over the course of the entire day, we often collected reports after the person had been viewing for a number of hours, perhaps lying on the couch with a book propped up on his or her stomach, that is, in the various ways that people actually view television.[1]

Experience Sampling Method Findings

Across nine different ESM studies, relative to other activities, when people are viewing television they report that they feel more passive, are less alert, and concentrate less than in almost any other activity. The main psychological reward reported is relaxation. Correlational studies strongly indicate that television is relaxing in large part precisely because it is easy to do and because it entertains us and distracts us with so little effort on our part.

Indeed, in ESM research, respondents are also asked to rate how challenging each activity is each time they are signalled. On a measure of perceived challenge, television viewing ranked lower than any other daily task. On a scale from 0 to 9, with 9 denoting high challenge, our respondents rated television 2.3. Even eating was rated higher at 3.0 ($p. < .05$). Television also ranked lower than any other daily activity on how much "skill" was required to do the activity as well as on how much was "at stake" in the activity.

If "viewers *work*," as Elihu Katz claims they do in the Wilbur Schramm Memorial Lecture (see Katz's chapter in this volume), why, Katz asks, does my research show that they perceive themselves as so unchallenged, unskilled, and with so little at stake when they view television? Clearly, one of the explanations has to do with how we construe the act of television viewing. By "work" Katz means that viewers *actively* decode and interpret what they view. And to be sure, some of the confusion that Katz points to early in this volume is purely semantic.

Indeed, we have never doubted that viewers enter into various interpretive processes when they partake of the media (Kubey and Csikszentmihalyi, 1990: 98–99). Unlike Katz, however, we do not equate any and all viewer interpretation as viewer activity. That such a definitional problem arises is of interest in itself, and helps highlight some of the problems in conceptualizing media effects as well as the audience.[2]

Although it is often useful—and accurate—in mass communication studies and research to conceptualize the television or film audience as active, I will try to make the case in the pages ahead that the reality of how audiences actually use and experience the media—and how the media affect audiences—is often lost sight of when a decided active audience orientation is chosen. Indeed, the active audience and uses and gratifications approaches often steer researchers away from apprehending important media effects.

Back to the research. Not only do our respondents report feeling relatively passive when they view, they also report feeling more passive after viewing than after other activities at home at the same time of day. I have called this a "passive spillover" effect. I have also shown that people are particularly likely to engage in a night of heavy viewing after reporting more negative moods and experiences earlier the same day. Heavier television viewers are particularly likely to report feeling bad when alone and when in unstructured situations, that is when they are between activities and when they have nothing explicit to do (Kubey, 1986). Such a finding may suggest a dependence on the medium for filling the void of solitude and open time.

The result may well be largely symptomatic; people who feel bad when alone or during unstructured time may gravitate to television, a medium that in the view of some is the cheapest and most readily available means of distraction and escape ever invented. It also seems possible that viewing four or five hours a day, almost every day, for ten, twenty, or thirty years could create in some people a dependence on television for structuring open time.

It is probably not coincidental that one of the few replicated findings cited in Anderson and Collins's massive research review, written for the U.S. Department of Education on the effects of television on scholastic performance, is that heavier young viewers of television, especially of violent television, exhibit poorer task perseverance and poorer impulse control (1988, 51). Similarly, two separate studies of self-labeled television addicts showed that heavier viewers scored higher on measures of mind-wandering, distracted thought, boredom, and unfocused daydreaming (McIlwraith, 1990; Smith, 1986). But, as always, it is extremely difficult to distinguish cause from effect.

Television Dependence

Let me suggest at least one way that a television viewing habit—or television dependence—is likely to develop. As noted, research shows that viewers feel relaxed when they view. This is true regardless of how long they view. Viewers also begin to feel relaxed shortly after they begin viewing. In short, the positive change in mood from some level of tension before viewing, to relaxation during viewing, and the quickness with which it occurs are likely to reinforce viewing. Viewers quickly learn—indeed even young children learn to use television to modulate their moods—to associate viewing with relaxation and escape. And because the viewer feels relaxed throughout viewing, the habit is strengthened.

My research also indicates, however, that viewers have to keep viewing to keep feeling relaxed, that is, after viewing, respondents do not report feeling as relaxed as they did during viewing. This is one of the reasons, along with "attentional inertia," to be discussed shortly, that people report difficulty in turning the set off.

Of course, television producers are past masters at encouraging people to view. Each of the commercial networks, local channels, and cable channels frequently promotes and "teases" what is coming next. Combined with how easy it is to view and the relaxation and diversion that television provides, the frequent invitation to view more can encourage people to view longer than they had originally intended.

I have suggested elsewhere that some features of some people's television habits may be akin to very similar features of substance dependence (Kubey, forthcoming; Kubey, 1990b; McIlwraith, Jacobvitz Kubey, and Alexander, 1991). In this work I have taken the diagnostic criteria for substance dependence put forth by the American Psychiatric Association (1987, 1994) in their *Diagnostic and Statistical Manuals III-R* (and in DSM-IV) and show how at least five of them could be applied to relatively common features of many people's viewing habits. In the rubric of DMS-III, for example, only three diagnostic criteria must be met to arrive at a diagnosis of "mild dependence."

The five criteria for substance dependence that I believe can be applied to aspects of some people's viewing habits are as follows (with the wording of DSM-III-R, pp. 166–167):

1. "Person uses the substance more than originally intended. Person may decide to take one drink of alcohol, but continues to drink till intoxicated."

   It is quite common for viewers to report sitting down to watch one program but winding up watching much more than planned. In a recent Gallup Poll, 42 percent of 1,241 adult Americans surveyed reported that they "spent too much time watching television" (Gallup and Newport, 1990).
2. "Person recognizes that substance use is excessive but has been unable to reduce or control it. Or person may want to reduce or control but has never actually made an effort to do so."

   As noted above, it is very common for people to express the belief that they view too much.
3. "Important social, occupational, or recreational activities are given up or reduced because of substance use. The person may withdraw from family activities and hobbies or use substance in private."

   Although there is a good deal of research that shows that television can bring family members together, there is also reason to believe that in some instances, viewing can reduce contact and/or the quality of family contact (Bronfenbrenner, 1973; Kubey, 1990a; Maccoby, 1951; National Institute of Mental Health, 1982). Not a few women feel neglected by husbands who watch numerous football games each weekend ("football widows"). And many people feel that they must regularly compete with television personalities for the attention of family members. Particularly disturbing is the suggestion that some children may be emotionally and/or physically neglected because their parent (or parents) is (are) too engaged in television programs to attend to their needs (Shanahan and Morgan, 1989).
4. "With heavy and prolonged use, a variety of social, psychological, and physical problems can occur and are exacerbated by continued use of the substance."

   As discussed, viewing can inculcate passivity and many viewers find it somewhat difficult to turn the set off, the longer they view. Activities that might have seemed relatively simple to do before viewing may be less likely to be tackled, the longer people view. Lack of exercise and weight gain constitute negative physical by-products of heavy television use (Dietz and Gortmaker, 1985). A vicious circle may occur where viewing perpetuates viewing.
5. "With continued use, characteristic withdrawal symptoms develop when the person stops or reduces intake of the substance. Symptoms vary greatly across classes of substances."

In *The People Look at Television,* Gary Steiner recounted his respondents' reports of behaviors attending the loss of a television set due to a technological malfunction. Three examples are of interest here: "The family walked around like a chicken without a head." "It was terrible. We did nothing—my husband and I talked." "Screamed constantly. Children bothered me and my nerves were on edge. Tried to interest them in games, but impossible. TV is part of them" (1963: 99). Twenty-five years after Steiner, Winick methodically studied families whose television sets were in repair. He writes:

> The first 3 or 4 days for most persons were the worst, even in many homes where viewing was minimal and where there were other ongoing activities. In over half of all the households, during these first few days of loss, the regular routines were disrupted, family members had difficulties in dealing with the newly available time, anxiety and aggression were expressed, and established expectations for the behavior of other household members were not met. People living alone tended to be bored and irritated. Over four-fifths of the respondents reported moderate to severe dislocations during this period. . . . By the second week, a move toward adaptation to the situation was common. (1988: 221–222).

That television dependence exists is highly plausible to most North Americans. In two different studies, one in Canada and one in the U.S., 65–70 percent of those surveyed reported believing that television could be addicting, although only 2–13 percent felt that they themselves were addicted (McIlwraith, 1990; Smith, 1986).

I prefer to use the word "dependence" precisely because the word "addiction" is misleading. But I do believe that many people in our culture have developed a psychological dependence on television. To be sure, "dependence" by no means must imply lack of "activity." Still, few active audience proponents are prepared to accept that a significant proportion of the viewing audience has developed entrenched viewing habits.

Varying Conceptualizations of Television Passivity

Some researchers have focused on the self-perpetuating nature of television viewing, and one can see how such processes can be equated with passivity. For example, "attentional inertia" refers to the observation and conclusion that "the longer people look at the television, the greater is the probability that they will continue to look" (Anderson, Alwitt, Lorch, and Levin, 1979: 339). A similar process refers to the "automaticity" of viewing (Hawkins and Pingree, 1986). Indeed, there is EEG evidence to suggest that some formal features of television command involuntary responses and that they may "derive their attentional value through the evolutionary significance of detecting movement" (Reeves, Thorson, and Schleuder, 1986: 271).

It should come as no surprise that passivity and television viewing have long been thought to be associated. In Steiner's (1963) classic work, 49 percent of respondents checked off TV as the activity that best fit the phrase, "Am I lazy!" In studies from the United States, Japan, and England over the past two decades, mild feelings of guilt that sometimes attend television viewing were found to be tied largely to the belief among viewers that they were spending their time passively and unproductively (Bower, 1973; 1985; Furu, 1971; Himmelweit and Swift, 1976; Kubey and Csikszentmihalyi, 1990; Steiner, 1963). Such guilt is most pronounced among middle-class viewers in each of these studies.

Katz is himself aware of such studies and writes that "most people, most of the time, are lounging in front of their sets, watching television, not programs" (Katz, this volume). But if viewers are truly working in their viewing, as Katz argues, why did the Corporation for Public Broadcasting (1978) find in a series of focus group studies with typical viewers that one of the primary reasons that PBS programming was not more popular was that viewers felt that they had to *work* too hard to understand and enjoy it? Among viewers' chief reasons for preferring commercial offerings is that they were much more likely to see familiar actors and characters in the usual episodic programming. On PBS, in contrast, they reported having to work much harder with an installment of "Masterpiece Theater," or the like, to understand and remember who the characters were. And rarely were the actors familiar. Inside the television industry, it is widely believed that a program such as "Hill Street Blues" reached a limited audience precisely because millions of viewers felt that it required too much "work" to watch (Bochco, 1983).

Katz also makes the curious suggestion that acceptance of the idea of viewer passivity might be synonymous with the idea that television and its audience are "unequipped to transmit or to receive ideological statements" (Katz, this volume). However, not only does television transmit ideology, but also the more assumptive the nature of the message, the more ideological it is likely to be. (Here we defer to Stuart Hall [1989]). By no means do we, or any others to our knowledge, believe as Katz suggests that passivity in viewers might somehow rule out television's political impact. On the contrary, the opposite view is quite common. Todd Gitlin, for example, concludes that viewing "flattens consciousness" (1972: 351). Similarly, Schiller points to a "diminution of mental activity" with viewing and a "pacifying effect on critical consciousness" (1973: 30).

On Looking for Activity and Finding Activity

Why does Katz find it difficult to reconcile our research with his own (e.g., Liebes and Katz, 1990)? His cross-cultural studies of the television program *Dallas* reveal audiences from one nation interpreting the program in very similar ways to one another but in different ways from audiences from other nations. When Katz characterizes Japanese viewers of Dallas as typically "bemused over seeming contradictions between episodes" and Russians seeing an "ideological threat in the program," one wonders just how *active* and individualistic each of his viewers really is.

One can view the glass as half-full or half-empty. Katz chooses to see the differences across nationalities as evidence of viewer activity. Looked at another way, one can marvel at the similarity in response within a given culture and ask why so many responses are so uniform.

The active viewer is uppermost in Katz's thinking. For example, he asks why such viewers do not "identify themselves to Kubey and Csikszentmihalyi" (Katz, this volume). Our answer is that perhaps they would have if we had gone out, as Katz has, looking for viewers to so identify themselves.

Obviously, part of the problem is related to the employment of different methods. As Greg Hearn has so aptly written: "It is clear that there is a tendency for the two opposite conceptions of the viewing audience to be buttressed by two different types of data. Passive conceptions tend to be built on behavioral data from diaries or meters; active conceptions tend to stem from self-report data via single-measure questionnaires or interviews" (1989: 858). One answer, then, is that different methods can sometimes elicit different results. Still, the ESM approach does not focus on one kind of daily activity any more than any other. Over the period in which data were being collected the media behaviors of our respondents were not singled out in any way. Of course, no one can tap into every aspect of viewing, but by collecting data from respondents, in the field and in the natural course of their viewing, we have gained a significant measure of validity (see Kubey and Csikszentmihalyi, 1990: 57–61).

Katz does the opposite. He typically focuses exclusively on television viewing, sometimes giving subjects specific programs to view, after which respondents are probed for their interpretations. Most revealing is that Katz explicitly states that "*what interests us, however, is not what people take from television but what they put into it*" (Katz, this volume; italics added). He admits to a decided bias and in deliberately ignoring media effects and in intentionally choosing only to look for his preferred phenomena, Katz virtually guarantees that he will find what he is looking for.

Katz frequently employs his highly intriguing research on how audiences partake in major media events for evidence of audience activity. During media events, he writes, "Audiences dress up to watch; they prepare their hearts—for the moon landings, for Kennedy's funeral, for the pope in Poland, for Sadat in Jerusalem. . . . They experience a sense of occasion, of 'communities'" (Katz, this volume; Dayan and Katz, 1992). But Katz is disturbed that such events are "unrecognizable" in our ESM research. Of course he is correct that we do not focus on such events, but he need not be so surprised as to our different results. Such instances of viewing rarely occur in our studies precisely because they *are* rare.

It is doubtful that in the 50,000-plus instances of viewing behavior we have studied with the ESM, that more than one or two major media events occurred. After all, Katz's classic examples come from once-in-a-lifetime—or once in *many* a lifetime—events: the moon landing and an Egyptian leader visiting Israel in a conciliatory gesture have each occurred *only once in recorded history.* Indeed, we discussed just such exceptional viewing occasions vis-à-vis the ESM over a decade ago when we wrote that "live reports of assassinations, moon landings, Watergate hearings, or sporting events. Such events *do* command our undivided attention and we respond almost as if we were there" (Csikszentmihalyi and Kubey, 1981: 326). Our research focuses on the television experience generally and we do not deliberately spend time on the extraordinary. In contrast, Katz frequently focuses on the extraordinary, spending less time on the ordinary or normative.

We have also made it clear that although people report viewing to be a relatively passive activity involving relatively little concentration, we are *not* claiming "that viewers are not thinking about and interpreting what they are viewing . . . or that they are not possibly engaged in a critical 'reading' of what they are viewing" (Kubey and Csikszentmihalyi, 1990: 98–99). Clearly, for an audience member to experience any stimuli—or message—sensory experiences must be processed through existing neurological apparatuses *and* through the prior experience of the individual (schemata, etc.). For us, these are *givens* of audience reception.

But by his own admission, Katz is not interested in what television does to people, only what people do with television. He intentionally looks only at particular aspects of the process and is then perplexed when other researchers report on the other, very frequent things that happen when people view. Ironically, in the same paper in which he expressly admits to intentionally ignoring an entire aspect of the viewing process, he suggests that "The first thing that must be done. . . is to agree on some rules of evidence, and the second is to do some research that will stand up to these rules, reliably and validly" (Katz, this volume).[3]

Methodological Limitations

We know that the ESM is not without problems. Nor are the methods often employed in ethnographic and cultural studies, especially when researchers engage in extended, in-depth interviews focusing on viewers' interpretations. In the view of Zillmann and Bryant (1985), one problem in such research is that most people are not particularly cognizant of why they choose what they choose to view, nor in their opinion are people generally able fully to explain what they derive from their media experiences. If nothing else, one can predict significant demand characteristics and considerable evaluation apprehension and other method reaction effects coloring the responses of an individual asked by a researcher from a university to unpack the meaning of what he or she watches on television.

This is a point of no small methodological significance and may partly explain why some researchers find viewers to be so active. Ang (1985) has argued, for example, that a viewer's immediate experience of a program may deviate significantly from the more cerebral and logical explanations often offered. Audience members who are interviewed about a television program may tend "to construct a more critical reading than they might do otherwise" (Wren-Lewis, 1983: 196). Elaborated meaning may largely be produced when viewers account for or reconstruct their experiences (Fry, Fry, and Alexander, 1988).

Zillmann (1985) is particularly concerned about attempts on the part of the respondent to impress the researcher.

In answer to Katz: *Yes*, agreement on rules for evidence would be helpful, but we can hardly expect many researchers to agree with Katz that the rules should be bent toward only observing what people "put into" viewing. Is this what James Halloran meant by the "Katztration of research" (1978: 122)?

Katz wonders why his active viewers do not identify themselves to us. *We* wonder why viewers such as the college English instructor interviewed by Marie Winn do not identify themselves to Katz:

> When the set is on, I cannot ignore it. I can't turn it off. I feel sapped, will-less, enervated. . . . So I sit there for hours and hours. . . . I remember when we first got the set I'd watch it for hours and hours, whenever I could, and I remember that feeling of tiredness and anxiety that always followed those orgies, a sense of time terribly wasted. . . . I remember feeling terribly drained after watching for a long time (1977: 21–22).

On Taking Leave (of Uses and Gratifications)

For Katz, even the television viewer who chooses to "take leave," that is, who uses the medium to escape reality, is active. Suddenly, viewer interpretation is no longer the sine qua non of viewer activity. In this instance, viewer activity is denoted only by the fact that the viewer has chosen to view. Indeed this has long been Katz's argument for viewer activity (Katz, Blumler, and Gurevitch, 1973/1974). From this perspective, viewing is necessarily active because the audience chooses to view. Given this approach, then, would not almost all human activity, including sleep, be active? Even if by some stretch of the imagination the viewer is not engaged in any interpretation whatsoever, he or she is active as long as he or she chooses to view. If instead of choosing, a person was forced to view a particular film, as was the central character in *A Clockwork Orange*, Katz would still consider the viewer to be active so long as he interpreted what he saw or, particularly, if he resisted the images displayed before him, as well he might. But in this example we can see that while choice is important, it is not at all a necessary condition for viewer activity. If one criterion for viewer activity does not apply, another is adopted.

Views from the Past

In thinking about the question of audience activity it may prove instructive to reexamine one of the early invocations of the term "uses and gratifications," by Joseph Klapper (1963). He entitled his article, "Mass Communication Research: An Old Road Resurveyed." (If this was an old road over three decades ago, one can only wonder what potholes we might encounter today.) Klapper wrote, "Viva los uses and gratifications studies, and may their tribe increase" (p. 517). But Klapper also issued a warning.

> If uses and gratifications studies are to achieve their potential they must, I believe, proceed further along the road on which many of them have stopped. They must consider not only the observed use, but the *consequence* of that use for the individual user, for social groups, and for society at large. . . . To lay claim to the functional analysis label . . . the consequences of the uses and gratifications must be stipulated. The stipulation, furthermore, cannot be a purely speculative manifestation of the researcher's personal values, for at that point he ceases to be a researcher and becomes a philosopher. If it is to merit the name of research, the stipulation of consequences must be made on the basis of scientific observation, or perhaps temporarily, in terms of hypotheses that may thereafter be put to scientific test (p. 520).

Klapper's demand for scientific observation may sound odd or dated in certain academic circles today. But to be sure, the *consequences* of mass media in uses and gratifications studies has been neglected just as Klapper warned they might. As with Katz's approach, one of the key problems with the uses and gratifications approach is that it is structured conceptually and applied in such a way as *not* to find—or in some instances to even look for—effects.

There can be no question that the uses and gratifications approach has contributed to our field. There can be no question that on any given evening of television viewing most viewers will spend some, if not a great deal, of their time bending the media to their needs. But they will also all spend a portion of their time being *affected* by what they view. Too many active audience and use and gratifications proponents lose sight of this. Katz has himself rejected much that this position once represented:

> Gratifications research has been through a long period of soul-searching and self-criticism. Too mentalistic, too empiricistic, too functionalistic, too psychologic in its disconnection from social structure—say the critical theorists . . . and they are largely correct. Early gratifications research has leaned too heavily on self-reports, was unsophisticated about the social origin of the needs that audiences bring to the media, too uncritical of the possible dysfunctions both for self and society of certain kinds of audience satisfaction, and too captivated by the inventive diversity of audiences uses to pay much attention to the constraints of the text (1987: 37–38).

But though the one-sidedness of earlier uses and gratifications positions have been largely abandoned, and though the approach may well have begun to take both the text and social structure more seriously, one wonders what the approach *now* stands for and whether cultural and critical studies of how audiences negotiate texts deservedly fall within, or even very near, the uses and gratifications rubric, as Katz would seem to wish. In fact, the question can legitimately be raised of whether there was ever anything particularly novel about the uses and gratifications approach. Indeed, was the much criticized "magic bullet/hypodermic needle" direct effects model cited so often by uses and gratifications exponents deliberately to set up a straw man in order to validate the importance of uses and gratifications? This is precisely the claim made by Chaffee and Hochheimer (1985), who report finding little evidence of any such application of direct media effects terms prior to the rise of uses and gratifications.

Furthermore, the post–direct effects "advancement" of the "two-step flow" concept holding that media effects on opinion and attitudes were transmitted first to "opinion leaders" and then from these leaders to the public (Katz and Lazarsfeld, 1955), was pioneered in actual practice much earlier by Sir Gilbert Parker, who very effectively directed British propaganda in America by first singling out "influential and eminent people of every profession" (Parker, 1918: 535) as the targets of propaganda in order to encourage American involvement in World War I (Lasswell, 1972: 155–156).[4]

More to the point, the idea that different people will use the same stimuli or entertainment in different ways was certainly not new when it was first proposed by uses and gratifications advocates. One hundred fifty years earlier, Goethe wrote in *Faust* (1808) that "Each loves the play for what he brings to it." And a hundred years later, the idea of film viewers actively projecting themselves into what they viewed was masterfully elaborated in considerable detail by psychologist Hugo Munsterberg in his 1916 book, *The Photoplay.* Warner and Henry (1948) conducted similar projective analyses of the radio soap opera listener three decades later. Certainly anyone familiar with projective tests such as the Rorschach or Thematic Apperception Test—developed in the early 1920s and middle 1930s—knows that different people see (or project) different things in response to the same ink blot, photograph, or drawing.[5]

But people also tend to see and experience very similar things, and it should come as little surprise that with much less ambiguous and prepotent stimuli such as radio, film, and television, there will be many similarities of response. This is precisely one of the key problems with the uses and gratifications and active audience approaches: there is little or no recognition of the degree to which the mass media bring about rather uniform and immediate responses in substantial portions of their audiences. Although it is true that moviegoers or television viewers come to the same film or television program with different backgrounds, needs, and expectations, these media are often powerful enough that once involved in an effectively produced drama, say, a suspense plot, most viewers will care at exactly the same time whether the hero survives, whether a victim is rescued, and whether the villain is vanquished (Comisky and Bryant, 1982).

And, to be sure, audiences for comedic and tragic fare typically laugh or cry at the same time. At his death, Noel Coward was unshakable in his confidence that an audience could be quite directly effected when he advised, "Coax it, charm it, interest it, stimulate it, shock it now and then if you must, make it laugh, make it cry, but above all . . . never, never, never bore the living hell out of it" (Simpson, 1992: 342). Nor is it a revelation that audiences can vary in response. Anyone with a fair amount of theatrical experience knows that no two audiences are exactly the same. A line that drew a huge laugh in the matinee might go flat during the evening's performance even when delivered in a nearly identical manner. But does that make audiences active, or simply idiosyncratic?

Consider also that if media "texts" do not result in some uniformity of response, how can we explain that new audiences gravitate to certain classic television programs and films year after year, whereas other fare is almost universally considered expendable? The popularity of Capra's *It's a Wonderful Life* or *The Honeymooners* cannot possibly be due to random chance. Rather, these productions were, and *are,* popular precisely because they reliably activate in their audiences relatively similar responses across time.

Just as certain films and television programs affect their audiences in similar ways, so, too, are television promotions and "teases" capable of peaking curiosity sufficiently that some viewers will watch the eleven o'clock news or the next scheduled situation comedy even though they had no intention of doing so when they first sat down. In such instances, to what degree is the person bending the media and to what degree are the media bending the person?

Furthermore, who could possibly disagree with the notion that the person's prior background, needs, and psychological makeup are critical to what happens when people use the media? Why does this negate media effects, as some active audience proponents would seem to think? Indeed, because prior instances of television viewing necessarily constitute a viewer's prior experiences and help determine future proclivities, prior television viewing necessarily helps to shape an audience's needs and expectations. But some uses and gratifications proponents seem to think that each television viewer is coming to television for the first time and has never been influenced by what he or she has seen previously.

Indeed, before the advent of the electronic media, people living in different parts of the country, or world, would necessarily have had fewer commonly shared direct cultural experiences. To the degree that one is shaped by prior experience, this means that the mass media may well have some similar socializing, homogenizing, or mainstreaming effects on viewers. If nothing else, exposure to television is certain to shape knowledge and therefore the needs, attitudes, and expectations we bring to new encounters with television materials.

The active audience approach also generally neglects to recognize that relative to print, certain cognitive and affective responses to television and film are much more likely to be uniform as a result of the pacing and pictorial nature of these media. There can be no doubt that people reading the book *The Wizard of Oz* without having seen a filmed or televised version will visualize each of the characters much more idiosyncratically than will people who are exposed to the more complete characterizations offered on the screen. Responses to film and television will also tend to be more uniform than those to print because the pace of information reception in film and television is dictated almost completely by the medium. By contrast, when one reads, one moves along at one's own pace. The reader is in much more active control of reception, and the resulting experience is likely to be more idiosyncratic.

Further Problems with Uses and Gratifications

As noted earlier, active audience proponents minimize media effects on the ground that people choose certain media products in seeking particular gratifications. Gratifications are not conceptualized as effects. For uses and gratifications, because a man chooses to use television to relax, television can no longer be thought of as relaxing him. Such a view misses the fact that there was a reason that the person chose television but did not choose to scrub the bathtub in order to relax. In other words, there must be something about how television characteristically affects the viewer that induces so many people around the world to pick television as one of their primary—if not *the* prime—means by which they relax. But many contemporary media scholars are completely unconcerned about what characteristic responses different media and media forms and genres tend to elicit, and indeed some actively resist any discussion of such phenomena.

The uses and gratifications approach also gives inadequate consideration to the fact that any given viewer is likely to have many different simultaneous needs that might be gratified by available media content. One individual might have a simultaneous need to be released from sexual tension; for escape from the family; for intellectual stimulation; and for information about both the stock market and a particular disease that he or she has just had diagnosed. In short, *other* needs may be stimulated by virtue of viewing. Likewise, various needs may be gratified that the viewer did not intend to have gratified at the outset. The idea that viewers set out to view or read with one or two precise needs set out from the beginning is ludicrous and rarely represents the way people actually use and experience the media.

If we follow the uses and gratifications/active audience lead and minimize television's effects on the basis that people choose to watch television in order to deliberately gratify this or that need would we not by necessity also have to minimize the impact that many *people* have on us, or for that matter, the power of *books,* or a *college education,* to influence how we think, feel, and behave? In other words, because we actively choose to enter a relationship, or read a book, or go to college, does that then necessarily mean that the person or the book or the education are less influential in our lives? Put another way, is there something inherently different about the nature of our interactions with people or books or education and our interactions with media? Surely there are important differences, and this is precisely my point. Overemphasis on the active audience can steer us away from building knowledge about, and a theory of, mass communication that reveals the nature of media effects generally, how they differ from other effects, or about the specific effects of a particular medium or genre.

Cultural Theorists Hold That the Audience Is Passive When It Suits Them Ideologically

Perhaps most perplexing about the active audience position is the frequent inconsistency in its application. Generally consistent with the active audience position is a respect for the media negotiation abilities of audiences—their intelligence and selectivity. Audience members are not easily duped, it is believed. But there are curious exceptions to this view. The noted British media literacy scholar David Buckingham and his colleague Julian Sefton-Green use the sociological term "moral panic" to describe the public outrage in England in 1993 over the idea that violent video games or the media might have played some role in the murder of a toddler, Jamie Bulger, by two ten-year-old boys. Buckingham and Sefton-Green marvel at how the public preferred to focus on the media's alleged role in the killing instead of understanding it "in terms of poverty and economic recession, or the erosion of leisure provision for young people" (forthcoming, 3). These were clearly the preferred explanations for Buckingham and Sefton-Green, and no doubt these critical facts *are* related to much crime. For them, these "facts" are givens—yet they provide no evidence that these factors played any role in the case in question. They do claim, however, that "the question of evidence" in support of the media effects thesis was "entirely ignored" in the ensuing British moral panic.

Buckingham and Sefton-Green are of the view that these moral panics are irrational and that they are indicative of misplaced public concerns. They may well be right. Curiously, however, nowhere in their argument do they ever mention that the boys were reported to have viewed a tape of the exceedingly violent movie *Child's Play 3* before the murder. As was the case with Jamie Bulger, in the movie, a lifelike young boy-doll is badly maimed while in the company of an older boy. The boy-doll is the victim of the worst violence in scenes involving a rollercoaster track. In reality, after the boys attacked Jamie Bulger, they left him on a railroad track to be run over by an oncoming train.

One cannot know whether the father of one of the boys, who reported their having viewed the video, invented this tale to protect his son, but if one or both of the boys did indeed view this film within a few days or even weeks of the murder, it is at least worth noting, particularly in trying to describe the reasons for the resulting moral panic. After all, ten-year-old boys, even those living in poverty and with few leisure opportunities, do not regularly abduct two-year-old children, beat them to death, and put their beaten bodies on railroad tracks. Yet the authors seem satisfied to explain such an outrageous, anomalous act in general sociological terms.

That Buckingham and Sefton-Green do not even mention that such a connection between the film and the act arose in the debate and in the trial, preferring to dismiss the public response as simply another mindless moral panic, is interesting to say the least. Nor do they mention that after hearing all the evidence in the case, the trial judge, while sentencing the boys, declared that she believed that the media (and film) had played *some* role in the murder and that she was disappointed that she could not do anything about it.

Curiously, the authors prefer to see the audience as active, but they do believe it is important that students be "encouraged to question received ideas about gender or 'violence'" (p. 22). Clearly, some of these "received ideas" are transmitted through and by the media. One of their solutions to this problem is to help students "to dissect the kinds of 'moral panics' discussed above, and to evaluate their underlying motivations." Though they say they respect the intellectual integrity of their students and want to encourage freedom of thought, it would seem that they believe that students need to be indoctrinated toward their own firmly held view that the public is merely scapegoating the media when moral panics break out.

Ironically, the only sort of media effect they are prepared to recognize is one that they believe misleads audience members to conclude that there are media effects.

A similar analysis occurred at the same conference in which this chapter and Katz's were first presented as papers. Ellen Seiter gave a talk on the views held about television by the parents in her parents' group (see Seiter's chapter in this volume). During her talk, not a few scholars in the audience laughed derisively at the reported assumptions of Seiter's respondents that television viewing needed to be controlled in their households. Seiter reported that many of the parents she studied ardently believed that too much television viewing, or the wrong kind of television content, could be harmful to young children. She ranked the seriousness of her subjects' concerns regarding the control of television viewing along with other, older parental concerns: "Television easily fit into an agenda of perceived problems facing parents, taking its place alongside feeding, potty training, and fire safety."

Several in the audience who normally hold to the view that the television audience's predilections are to be respected cackled in delight at the characterization of parents' being seriously concerned about their children's television viewing. So obvious was their reaction that Seiter chose to make a side comment, noting that the kind of "passive" characterization of the audience that I had presented the previous day, which is discussed earlier in this essay, was shared by many other television researchers as well as by the parents in her group.

The question arises as to why active audience proponents who voice a healthy respect for the audience and its predilections laugh at and deride audiences when their views are contrary to their own? I posed this same question to a few active audience proponents after Seiter's talk. The view generally espoused was that the parents in Seiter's were simply wrong that children could be harmed by watching television. Their stated view was that the parents in Seiter's group were essentially passive media recipients. Without thinking, the parents bought into the mass media's frequent presentation of psychologists, pediatricians, and other "experts" who pass on the received wisdom that children's television viewing should be limited or supervised.

Put another way, in their view, the dominant, mass-mediated ideology about television viewing's value for children had been passively adopted by Seiter's respondents. It never seemed to occur to the scholars with whom I spoke that any of these parents might have come to their conclusions more independently (or *actively*).

I asked the active audience proponents why they thought Seiter's respondents were so passive, why they were prepared to buy so completely into what they believe are the culture's misguided ideas about the value of television? If people are not influenced by the media, why did they believe that Seiter's respondents were so influenced? Why were they not resistant? Why were they not more active?

One scholar answered that in time, when the cultural studies approach began to receive more media attention, more parents would begin to understand that television was an essentially harmless source of entertainment and that their children could easily negotiate the medium.

But wouldn't that be a media effect, I asked? Wouldn't the parent's adoption of a new and different media-transmitted notion of how their children use the media still be a media effect? My partner in conversation was stymied. Clearly, like so many who adopt the cultural studies perspective on active audiences, he was very much taken with the intriguing insights made possible by a poststructuralist deconstruction of media and culture. He was very good at espousing the dominant cultural studies ideology of active audiences and the lofty "that's the wrong question" response to inquiries about media effects. But when momentarily brought back to earth by a few simple questions he was as pedestrian in his thinking as the rest of us. Down deep, he still believed in media effects even though it violated his ideological position to admit as much. He clearly still thought the media could alter the attitudes and beliefs of its audience.

Notes

1. One of the reasons that some laboratory researchers claim that television viewing is psychologically and physically activating may be because the viewing situation in laboratory studies is necessarily contrived. It is my contention that in being seated in a university research room by a researcher and asked to watch something on television—or even when left to their own devices to watch whatever is available at the time—research subjects will view differently, probably with greater alertness and more "actively," than when they view in the privacy of their own homes. Many laboratory research subjects rightly expect that the researcher will return later to ask questions about what they have viewed. Thus, the social psychology of the laboratory media study experiment often interacts in a problematic way with one of the key phenomena under study, namely, the level of the viewer's psychological activity. Furthermore, laboratory respondents are often not permitted to make their own viewing choices, and data are often collected after the person has viewed for only a few minutes or, at best, an hour or so.

2. One explanation for some of Katz's confusion about our findings is that he appears to have misinterpreted some of them. Contrary to Katz, we have never reported or suggested that "viewers are nearly asleep," and nor have we reported that measures of "variables such as alert, excited, and happy . . . register very low in television viewing" (Katz, this volume). Yes, television ranks low *relative* to other activities on these self-report variables but the actual mean scores are not so low. It is true that relative to other activities television viewing is reported as less activating, as more passive, and as involving less concentration. But in terms of the raw reports from research subjects, television viewing was reported as falling midway between high activation and low activation and high and low concentration (Kubey and Csikszentmihalyi, 1990: 82). Indeed, in the case of our measure of self-reported "happiness," we have made this quite explicit in writing that "People frequently report feeling slightly happy during viewing but on average, *not* any happier than usual" (p. 101). On our "activation" measure—a self-report cluster variable made up of active-passive, alert-drowsy, excited-bored, and strong-weak—our respondents rated themselves at an average of 4.29 during viewing (p. 83), with 4 as the midpoint between a high of 7 and a low of 1. But though the mean television viewing activation score was above the median, television viewing came in fifteenth out of sixteen activities ranked. The only "activity" reported as less activating than television viewing was "resting." Similarly, our respondents' mean score on the alert-drowsy semantic differential was 4.76 (p. 82). The variable was set on a seven-point scale with 7 denoting high alertness and 1 the highest drowsiness, and 4 denoting a state midway between alert and drowsy (the actual instrument used is presented on pp. 54–55).

The score of 4.76 means that on average, our respondents reported feeling more alert than drowsy when they viewed. Still, television viewing was reported rather low on alertness *relative* to the other activities in the comparison. One would not know it from reading Katz's characterization of our work that we have reported that a group of black African graduate students living in the United States reported significantly greater concentration and alertness when they viewed television than did black U.S. graduate students (Kubey and Csikszentmihalyi, 1990: 77–78). We know also from Gavriel Salomon's (1979) work that Israeli children take television more seriously than do U.S. children.

3. Although I by no means believe that the method that I have employed is the only way, or the best way, to tap viewer behavior, at least it fulfills Katz's call for rules of evidence—and we have reported regularly on the reliability and validity of the method (Kubey and Csikszentmihalyi, 1990: 57–61). Furthermore, the findings emanating from this method—or *rules*—if you will, have been replicated in numerous settings, with different types of subjects, of different ages and different nationalities.

4. I wish to thank Ben Fisher for bringing this last point to my attention.

5. Nor is the idea that the television audience sometimes resists content particularly new either. It has been known for at least three decades, for example, that much of the audience resents and resists television ads (Steiner, 1963). Over 40 percent of Steiner's 2,400-plus U.S. research subjects reported that they would prefer television without commercials, and roughly 40 percent found ads "in poor taste and very annoying." Sixty-three percent found ads to be too long. Steiner also reported that nearly half of his respondents believed that advertisers had the most say about what kinds of programs were on the air (p. 221). An active dislike for television commercials and a belief that television is dominated by commercial interests may not qualify as resistance for some. For this writer, it is at least worth noting.

References

American Psychiatric Association (1994). Diagnostic and Statistical Manual of Mental Disorders: DSM-IV. 4th edition. Washington, D.C.: American Psychiatric Association.

American Psychiatric Association (1987). *Diagnostic and Statistical Manual of Mental Disorders.* 3rd edition, revised. Washington, D.C.: American Psychiatric Association.

Anderson, D. R., L. F. Alwitt, E. P. Lorch, and S. T. Levin, (1979). Watching children watch television. In G. Hale and M. Lewis (eds.), *Attention and the Development of Cognitive Skills* pp. 331–361. New York: Plenum.

Anderson, D. R., and P. Collins (1988). *The Impact on Children's Education: Television's Influence on Cognitive Development.* Washington, D.C.: U.S. Department of Education.

Ang, I. (1985). *Watching "Dallas": Soap Opera and the Melodramatic Imagination.* London: Methuen.

Bochco, S. (1983). Interview. *Playboy,* October, pp. 157–164.

Bower, R. T. (1973). *Television and the Public.* New York: Holt, Rinehart and Winston.

———. (1985). *The Changing Television Audience in America.* New York: Columbia University Press.

Bronfenbrenner, U. (1973). Television and the family. In A. Clayre (ed.), *The Impact of Broadcasting.* London: Compton Russell.

Buckingham, D., and J. Sefton-Green (forthcoming). Multimedia education: A curriculum for the future? In R. Kubey and B. Ruben (eds.), *Information and Behavior,* vol. 6, *Media Literacy in the Information Age.* New Brunswick, N.J., Transaction Books.

Chaffee, S. H., and J. L. Hochheimer (1985). The beginnings of political communication research in the United States: Origins of the "limited effects" model. In E. M. Rogers and F. Balle (eds.), *The Media Revolution in America and Western Europe*, pp. 267–296. Norwood, N.J.: Ablex.

Comisky, P., and J. Bryant (1982). Factors involved in generating suspense. *Human Communication Research* 9: 49–58.

Corporation for Public Broadcasting (1978). *A Qualitative Study: The Effects of Television on Peoples' Lives.* Washington, D. C.: Corporation for Public Broadcasting.

Csikszentmihalyi, M., and R. W. Kubey (1981). Television and the rest of life: A systematic comparison of subjective experience. *Public Opinion Quarterly* 45: 317–328.

Dayan, D., and E. Katz (1992). *Media Events: The Live Broadcasting of History.* Cambridge, Mass.: Harvard University Press.

Dietz, W., and S. L. Gortmaker (1985). Do we fatten our children at the television set: Obesity and television viewing in children and adolescents. *Pediatrics* 75: 807–812.

Fry, D., V. H. Fry, and A. Alexander (1988). *The relative importance of primary and secondary contexts in the constitution of textual meaning.* Paper presented at the meeting of the International Communication Association, New Orleans, May.

Furu, T. (1971). *The Function of Television for Children and Adolescents.* Tokyo: Sophia University Press.

Gallup, G., and F. Newport (1990). Americans love—and hate—their TVs. *San Francisco Chronicle,* 10 October, p. B3.

Gitlin, T. (1972). Sixteen notes on television and the movement. In G. White and C. Newman (eds.), *Literature in Revolution.* New York: Holt, Rinehart & Winston.

Hall, S. (1989). Ideology in communication theory. In B. Dervin, L. Grossberg, B. O'Keefe, and E. Wartella (eds.), *Rethinking Communication Theory,* vol. 1, *Paradigm issues.* Newbury Park, Calif.: Sage.

Halloran, J. D. (1978). Further development—or turning the clock back. *Journal of Communication* 28: 120–132.

Hawkins, R. P., and S. Pingree (1986). Activity in the effects of television on children. In J. Bryant and D. Zillmann (eds.), *Perspectives on Media Effects,* pp. 233–250. Hillsdale, N.J.: Erlbaum.

Hearn, G. (1989). Active and passive conceptions of the television audience: Effects of a change in viewing routine. *Human Relations* 42: 857–875.

Himmelweit, H., and B. Swift (1976). Continuities and discontinuities in media usage and taste: A longitudinal study. *Journal of Social Issues* 32: 133–156.

Katz, E. (1987). Communication research since Lazarsfeld. *Public Opinion Quarterly* 51: 25–45.

Katz, E., J. G. Blumler, and M. Gurevitch (1973/1974). Uses and gratifications research. *Public Opinion Quarterly* 37: 509–523.

Katz, Elihu, and Lazarsfeld, P. F. (1955). *Personal Influence.* Glencoe, Ill.: Free Press.

Klapper, J. (1963). Mass communication research: An old road resurveyed. *Public Opinion Quarterly* 27: 515–527.

Kubey, R. (forthcoming). Television dependence: Toward diagnosis, treatment, and prevention. In T. M. Williams (ed.), *Television and Children.* Newbury Park, Calif.: Sage.

———. (1990a). Television and family harmony among children, adolescents, and adults: Results from the experience sampling method. In J. Bryant (ed.), *Television and the American Family,* pp. 73–88. Hillsdale, N.J.: Erlbaum.

———. (1990b). Psychological dependence on television: Application of DSM-III-R criteria and experience sampling method findings. Paper presented at the annual meeting of the American Psychological Association, Boston, August.

Kubey, R. W. (1986). Television use in everyday life: Coping with unstructured time. *Journal of Communication* 36(3): 108–123.

Kubey, R., and M. Csikszentmihalyi (1990a). *Television and the Quality of Life: How Viewing Shapes Everyday Experience.* Hillsdale, N. J.: Erlbaum.

Lasswell, H. D. (1972). *Propaganda Technique in the World War.* New York: Garland (originally published 1927).

Liebes, T., and E. Katz (1990). *The Export of Meaning: Cross-Cultural Readings of "Dallas."* New York: Oxford University Press.

Maccoby, E. (1951). Television: Its impact on school children. *Public Opinion Quarterly* 15: 421–444.

McIlwraith, R. D. (1990). Theories of television addiction. Talk to the American Psychological Association, Boston, August.

McIlwraith, R. D., R. S. Jacobvitz, R. Kubey, and A. Alexander (1991). Television addiction: Theories and data behind the ubiquitous metaphor. *American Behavioral Scientist* 35: 104–121.

Munsterberg, H. (1916). *The Photoplay: A Psychological Study.* New York: D. Appleton.

National Institute of Mental Health (1982). *Television and behavior: Ten Years of Scientific Progress and Implications for the Eighties.* Vol. 1. Rockville, Md.: U.S. Department of Health and Human Services.

Parker, S. G. (1918). The United States and the war. *Harper's Magazine* 136: 521–531.

Reeves, B., E. Thorson, and J. Schleuder (1986). Attention to television: Psychological theories and chronometric measures. In J. Bryant and D. Zillmann (eds.), *Perspectives on Media Effects,* pp. 251–279. Hillsdale, N.J.: Erlbaum.

Salomon, G. (1979). *Interaction of Media, Cognition, and Learning.* San Francisco, Jossey-Bass.

Schiller, H. (1973). *The Mind Managers.* Boston: Beacon Press.

Shanahan, J., and M. Morgan (1989). Television as a diagnostic indicator in child therapy: An exploratory study. *Child and Adolescent Social Work* 6: 175–191.

Simpson, J. B. (1992). *Webster's II. New Riverside Desk Quotations.* Boston: Houghton Mifflin.

Smith, R. (1986). Television addiction. In J. Bryant and D. Zillmann (eds.), *Perspectives on Media Effects,* pp. 109–128. Hillsdale, N.J.: Erlbaum.

Steiner, G. (1963). *The People Look at Television.* New York: Alfred A. Knopf.

Warner, W. L., and W. E. Henry (1948). The radio day time serial: A symbolic analysis. *Genetic Psychology Monographs* 37: 3–71.

Winick, C. (1988). The functions of television: Life without the big box. In S. Oskamp (ed.), *Television as a Social Issue,* pp. 217–237. Newbury Park, Calif.: Sage.

Winn, M. (1977). *The Plug-In Drug.* New York: Viking.

Wren-Lewis, J. (1983). The encoding/decoding model: Criticisms and redevelopments for research on decoding. *Media, Culture, and Society* 5: 197–198.

Zillmann, D. (1985). The experimental exploration of gratifications from media entertainment. In K. Rosengren, L. A. Wenner, and P. Palmgreen (eds.), *Media Gratifications Research: Current Perspectives,* pp. 225–239. Beverly Hills, Calif.: Sage.

Zillmann, D., and J. Bryant (1985). Affect, mood, and emotion as determinants of selective exposure. In D. Zillmann and J. Bryant (eds.), *Selective Exposure to Communication,* pp. 157–190. Hillsdale, N.J.: Erlbaum.

PART III

THE POLITICS OF AUDIENCE STUDIES

The Politics of Producing Audiences

13

Martin Allor

Since the mid-1980s there has been a growing body of literature(s) from across the range of research traditions within media studies that has returned to questions concerning audiences. The reasons for this are complex and would make an interesting case study in a conjunctural epistemological analysis—one that would advance beyond the limitations of either a Kuhnian model of paradigms or Feyerabend's version of "ad hocing."[1] My project in this chapter is a more modest one. For several years I have been investigating the theoretical entailments of the various versions of "audience" as concept within recent media studies and cultural studies. Thus the kind of audience research that I have been engaged in has been focused on the gains and liabilities of the utilization of the concept of audience as a ground of theorization and in the designation of objects of research.

My thesis is a simple one. Rather than discovering the truths of the social practices of individuals (their sense-making) or the effectivity of media, recent research that begins with audience as a fixed analytic category works to *produce* audiences as objects of knowledge and intervention. As do other public intellectuals (advertising copywriters, pollsters, program producers, regulators), we work to produce figures of audience—figures that work to ground the claims of our research; figures that often work to condense contradictory assumptions about the social field and that often occlude the political implications of our work.

In simple terms, then, I want to argue against the utility of "audience," as an abstract conception of a totality (like "population" or the "public"), for the necessarily political project of studying the structures and practices of media reception. Moreover, I will argue against the possibility of any comprehensive theory of audience and argue in favor of something more modest, more appropriate to the kinds of interventions that we as public intellectuals can effectuate. In my analysis I will follow the example of the historian of science Georges Canguilhem (1980) in his posing of the question, "What is the subject of psychology?". Canguilhem rejected the internal claims to unity across methodological divides—a general theory of conduct—and instead focused on the project or the conceptual work of the different traditions in the field and, most importantly, in their anchoring in different neighboring and antecedent disciplines. Like Canguilhem, I want to reject the starting assumption that the various traditions of audience research are linked by a common object of inquiry. I will focus instead on the commitments and the possibilities of producing conceptions of audience activity. This will involve the consideration of the relations between objects of inquiry and logics of intelligibility, both in terms of the production of understanding (a scientific imperative) and intervention (the necessarily political concomitant to intellectual inquiry).

Figures of Audience

In framing my consideration of conceptions of audience through the concept of figure, I mean to foreground the necessary work that analytic concepts accomplish. That is, I believe that our theoretical tools need to be interrogated for both the ways in which they discursively produce claims to facticity (anchoring the claims to the validity of a given analysis of the social field), and the ways in which they articulate research problems to particular logics of inquiry.[2] It is precisely this doubled work of the discursive production of analytic concepts that often leads to the tensions and contradictions that undermine projects of analysis. The figures of audience produced in analytic projects often overlap with those produced by other public intellectuals, in the sense that they too attempt to make sense of the shifting social relations enacted by the apparatuses of public media. Thus, for much of this century public figures of the problematic social individual have centered on the relations between people and the machinery of communicative mediation.

I will cite two ideal-typical examples of this wider public discourse in order to specify the work of figures and to identify the dual axes of questioning that underpin both the general public discourse of audience and the analytic projects of media research. The first example marks a key moment in the successive moral panics attending the introduction of new media—the first decade of the commercial exploitation of film in urban North America. It is exemplary of the shift from social fears involving emigration and urbanization to a concern with film as a new machinery of representational mobility. Thus, the New York magazine *The Outlook* argued in the summer of 1916, in an article entitled "Movies, Manners and Morals":

> But it is the psychology—or rather the total absence of it—in the average moving picture play that constitutes its greatest danger to the growing mind. Especially is this injurious to the more or less rudderless being whom we must educate into a good citizen, the child of alien parents who too often is contemptuous of the habits and maxims of his parents and ignorant of anything American but the hybrid pavement life of a polyglot city. The version of life presented to him in the majority of moving pictures is false in fact, sickly in sentiment, and utterly foreign to the Anglo-Saxon ideals of our Nation (*Outlook*, 1916: 695).

What is most remarkable about the figure produced here is the way in which it condenses together the immigrant/the adolescent/the spectator as the problematic social body. In addition, it explicitly positions these pressures and influences on the problematic social body within the social trajectory of the production of citizens—that is, within the terms of a generalized social epistemology: the citizen as bearer of "Anglo-Saxon ideals."

The second example that I will cite exemplifies the range of public discourses framing the experiential or praxical adjustments of individuals to new leisure technologies. Here it is less a question of the production of the citizen than of the construction of a social imaginary framing the practices of mediated reception. Thus, one of a series of articles in the early 1920s working to position and popularize radio broadcasting ("The great audience invisible") centered precisely on the construction of this imaginary.

> Another difference that marks this vast company of listeners is that they do not sit packed closely, row on row, in stuffy discomfort endured for the delight of the music. The good wife and I sat there quietly and comfortably alone in the little back room of our own home that Sunday night and drank in the harmony coming three hundred miles to us through the air. How easy it is to close the eyes and imagine the other listeners in little back rooms, in kitchens, dining-rooms, sitting-rooms, attics; in garages, offices, cabins, engine-rooms, bungalows, cottages, mansions, hotels, apartments; one here, two there, a little company around a table away off yonder, each and all sitting and hearing with the same comfort just where they happen to be (Scribners, 1923: 411).

This domesticating imaginary works to install the radio receiver in the home at the same time that it provides a figure of and for the practice of listening in that space. Thus, the figure works by locating the technological apparatus in the private sphere as it simultaneously registers the practice of listening in relation to a public of "*semblables.*" In this way it can be seen to be an enactment of a social ontology of reception, turning around the question of the kind of "being in the world" embodied in radio listening.[3]

These two axes of interpretation (and interpellation) underpin public discourse about the relations between the media and the people. The first explicitly frames this relationship as a representational one, as a question of social knowledge, of social learning—the audience as the endpoint and guarantee of the stability (or instability) of the citizenry or the social itself. The second frames the relationship as one of enacted practice and experience; it links the specificities of particular practices of reception with larger formations of agency and the articulation of public and private spheres.

As researchers and theorists, we too are public intellectuals. The figures of audience that we produce have a wider public circulation that has a political effectivity outside of our particular analytic goals. At the same time the figures that we produce are different from others in that we demand that they tell us something grounded in facticity. We produce these figures (these concepts) within epistemological regimes that condition the ways in which we structure the engagement between concepts and empirical analyses. Our work, then, is conditioned both by the more general discourses figuring the audience (representation and enacted agency) and by the specific logics of inquiry and interpretive rules of particular research traditions. This doubled set of determinations (conjunctural and epistemological) is a condition of possibility of all social research, but audience research is particularly compromised by its tensions.[4]

In a recent essay, Pierre Bourdieu (1988) offers a cogent critique of the consequences of the continuation of theoretical orthodoxies in social theory. More specifically, he constructs an important argument against the functions of the underlying epistemological figures conditioning the differentiations of the domain of social theory and research. Thus, for Bourdieu, the many antonymous figures of "antagonistic pairs" ("*couples ennemis*"), for example, between theoreticism and empiricism, that structure theory construction and debate work more often to occlude the development of social knowledge than to aid it. We are all familiar with these pairs: micro/macro; agency/structure; subject/object; materialism/idealism; qualitative/quantitative; theory/methods. Such typologies are, of course, important tools for thinking with. And, taken as ideal types, they can function to spur the process of theoretical specification. More often, unfortunately, they have tended to become fixed and hypostatized. Instead of opening up a process of elaboration, they work to overdetermine the goals of research, either in framing the limits of the social ontology underlying research approaches, or in literally providing the subtending "contents" for specific research concepts.

The range of research concerning audiences is conditioned in various ways by these antonymous figures. In addition, however, our debates are framed by another set of oppositions (which are in evidence in this volume)—that between passivity and activity and that between the audience member conceived as part of a public and the audience member as a social agent. These particular dichotomies are the specification within social research of the more general public discourses on the media and the people. They specify the terms in which the questions of social epistemology and social ontology are brought to bear on questions of audience.[5] These two sets of oppositions operate on different levels of abstraction. The former grounds analytic interpretations of the practices of individuals (and groups), and the latter specifies the kind of social being that "audience" is put in relation with.

These differences of level, or register, often lead to confusion in theoretical debates when one level is condensed into the other in the support of a particular argument. For example, this is often the case when the claims of the political ramifications of particular text/subject relations are grounded in underlying (and unquestioned) assumptions about activity or passivity. In addition, both sets of dichotomies involve terms that operate at both a high level of abstraction and with a high level of generality. The values of activity or passivity can only be supplied by particular psychological or sociological assumptions; the logic of relations between the social agent and the social field can only be specified within particular models of the social formation.

However, it is the instability of the concept "audience" itself that most exacerbates the development of coherent research approaches and productive debate between different theoretical formations. Like many other analytic terms essentially adopted from ordinary language, the term "audience" is always already implicated in a complex (and sometimes contradictory) semantic/discursive field. Without going into its specific etymology, it can be seen in different cases to designate a range of traits and levels, including (but not limited to): the interpretive competencies of individuals; the co-presence of individuals in a reception situation; an active social relation of collective interpretation; the "market" for a particular cultural commodity; the "imaginary" constructions of cultural creators; the "public" of a particular genre; and the totality of potential receivers of a given media form. Audience, then, has discursive links with other nineteenth-century terms designating collectivities—the mass, the crowd, the popular, and the public. At the same time, it is linked to concepts emerging from within twentieth-century disciplines centering on the individual-psychology, social-psychology, psychoanalysis. Moreover, it is increasingly linked to a new series of terms that are developing within emerging technological apparatuses and the cultural industries themselves—the user interface, the lifestyle blocs of marketing research, or the "demos" of broadcast research and marketing.

All of which is to suggest that the relative abstractness of the term audience is always filled in by the underlying traits and discourses that it mobilizes. At the same time the continued utilization of audience as the organizing term for research on social relations of reception tends to occlude the specific origins of the traits that specify the term. The relative polysemy of audience also links, in problematic ways, with the sets of antagonistic pairs that subtend theories about media reception. The active/passive dichotomy and the individual viewer/member of a public dichotomy operate as epistemological horizons. Like all horizons, they function as hypothetical limits, marking boundaries of inquiry. As an abstract conception of totality, audience either collapses the epistemological terrain of the dichotomies through being linked exclusively with underlying traits pertinent to one pole or the other, or it oscillates problematically between the poles, for example, in making claims to validity about social being based in data derived from individual practices.[6]

It is this polysemy of the term audience that undermines any attempt at the construction of a unified theory. In fact, there is no such underlying compatible object of analysis around which a field theory could be constructed. Any attempt to do so inevitably leads to a set of reductions that effectively rule key issues out of the boundaries of analysis.

In recent years social theorists in different domains have begun a process of theory elaboration that attempts to overcome the liabilities of abstract terminologies that attempt to name an object domain. At the same time, these terminologies also work to overcome some of the liabilities of the antagonistic pairs structuring social theory. For example, Anthony Giddens's (1979) conception of structuration is part of an explicit theoretical strategy aimed at overcoming the dichotomy between agency and structure and subject and object. Its work as a concept is to displace competing conceptions that function by naming an object domain and to replace them with a term designating a social relation. The advantage of this strategy is that it makes explicit the underlying social ontology framing theory development. However, because Giddens's conception of structuration functions directly on an ontological plane, it is less immediately useful in the construction of particular objects of inquiry, as it tends to identify the same ontological status across practices and structures.

A complementary strategy is adopted by Bourdieu himself in the analytic distinction that he has developed between habitus and field.[7] These concepts also attempt to frame the understanding of the relationship between agency and structure (or practice and institution) outside of a logic of subject/object dualities. Bourdieu's strategy is to interpret action and experience (agency) as the structured (habituated) and structuring (sedimenting) motor of the development and differentiation of institutionalized domains (fields) of the social formation—"the social agent in his [her] true role as the practical operator of the construction of objects" (1990: 13). And, while these terms designate a recursive ontology similar to Giddens's, they also provide more specific analytic tools that link this ontological move to particular domains of inquiry. That is to say that the habitus/field distinction is mobile and capable of rearticulation; it explicitly calls for the exploration of the different relations within various fields of social relations.

Regional Explication

These kinds of analytic distinctions should be seen as one of the strategies through which more precise and productive approaches to the study of media reception might be articulated. What they suggest is a more "regional" approach to theory, research, and intervention. Rather than seeking to replace the abstract concept "audience" with another seemingly more suitable term, we should be working at building analyses around the specificities of particular aspects or moments of the social relations of media reception. This kind of approach begins from the assumption that relations of reception involving different media, different kinds of social agents, different texts, and different conjunctural contexts might indeed involve a range of differentiated practices. In the recent development of audience research in media studies, two key orthodoxies have tended, across theoretical and methodological divides, to further focus and limit inquiry. The first has been the predominance of broadcast reception, in particular television, as both the model for the relations between social agents and media, and as privileged domain of research. The second has been an underlying epistemology of reading/meaning-making/text-decoding as the dominant model of the reception process.

Both orthodoxies have narrowed the scope of the issues under study. The privileging of broadcast reception has reinforced atomistic models of the activity—privatized reception by lone viewers or the family unit. The underlying epistemology of reading has worked virtually to exclude other aspects of social practice connected to the relations between media and people, focusing (whatever the method) on the semiotic relations of the constructed text and the (re)constructed text of reader/viewer/spectator accounts. Both enact a psychologistic logic of inquiry in that they link reception to the privatized (re)production of sense by social agents.

In order to open up a process of regional theory construction and specification, it is necessary to break with the centripetal pull of these orthodoxies and pose the question of what are the boundaries of researches bearing on relations of reception. Put in other terms, what kinds of practices in relation to what aspects of cultural (re)production should we frame as aspects of audience researches? As soon as one backs away from the inclusive psychologism of text decoding models, it becomes clear that reception itself may not be an adequate analytic frame for analysis. A concept of reception poses the question of social action in relation to communicative mediation within an epistemology of re-action. Whether or not particular models view that action as active or passive, logics of reception prioritize the medium, the text, or discursive structures. In this way the practices of people and the network of social relations that frame them are always already rendered secondary, determinate, or reactive.

Perhaps the most important recent trend in approaches to "audience" research has been the explosion of ethnographically influenced strategies. These developments can be read both positively and negatively. In focusing on situated, contextualized practice, fully ethnographic research engages with the complexity of the relations between agents and cultural mediation. It necessarily encounters the embeddedness of practices of "reception" within other social relations, other mediations, with levels of engagement that are transversal to processes of textual decoding. More often, however, qualitative research that is less depth-oriented than full ethnography (interviews, focus groups, limited participant observation) has taken experiential data of limited scope as a demonstration of traits of global models of audience. This research, then, has taken the tangible talk, practice, or bodies of research subjects as the guarantee, within a strong epistemological realism, of the claims of general theoretical positions.

Put in other terms, ethnographic research regimes, in the praxical materialities of their objects of inquiry, encounter one of the epistemological determinations framing a regional reconstruction of approaches to audience. Depth ethnography, in its focus on the embeddedness of strips of action within the frames of the everyday, tends to evacuate the specificity of research questions centering on mediated cultural practices. For example, Janice Radway's (1988) call to replace text-based audience studies with a general ethnography of the everyday sketches the limit position of an ethnographic approach to reception study. In displacing the text-reading logic of inquiry through a strategy of studying the everyday transversally, the position effectively overturns the underlying assumptions of reaction-framing audience ethnography. At the same time, it also displaces any simple conception of audience or reception as an object of inquiry. In this way it enacts a social ontology that refuses to privilege particular strips of interaction outside of the analysis of their embeddedness in the everyday. Radway's argument, then, develops the ethnographic strategy at the horizon of audience study. That is, both the questions of the representational and the social being aspects of audience are subsumed in a contextual ontology of the everyday, following the logic of intelligibility of ethnography to its limit.

The radical contextualism of this approach poses the critical question of the specification of political projects of analysis. To use Bourdieu's distinction, it poses the problem of defining the pertinent field(s) (the social and institutional relations of power) through and in which one grounds the analysis of the habitus.[8] This issue of specification, of identifying the particular stakes of analysis, works within ethnographic research regimes to undercut any tendency to a strong epistemological realism or foundationalism in approaches to audience. That is, the more closely one specifies the nature of the social relations framing practices of media consumption in particular contexts, the more one recognizes the impossibility of discovering the truth of an embodied "audience" through ethnographic inquiry.

Rather than delivering the ontological truth of an abstraction called audience, ethnography should be seen as providing one of the conditions of possibility of the study of the social relations of media consumption. It points to the necessarily praxical moment of any general understanding of the relations between media and people. At the same time, in its contextualism, ethnography specifies the necessarily social relations of inquiry, and, therefore, it designates the necessarily partial nature of ethnographic accounts.

The general strategy of regional explication (and political intervention) follows from this insight. The recentering of particular aspects of the mediations of relations of reception can be articulated around the twinned recognition of the partiality and institutional embeddedness of ethnographic accounts. The most promising recent research developments in approaches to audience derive from this insight. Rather than beginning from the ontological ground of a certain social subject (either active or passive), regional research strategies begin from the construction of levels of analysis and objects of inquiry derived from the specification of models of mediation.

The epistemological resources for the models range from Foucaultean conceptions of social apparatuses (*dispositifs*), through feminist models of institutional relations of engendering, to materialist conceptions of the cultural industries and domestic technologies. In different ways, they work to construct regional planes of determinations and agency that specify both the specific social relations of mediation under analysis and the aspects of human agency that serve as the political horizon of ultimate intervention. At the same time, these models provide the analytic resources for transversal interpretations of relations of reception. In different ways, they map the relations between agency and determination in a manner that avoids either the evacuation of the practices of viewers/readers/visitors or the simple universalization of models of the social determinants bearing on relations of reception.

Thus Foucault's conception of the *dispositif* offers a model of the relations among institutions, discourses, and agency that privileges neither subject nor object. Rather, the analytic and political focus is on the anaclitic productivity of power/knowledge relations within bounded institutional domains. As Gilles Deleuze recently put it:

> The components of [Foucault's] *dispositifs* are lines of visibility, of enunciation, lines of force, lines of subjectification, lines of fissure and fracture which intersect and tangle up with each other. Where certain lines supply the others, or instigate still others, across variations or even mutations of disposition or organization (Deleuze, 1989: 188; my translation).

Deleuze's point here is to emphasize the transversal nature of the analytic model. More significantly for the study of relations of reception, Deleuze specifies the components of the terrain of the *dispositif* as articulations of discursive principles of inclusion and exclusion (visibility, enunciation), institutional power (force), and of individuated or singularized agencies (subjectification). The *dispositif,* then, frames a research strategy for relations of reception that foregrounds the intercalation of practices of subjectification, institutional relations of power, and discursive mediations without privileging any single level as ground or guarantee.[9]

Recent debates within feminist sociological theory also offer models for the regional reconstruction of reception study. For example, Dorothy E. Smith's (1987) recent work elaborates a sociology "from the standpoint of women" that subsumes an ethnographic or ethnomethodological logic of inquiry within an encompassing critique of the institutional relations of power that frame both the silencing of domains of women's practice and the problematic power relations troubling neomarxist and feminist social inquiry. In this way, her work attempts an inquiry into the experience moment of social agency within the terms of a model that foregrounds the institutional and discursive relations of power that subtend and "trouble" any naturalistic account of the experiential.

Recent feminist researches into media reception have entered into and rearticulated this debate both methodologically and politically. Elspeth Probyn and Janice Radway have, in different ways, rethought the political implications of the social relations inherent in ethnographic inquiry in reception study.[10] In this work conceptions of gender do not function as abstract universals. Instead, relations of gender or "sexage" are taken to designate the differentiations and specifications of social power that cut across, and articulate, the production, circulation, and reception of social texts. Following this logic, then, gender relations are investigated as structuring determinations that can be traced across the locations (located conditions) of production and reception as much as in the practices of the agents of cultural consumption.

In a related way, recent reevaluations of models of mass consumption and the cultural industries have pointed to a third direction in regional approaches to reception. These newer approaches treat the practices of commodity consumption and domestic technologies as moments of engagement rather than as instances of reification. At the same time, they refuse simply to treat commodity consumption under the sign of free practice or creativity. Cultural commodities and domestic technologies are framed precisely as problematic objects of inquiry, productive of surplus value and appropriated into localized networks of significance and distinction.[11]

The research projects led by Roger Silverstone and David Morley (both individually and jointly) on the relations between domestic technology, family structures, and the localized reproduction of the "national" are important indices of this trend. In this work, regional or localized abstractions—the moral economy of the household, for example—work to specify the objects of inquiry, levels of analysis, and the political stakes of inquiry.[12] Here again, the analytic terms "consumption," "the domestic," and "family" are not taken to be universals. They function to specify a field of inquiry—a set of specified practices, structures and relations that are viewed as instances of relations of reception rather than the site of "the" audience.

In different ways, these three trajectories of recent research offer models for the regional reconstruction of audience study. Each displaces the search for the audience into a specified analysis of particular social relations of media reception. Each accomplishes this specification within the framework of a politics of research that foregrounds the necessary moment of intervention. That is, the stakes of analysis are seen from the beginning as involving more than simple explication or understanding. The politics of audience research is seen here as necessarily involving a public production of knowledges about specific social relations for the amelioration of particular relations of social power. This is to say that this work rejects the search for a general theory of the audience in favor of a practice recognizing our role as public intellectuals in the production of altered relations of reception.

Notes

1. For one important version of this conjunctural epistemology of the return of the "active media audience," see Lawrence Grossberg, 1989.

2. See Michele le Doeuff, *L'Etude et le rouet,* for an analysis of the work (*le faire*) of analytic images. My introduction to this work has in turn come from the work of Elspeth Probyn (1991) and Meaghan Morris (1981/1982).

3. Cf. Lesley Johnson (1988) and Lynn Spigel (1988) for analyses of the relations of these discourses to the rearticulation of gendered domestic space.

4. Cf. Jean-Michel Berthelot (1990) for a more general discussion of logics of intelligibility in the social sciences.

5. Thus, to take ideal-typical examples from work presented in this volume, John Fiske and Robert Dawson's research could be mapped onto the activity/social agency poles, and Robert Kubey's onto the passivity/public poles. That is to say that Fiske frames the question of audience in relation a political social ontology and that Kubey frames his research in relation to an underlying social epistemology of social learning.

6. The conceptual and theoretical problem is mitigated in other languages where there is no tradition of single terms that cover both the pole's of this terrain. Thus, in French, *public* is used to designate the collective aspects of the domain, while more specific terms—*lecteur(trice), telespectateur(trice)*—are used to designate the processual or praxical aspects. In addition, this conceptual division itself undermines any claims to universal claims.

7. Cf. Bourdieu (1985) and the essays in *In Other Words* (1990). There are problems with Bourdieu's overall theoretical position that are outside the scope of the present discussion.

8. Radway describes this field in her proposal as the relations of taste articulated across the public and private worlds of the institutions of the family, the school, and leisure. In this sense she has proposed a transversal model of both the habitus and fields.

9. For two very different examples of this discursive model, see Ien Ang (1991) and Lawrence Grossberg (1988).

10. See, for example, their respective essays in the Cultural Studies/Ethnography issue of the *Journal of Communication Inquiry*, vol. 13, No. 2, 1989.

11. See, among others, Mike Featherstone (1981) and Daniel Miller (1987).

12. See, for example, Morley and Silverstone (1990).

References

Ang, Ien (1991). *Desperately Seeking the Audience.* London: Routledge.

Berthelot, Jean-Michel (1990) *L'Intelligence du social.* Paris: PUF.

Bourdieu, Pierre (1985). The genesis of the concepts of habitus and field. *Sociocriticism* 2(December): 11–24.

———. (1988). Vive la crise! For heterodoxy in social science. *Theory and Society* 17: 773–87.

———. (1990). *In Other Words.* Cambridge, Eng.: Polity.

Canguilhem, Georges (1980). What is psychology. *I&C* 7: 37–50.

Deleuze, Gilles (1989). Qu'est-ce qu'un dispositif? In *Michel Foucault philosophe: rencontre internationale, Paris, 9, 10, 11 janvier 1988.* Paris: Seuil.

Featherstone, Mike (1991). *Consumer Culture and Postmodernism.* London: Sage.

Giddens, Anthony (1979). *Central Problems in Social Theory.* Berkeley: University of California Press.

Grossberg, Lawrence (1989). The context of audiences and the politics of difference. *Australian Journal of Communication,* 16: 13–35.

———. (1988). *It's a Sin: Essays on Postmodernism, Politics, and Culture.* Sydney: Power Publications.

Johnson, Lesley (1988). *The Unseen Voice: A Cultural Analysis of Early Australian Radio.* London: Routledge.

Le Doeuff, Michele (1989). *L'Etude et le rouet.* Paris: Seuil.

Miller, Daniel (1987). *Material Culture and Mass Consumption.* Oxford: Basil Blackwell.

Morley, David, and Roger Silverstone (1990). Domestic communication: Technologies and meanings. *Media, Culture, and Society* 12(1): 31–55.

Morris, Meaghan (1981/1982). Operative reasoning: Michele le Doeuff, philosophy, and feminism. *I&C* 9: 71–102.

Outlook (The) (1916). Movies, manners, and morals.

Probyn, Elspeth (1989). Take my word for It: Ethnography and autobiography. *Journal of Communication Inquiry* 13(2)___.

———. (1991). This body which is not one: Technologizing an embodied self. *Hypathia* 6(3):___.

Radway, Janice (1989). Ethnography among elites: Comparing discourses of power. *Journal of Communication Inquiry* 13(2)___.

Scribners (1923). The great audience invisible.

Smith, Dorothy E. (1987). *The Everyday World as Problematic: a Feminist Sociology.* Toronto: University of Toronto Press.

Spigel, Lynn (1988). Installing the television set: Popular discourses on television and domestic space, 1948–1955. *Camera Obscura* 16: 11–48.

Power Viewing: A Glance at Pervasion in the Postmodern Perplex

14

John Hartley

What follows is an attempt to take television audiences seriously as fictional constructs, their fictionality not being taken as a disqualification from, but as a demonstration of the social power (even truth) of fictions. The energy with which audiences are pursued in academic and industry research bespeaks something much larger and more powerful than the quest for mere data. The television audience is pervasive but perplexingly elusive; the quest for knowledge about it is the search for something *special;* literally, knowledge of the *species.*

The question of the power of the visual media is traditionally posed in terms of their power over the populace and the power of economic or political elites over them. The question of media power is thus a social question. But the media are encountered by those over whom they are supposed to hold sway, the audience, only in textual form. Thus to study television is to study the relationship, if any, between social and textual power. If the media exert power and influence over their audiences—that is, socially—how is it done textually? And if media texts exert power, what is the place of meaning in the analysis of power?

If politics, the economy, and society are systematically meaningful, how can meaning be investigated? Despite the current corporate enthusiasm for new technologies, broadcast television remains one of the most fully developed and globally pervasive phenomena for the production and distribution of meaning. To investigate it culturally is to ask how *humanity* makes sense, not as primitive or prehistoric species, but here, now. As I have argued elsewhere, television is the "power of speech" as transformed and developed in relation to material forces and social power (Hartley, 1992, chapter 2).

From this cultural perspective, which is close to the anthropology of meaning of Marshall Sahlins (1976), the television audience can be theorized quite simply, not by reducing it to a false unity, but precisely because it is *not* a specific group. It represents a way of conceptualizing not a group in society but what *constitutes* society as a whole in its specifically sense-making mode.

Watching television is not merely useful (economic), it is meaningful (cultural); and those for whom television has meaning are the audience. But from a cultural perspective the audience is not made of "individual viewers." Individual people are not merely individuals, opposed to or distinct from society, they are the totality of society; as Marx put it, the "ideal totality—the subjective existence of thought and experienced society present for itself" (Marx, 1961: 105). What this helpful insight means in practice, however, is that the notion of the "individual" is far too complex to deal with as an analytical unity ("writing" individuals as social totality has long been the province of fiction and drama, not least on television, but is not common in academic analysis). Hence I am reluctant to equate "audiences" with "individuals," for individuals are not always audiences—and the point of Marx's statement is that though individuals are particular they are totally social at the same time. Instead of trying to decide between them, cultural analysis might do better to show how the individual and the social are in fact connected. One way of doing this is to look at television as meaningful, as a system for circulating social meaning among particular individuals in which the connection is textual.

The textual and discursive forms of television are by no means unique to it, and they are not confined to the screen, so by text I mean more than a program or even "supertext" (Brown, 1987). Television's textuality includes the industrial, regulatory, and critical discourses that literally channel what is on the screen, without ever appearing on it; and of course it includes the dialogic participation of its viewers, both during and beyond viewing time. However, I do think that such texts and discourses, representations and meanings are real; they can even be observed in their "exosomatic" or empirical form, where they display the properties of what Karl Popper called "objective knowledge" (Popper, 1983: 72). I take audiences to be textual (rather than somatic—though of course bodies are both textualized and legible) too. It is not just a matter of finding fictive representations of audiences in other texts (addressed to them or about them), but also I think the practice of "being" an audience cannot be understood except textually. Certainly the act of making meaning cannot be observed directly. Our "behavior" as audience is an act in the performative sense; the role we play is rehearsed and our lines have been scripted, and the scene we are in is a dialogue with the social totality. As a matter of fact I have always been impressed by how difficult it is to think, experience, or know anything, as an individual, without discovering that it has already been done. Buckminster Fuller once wrote a quirky little book called *I Seem to be a Verb* (1970); well, my autobiography should be entitled *I Seem to be a Quote*—but then, so should everyone's.

The fundamental attribute of television viewing is pervasion. It is global, insistent, and quite indifferent to any demographic boundary you can name. Give people a chance and they will buy television sets; give them time and they will watch TV. Not even totalitarian governments can stop the will to watch; it seems to belong to the species. At the same time, no one *just* watches television, as far as I know. Just as television is pervasive, so it is pervaded. Its meanings circulate in a context of talk, other media, and the myriad semiotic systems, from clothing and housing to industrialized production itself. To understand television's textuality, we need to take some notice of such concomitant goings-on; it is very hard to see how "family television" makes sense (to itself or to an observer) without an account of the meaning and history of housing, the domestic ideology that turns houses into homes, and a productive industry that depends on the consumption of symbols in the form of commodities. Television makes sense in such a context, and I think, given time, that the "totality" of the society that is "present for itself" in a family watching television could be described. But it would be a big project, and who could do it justice but a playwright or novelist?

Textual and social power, connected in the form of meaning, can form a proper object for television studies, novelists notwithstanding. Power is not merely sovereignty or the power to command; after Foucault, it has been rethought as occurring in all the myriad transactions and networks of everyday life (Foucault, 1977). Although television is radically different from the classic Foucaultian disciplinary institutions in that watching it is strictly voluntary, it is certainly not exempt from its own internal power relations, which are much better understood through a Foucaultian notion of power than by means of earlier versions—no one has the "right" to watch, produce, or appear on television, so a notion of power as sovereignty (which generates rights), for example, does not apply. Nevertheless, television's institutional power relations are constantly worked out among its participants on both sides of the screen.

Within the repertoire of viewing practices, the power of the viewer escapes that of the institution, its controllers, regulators, and textual regimes; the disciplinary apparatus does not only circulate to the viewer from the screen but also in the reverse direction. Watching television is exercising the power to turn one's own disciplinary gaze or glance on and through the screen, using the act of looking to keep an eye on the social and discursive organization of the world at large, and to make judgments and take actions that are themselves exercises of power, often enough of a directly political kind, over which the forces of disciplinary domestication have much less control than they or their critics would like to think. Television does exercise social power, but half of the equation has historically been ignored: the social power of surveillance exercised by audiences in the meaningful use of television as a cultural resource of their own.

Although it would be a simplification to argue this too strongly, it is nonetheless tempting to suggest that while Foucaultian notions are being applied to the terrain previously occupied by questions of television's social power, questions of textual power have been abandoned to the postmodernists. Taking their cue from such writers as Derrida, Baudrillard, and Lyotard, the postmodernists rode into town looking not for power but for difference. They found it in the act of reading. It seems that television texts are at their viewers' command and mean whatever the viewers decide. However, those who read television texts in the name of postmodernist techniques tend to be less interested in the social and textual power of television than in performing feats of postmodern criticism, a form of textual display that celebrates fragmentation, difference, the dissociation of sign from reference, text from readership. Far from producing exemplary readings on behalf of less astute audiences (the traditional literary method that has its roots in biblical exegisis), postmodernist readings begin from the position that texts are not in the first place acts of communication; they do not have addresses written on them, or even if they do, they may never reach their destination (Derrida, 1987). However, television and other media texts are still readable, even if they do not have an interlocutor, for what they reveal to postmodernism is something like this, and I quote:

> Once upon a time our world was enslaved by teleological meta-narratives like History which told our story for us in terms of an ending which never occurs but is endlessly deferred, while in the meantime we spend our intellectual energies trying to prove that the story is true and the world is real, authentic, beyond discourse and in a state of existence that is not only separate from our knowledge of it but capable of being known directly. But now, instead of using such totalizing fictions to prove how authentic our picture of reality is, we have taken the path of the computer animation artist, whose claim to fame is that the object revolving in front of our eyes is *completely simulated*, despite the fact that

1. It looks more real than the object it simulates ... and
2. It is now doing things that cannot really be done ... and
3. The things that cannot be done are in fact being done in front of our very eyes, so they're real ...

> Which means that the simulation is the real, that there's nothing outside the text, and that the authentic icon of postmodern art is the fake. Where this leaves television is uncertain, for television promotes itself on its realism while cashing in on computer animation; it rhetoricizes the world even while telling the truth, and its audiences are drawn to, held by and glory in its trashiest fakery while simultaneously asserting an earnest commitment to realistic portrayals of real life, including those performed by glove puppets purporting to be alien life forms posing as cute emotive child-pets.

Unquote. This quotation is of course a fake. The postmodernist perplex is not fazed by any of this.

Recent intellectual currents have proved exciting but somewhat frustrating for television studies. In spite of Foucaultian, postmodernist, feminist, and other interventions, or perhaps because of them, it seems as hard as ever to explain the link between textual and social power. There are contending and incommensurable theoretical approaches, and there are real difficulties in isolating a coherent object of study; the act of watching television is itself a widely dispersed and variable cultural practice, undertaken by a community (the audience) that is never encountered as such, while on screen television is what I have called a "blivit"—that is, a "dirty" textual form in which fact, fiction, faking, and fabrication are all mixed up like "two pounds of shit in a one-pound bag," in Kurt Vonnegut's phrase (Hartley, 1992, chapter 3).

Is it possible to supersede the variety, difference, "blivitousness," and elusive complexity of both television texts and viewers, not to mention theories, encompassing the whole lot into a general framework that can say *how it works* without having to bother too much about *what it is?* Or is it proper to abandon general frameworks as fictional and imperializing totalizations, and to content ourselves with local analysis, done in the flux of the "ceaseless reformulation of symbolic relations within the national social life," (Sahlins, 1976: 217), which not only form our object of study, but our subjectivity too, and also constitute the raw materials for the creative personnel who make symbols saleable in the industries we analyse? I think cultural studies, at least my own, retains a commitment to both of these positions, the global and the local, illogical though that may be. For one thing, a recognition of the textual productivity of contemporary culture requires both engagement and analysis, while skepticism about general frameworks does not mean they are not ceaselessly produced and deployed, comprising a major component of textual and social power, and a major component of academic research too.

Meanwhile, the focus of critical attention in cultural studies switched from ideology and its effects toward audiences or readerships, since it is at this point that meanings generated in and by media discourses actually go live socially, where textual and social power intersect, and where the distinction between them is meaningless. Ethnographies of reading ask "what (in fact) *do* audiences do with the media?" Meanwhile critical readings, mindful that the polysemic qualities of texts open them up to any reading that may plausibly be brought to bear upon them, ask "what (in principle) *can* readerships do with the media?" The proof is sought in the eating, not in the pudding.

But the conceptual gap between social and textual power is if anything widening in these moves, because instead of seeking to analyze its intersections, these approaches reproduce the gap. Ethnographies of reading must presuppose that they will reveal something in the "other" that can be observed—the act of reading, the real or "natural" audience. Thus ethnographic research runs counter to textual theory, which holds that nothing outside meaning exists for humans, that discourses organize practices, and that when you ask people what they think of texts, or how they read, even when you go into their very lounge-rooms to do it, you do not end up with the real, but with more text, requiring just the same sort of critical reading as is given to television texts themselves. The direct reflections of participants in a process are *reflections,* already mediated, theorized, fictionalized, and selecting them for analysis is a further act of textual creativity, as any television documentarist will tell you. The return to audience studies might look like a welcome return to common sense after some rather bizarre theoretical excursions, but if it is, then it has already forgotten that common sense is an effect of texts, not a cause.

The Oxford English Dictionary derives "ethnic" from the Greek *ethnos,* "nation," making clear that its arrival in English was associated with biblical translations of the Hebrew "goyim," "the nations," that is, Gentiles. In fact, early lexicographers thought "ethnic" to be the same as "hethnic"—heathen. In other words, "ethnic" referred to non-Christian or non-Judaic nations; it meant heathen, pagan, goyim (first citation 1470). What, then, is "ethnography"? The OED cites this 1834 description: it is "the scientific description of nations or races of men, with their customs, habits, and points of difference;" it is an imperial discourse, rendering difference into knowledge, knowledge into control.

In the 150 years since then, ethnography in its home discipline of anthropology has changed, not least in its sensitivity to the politics of the "other," of the relationship between researcher and subject, researcher and knowledge, researcher and his or her own self—including *its* other, the "heathen or infidel" *within the self* (Fiske, 1990). How ethnography came to be applied to television audiences is of course a matter of institutional history, not etymology. But what worries me about the recent upsurge in the use of "ethnography" to describe how viewers may be brought into the purview of cultural studies is that history may be repeating itself as farce; a careless reinstatement of the old distinction between believer and heathen, between the imperial colonizer and other nations, but now in the ludicrous form of a distinction between TV researchers and TV viewers, as if they are not one and the same.

Ethnography must be founded on *distance from self;* the object of study must be "other" in order to be written. But ethnographies of audiences ought not to be *orientalisms,* in Edward Said's sense (1978), "heathen-ography" if you like, descriptions of the other, imagined by the research project that nevertheless pretends exemption from its own otherness. The otherness of the television audience has historically taken many forms, not only of ratings but also of deviant, dissident, disadvantaged, and politically disabled groups (children, women, working class men, adolescents, racial minorities). Social and psychological scientists tend to look for "others" who are in need of correction or protection—watching television causes violence and passive behavior (all at once!)—while oppositional intellectuals look for political resistance in "others" who will vindicate the intellectual stance of opposition.

Such tendencies indicate to me that there is a need to "decolonize the audience." But this means decolonizing *our* "other;" it does not mean looking for new ways to capture a purer, less partial conception of the audience, for example, by doing more sophisticated research into what the "natural" audience really does. Indeed, the "natural" audience of academic researchers comprises academic readers. Research is conducted along lines, and in pursuit of questions, that are agreeable to colleagues, editors, conference organizers, students. It is not disciplined by television audiences; what they think about it is normally neither here nor there.

Given that we (academic intellectuals) are what we study, I would like to find a different metaphor for power, one that is appropriate to television, while recognizing the public's capacity and willingness to don the garb of an audience and our ability to wear it for our own purposes in our own style.

If acting as audience means playing a role, then the fashion industry can supply a description of the costumes people are wearing for the part. Marshall Sahlins, in his analysis of American clothing, "*La Pensée Bourgoise:* Western Society as Culture," concludes with an insight into the function of the clothing system that is suggestive to my own project—the power of looking:

> "Mere appearance" must be one of the most important forms of symbolic statement in Western civilization. For it is by appearances that civilization turns the basic contradiction of its construction into a miracle of existence: a cohesive society of perfect strangers. But in the event, its cohesion depends on a *coherence*

> of specific kind; on the possibility of apprehending others, their social condition, and thereby their relation to oneself "on first glance" (1976: 217).

Sahlins claims that this kind of looking has logical precedence over rationality and indeed forms the base on which reason's superstructure is constructed:

> This dependence on the glance suggests the presence in the economic and social life of a logic completely foreign to the conventional "rationality." For rationality is time elapsed, a comparison: at least another glance beyond, and a weighing of the alternatives (1976: 203–204).

At this point I will direct only the most fleeting glance toward my own suggestion that the pleasure of television is akin to *frottage:* "a glimpse, a frisson of excitement provoked by taking private pleasure from public contact" (Hartley, 1990, chapter 12), the pleasure of watching television to keep "in touch" by means of "the brush, not with skin . . . but with clothing, surfaces, textures, furtive appropriations of the *look of otherness.*" . . . "Prime time presents Americans with the clothed body of America" (Hartey, 1987: 256).

Be that as it may, I am interested in pursuing the idea that "mere appearance" and "the glance" may be neglected or undervalued concepts in cultural studies. I am also interested in the idea that fashion can help to identify some larger issues. In fact, I am searching for a metaphor that will unite two kinds of power, textual power and social power, in the name of a "comprehensive theory of the audience" (which of course was the theme of the conference that gave rise to this volume). But rather than searching through time-elapsed, weighty textual and social theory for my rationale, I turn to something immediately at hand as I write; a fashion feature in the weekend magazine of *The Australian* newspaper (10–11 March 1990): "Each decade within reasonable memory has had a distinguishing fashion feature: 1920s, the Little Black Dress; 1930s, the Bias Cut; 1940s, the New Look; 1950s, the A-Line; 1960s, the Mini; 1970s, the Big Look; 1980s; Power Dressing." It is tempting to match these changes in the history of looks with a history of looking. The 1930s of the Bias Cut was certainly also an era of biased looking, what with totalitarian propaganda, the Frankfurt School, and F. R. Leavis's journal *Scrutiny,* and I suppose the New Look/New Deal, A-Line/A-Bomb decades show some connection between high fashion and high politics. The New Look 1940s also marked the cultural (if not technical) debut of new-look TV, and the A-Line 1950s was the period when television viewing was cut most simply; half the nation plugged into one show, ideology and domesticity harmoniously balanced while the eye was drawn uninterruptedly from hemline to head, or from suburbia to centrality, from *I Love Lucy* to "I like Ike." The 1960s, decade of the mini, was also the decade of the little screen's displacement of the big screen, when television, like miniskirts, used less but showed more, abandoned the high-definition artistic pretensions of cinema and flaunted the greater social (but not sexual) equality of its apparent freedom. The 1970s and 1980s coincide with questions of social and textual power very neatly. The era of the Big Look and Power Dressing in fashion—here's a useful metaphor for the politics of looking in the popular media. The Big Look and Althusserian Grand Theory of the 1970s gave way in the 1980s to a new phase in the history of looking and a new type of audience participation in the forum of media citizenship. The Couch Potato was wearing shoulder pads and going in for the ocular equivalent of Power Dressing; suddenly, around the world, the new look was *Power Viewing.*

The *Weekend Australian*'s laudable acknowledgement of historical change leads it to predict that the 1990s will be "an era of fashion driven by consumers, not designers." If so, the prediction for 1990s television is that the fashion in looking will be for audiences to "drive" programming. But the lines have softened and the look is fragmented; power viewing continues, but the shoulder pads are giving way to their bodily simulacrum—power bodies decked in sportive fashion, fashion decked in the prowess of sport.

Power dressing is in principle available to everyone, but as in fashion, so in looking; not everyone is dressed to kill. So who is wearing the new garb of power viewing—who is willing to be seen in public flaunting a style of looking that to other eyes appears ludicrous, even dangerous? Fashion-conscious people will often reserve their most powerful statements for clubs, which in turn are apt to restrict entrance to those who dress appropriately. Such places are public but private, open but restricted, village communities flickering to life here and there, part of but distinguished from the social terrain they both camp within and illuminate. Similarly, power viewers will not be distributed evenly throughout a given population; they often gather into clubs too, where they can flaunt their excessive obsessions in public but in safety. These clubs have a name. They're called universities.

Robert Park, writing in 1916, compares modern politics to primitive society; "we" are groups of insiders known to one another, scattered across an urban landscape, for whom the rest of the city, the outsiders, are not quite human, not quite alive. The relations between insiders is personal and feudal, based on fealty; the relations between outsiders is mediated and modern, based on publicity. In the latter groups and in the city, "fashion tends to take the place of custom, and public opinion … becomes the dominant force in social control." Hence, for Park, "the medium of the press, the pulpit, and other sources of popular enlightenment" are the disseminators of public opinion, which itself is a creation of agencies of publicity, not of the public *per se,* since "members of the public are not as a rule personally acquainted" (Park, 1969: 121–124).

So power viewing coexists with other kinds of looking, as insiders coexist with others, while all are connected via the communicative apparatus of fashion and the media. There are no doubt citizens who watch television in more traditional ways, uncritically and sympathetically, in cardigans and carpet slippers, and there are those who watch critically but unsympathetically—dressed like Margaret Thatcher and keeping a beady eye on television in order to find fault, finger poised for wagging. There are cross-cultural differences between cultures of ethnicity, nation, class, gender, age-group, and so on, and doubtless the way people look is differentiated along such lines too—though looking also transgresses those cultural divides. No doubt looking is not only characterized by historical change, but is also synchronically very complicated, displaying many modes at once. One piece of evidence for this is that in Mandarin the very word for television is—almost—"power viewing:" that is, "dianshi" or "electric looking." And there is evidence that power viewing is not confined to academics, but is already being reworked into a new generation's cultural politics:

> Some young people even self-consciously consider themselves as beyond postmodernism. Thus, one young local media activist from Amsterdam who calls himself an "illegal intellectual" says, shrugging his shoulders, "Postmodernism is for older people, for people who are still struggling with their pasts. We don't have that problem. We are sovereign" (Ang and Morley, 1989: 137).

However, before age or utopianism overtake me, I want to backtrack and show how the connections between Park's personal and mediated, feudal and modern communities might be thought of, not only in terms of fashion and publicity, but also in terms of power viewing. For insiders like legal and illegal intellectuals, sovereignty is achieved within their own communities, and their textual productivity may be mediated more widely by the "sources of public enlightenment." These sources, television most of all, are taken to be forces of social control, especially by those who claim sovereign exemption from their sway. But what is the situation for the general audience, the public that is "not personally acquainted?" How is Sahlins's "miracle" of a "cohesive society of perfect strangers" achieved?

I think one answer to that is hinted at in Sahlins's own observation of the power of "mere appearance," the power of the look, of the glance, of the "first glance," which is *constitutive* of social cohesion, of social totality in a particular individual. Such looking, glancing at appearances, is commonly taken to be an attribute of television viewing; the unengaged glance, the distracted look, the uncritical apprehension of an unordered surface. John Ellis (1982), for instance, makes much of a distinction between television and cinema, which he locates in the difference between glancing and gazing (where gazing is intense looking). I do not think such a distinction is valid to separate television from cinema. Indeed I think Ellis's position is just one more example of the pervasive binarism of Western intellectual thought, which constantly reinvents a Veblenesque distinction between the *honorific* and the *serviceable* (Velben, 1953)—between prestigious cinema and humble television. But I do think glancing and gazing are different modes of looking—both of which can be applied to television.

Where gazing may be favored as honorific by critics schooled in rationality (the second glance), glancing—the apprehension of social totality through appearance—seems to me to be the serviceable version of power viewing. But, like Veblen and Sahlins, I am impressed by social productivity as well as individual skill, and the power of the glance is not that of individual skill but of anthropological pervasion. Its analogy in natural attributes is the "power of speech;" its analogy in the historical world of material production is fashion. Furthermore, even in Western political mythology, glancing is the very cornerstone of politics.

Aristotle is definite about the fundamental importance of the agora to politics and to the state, concluding that the optimum size of a state is limited by technologies of looking: "Clearly, then, the best limit of the population of a state is the largest number which suffices for the purposes of life, and can be taken in at a single view" (*Politics* vii. 4–14; cited in Wirth, 1969). The Aristotelian "single view," where citizens can judge each others' merits "at a (Sahlinsesque) glance," is an apt description of contemporary politics on television. However, given that television is literally a technology for *distant* looking (which is what the word means), it is clear that such mediated appearances are not the same as a jostling crowd in the agora. But before we dismiss the idea that a marketplace like television cannot be used for politics, and for philosophy too, it is worth noting where our (i.e., Western) philosophy, as well as our political terminology, came from.

According to the political philosopher Thomas Hobbes, writing in 1651, the original occasion for philosophy was leisure: "*Leasure* is the mother of *Philosophy*" (1968: 683). Philosophy flourished as a form of conspicuous consumption, a leisure entertainment for the underemployed youth of a city got wealthy by conquest and dominion. "They that had no employment, neither at home, nor abroad, had little else to employ themselves in, but either . . . *in telling and hearing news*, or in discoursing of *Philosophy* publiquely to the youth of the City." Says Hobbes:

> From this it was, that the place where any of them taught, and disputed, was called *Schola*, which in their Tongue signifieth *Leasure;* and their Disputations, *Diatribæ*, that is to say, *Passing of the time* (p. 684).

It is worth noting in passing that the OED gives as its first historical instance of the word "agora" in English this highly pejorative view from the year 1800: "The agora or forum was the resort of all the idle and profligate in Athens." It seems that the agora-forum was not anglicized as a respectable concept (it is remembered as Aristophanian, not Aristotelian)—small wonder, as the root causes of Western philosophy turn out to be leisure, underemployment, and what Hobbes calls *loytering and prating* (p. 685), not to mention attending to the news. Furthermore, philosophy was from the beginning pedagogic, dedicated to "passing of the time" in the pursuit of teaching through discourse.

I mention all this not only because I think television deserves a better reputation—and what better than a classical pedigree—but also because I am aware that my emphasis on power *viewing* has ignored television's propensity for *talk*. The model supplied by Hobbes and Aristotle suggests that power viewing is not just personal display but also dialogic, and that talking, teaching, and watching are the attributes of today's "idle and profligate" medium of television. If we give it a second glance, may we see in television the beginnings of some new philosophy, some new politics—the miracle of social cohesion in the era of an urban public whose members are "not personally acquainted" but span the world?

The audience as tribal other, or as some kind of god. If you think these are far-fetched metaphors, I should point out that they are not of my own invention. Social scientists and television providers (producers, executives, presenters) are not known for their flights of fancy; on the contrary, they pride themselves on prosaic, no-nonsense understandings of the world and often espouse some pretty uncomplimentary theories about viewers. But within the industry and among its observers there circulates a view of the television audience that amounts to a secularization of *medieval* conceptions of God. That is, for those whose livelihoods depend on its power, the television audience is ubiquitous, unknowable, omnipotent, mysterious, capricious, benign, and cruel. It needs constant propitiation, endless offerings which it may or may not deign to accept.

Furthermore, monastic sects have arisen that are dedicated to the contemplation of this mighty but elusive being, erecting a fantastic edifice of writings, knowledges, methodologies, and metaphors to encompass its myriad manifestations. These learned but unworldly clerics are withdrawn from the contagions of everyday life behind the walls of secluded university campuses. They are divided into various orders, like their medieval counterparts the Dominicans, Franciscans, or black Benedictines; and as with them, there is intense rivalry between and within the sects of psychologists, social scientists, political economists, and culturalists, not to mention ratings agencies, public opinion pollsters, market researchers, and media analysts. Charges of heresy are common. Each sect claims privileged knowledge of and access to the only thing that unites them, namely their belief in the existence of something that can never be observed directly, their faith in a being that is never present but everpresent, pervasive but perverse.

I suggest that the television audience ought not to be imagined in terms of individuals with identities, experiences, motivations, or personalities, for in that direction lies the trap of deification, turning the audience into the ultimate "other." Instead, I have characterized television viewing in terms of "power viewing," invoking fashion and the power of the glance—the capacity of our species to apprehend social totality through mere appearance at a single look. Power in this context is not conceived as moral or political power—the power to select authoritatively or correctly what has been judged good or right. Power is not dominion over others. Power is not even sovereignty within a system of differentiated domains. In a cultural theory of television, power is pervasion—the power of the species to do what the species is capable of doing, the power of Veblen's "matter-of-fact knowledge of mankind and of everyday life" (1953: 252).

My analogy with the fashion industry and my emphasis on the anthropology of looking does lead to a cautionary note for those who would undertake audience research. In the case of fashion, attention is focused on the clothes, not exclusively on the model. The human body is essential to fashion, as it is to television, but fashion is strictly speaking exosomatic—outside the body. Here it differs radically from most audience research, which is dedicated to looking *inside* the body, interpreting inner states from outward signs, whether it uses electrodes and galvanometers or ethnography and participant observation to do it. What would you make of a study of the fashion industry that took off everyone's clothes and tried to work out the meanings of fashion by gazing intently at the models' naked bodies? It is my feeling that audience research is somewhat too interested in what is going on beneath the surface. My own modest proposal is that the television audience ought not to be treated like some naked, tribal, fetishized body, but rather that it should be imagined as the bearer of fashioned meanings that, like clothes, are at once public and private, personal and economic, real and rhetorical.

Finally, like fashion, "power viewing" can also be judged on its merits. Those who understand clothes will know at a glance whether a particular ensemble is good or bad, taking into account who is wearing it, on what occasion, for what purpose. Television audiences are likewise more or less successful in their appropriation of the textual and social power of television, and once they are understood as wearers of meaning rather than as natives, "others," or gods, the proper business of television studies can begin—teaching the public (our other self) how to get the most out of the available resources; where to find, how to fit, and when to wear the most appropriate outfit.

While television audiences are understood as persons, they will paradoxically be imagined as both sub- and superhuman (tribal others and god), under- and overvalued at once in the classic Levi-Straussian manner. The alternative is to develop a theory of television audiences that is based on the anthropological pervasion of the human species' power to make sense "at first glance," and to "read" television audiences within a history of looking based on fashion: Power Viewing.

References

Ang, Ien, and David Morley (1989). Mayonnaise culture and other European follies. *Cultural Studies* 3(2): 133–144.

Brown, Nick (1987). The political economy of the television (super) text. In Horace Newcomb (ed.), *Television: The Critical View,* 4th ed., pp. 585–599. New York: Oxford University Press.

Derrida, Jacques (1987). *The Postcard: From Socrates to Freud and Beyond.* Chicago: University of Chicago Press.

Ellis, John (1982). *Visible Fictions: Cinema, Television, Video.* London: Routledge & Kegan Paul.

Fiske, John (1990). Ethnosemiotics: Some personal and theoretical reflections. *Cultural Studies* 4(1): 85–99.

Foucault, Michel (1977). *Discipline and Punish: The Birth of the Prison.* Translated by Alan Sheridan. Harmondsworth: Penguin.

Fuller, Buckminster R., with Jerome Agel and Quentin Fiore (1970). *I Seem to Be a Verb.* New York: Bantam Books.

Hartley, John (1987). Been there—done that: On academic tourism. *Communication Research* 14(2): 251–261.

———. (1992). *Tele-ology: Studies in Television.* London and New York: Routledge.

Hobbes, Thomas (1968). *Leviathan.* Edited by C. B. Macpherson. Harmondsworth: Penguin (originally published 1651).

Marx, Karl (1961). *Economic and Philosophical Manuscripts of 1884 [The Paris Manuscripts].* Moscow: Foreign Languages Publishing Service (originally published 1844).

Park, Robert (1969). The city: Suggestions for the investigation of human behavior in the urban environment. In Richard Sennett (ed.), *Classic Essays on the Culture of Cities,* pp. 91–130. Englewood Cliffs, N.J.: Prentice Hall. Originally published in *American Journal of Sociology,* vol. 20, 1916.

Popper, Karl (1983). Knowledge: Subjective versus objective. In David Miller (ed.), *A Pocket Popper,* pp. 58–77. London: Fontana (originally published 1967).

Sahlins, Marshall (1976). *Culture and Practical Reason.* Chicago: University of Chicago Press.

Said, Edward (1978). *Orientalism.* London: Routledge & Kegan Paul.

Veblen, Thorstein (1953). *The Theory of the Leisure Class: An Economic Study of Institutions.* New York: New American Library (originally published 1899).

Wirth, Louis (1969). Urbanism as a way of life. In Richard Sennett (ed.), *Classic Essays on the Culture of Cities,* pp. 143–164. Englewood Cliffs, N.J.: Prentice Hall (originally published 1938).

The Hegemony of "Specificity" and the Impasse in Audience Research: Cultural Studies and the Problem of Ethnography

15

Janice Radway

In an astonishing poem entitled "Portrait (2)" published in 1940, the American poet Kenneth Fearing tellingly and presciently limned the profile of an individual now easily recognizable to us as the postmodern subject. "The clear brown eyes, kindly and alert, with 12–20 vision," he wrote, "give confident regard to the passing world through R. K. Lampert & Company lenses framed in gold; His soul, however, is all his own: Arndt Brothers necktie and hat (with feather) supply a touch of youth." "Love's ravages," Fearing continued, "have been repaired (it was a text book case) by Drs. Schultz, Lightner, Mannheim, and Goode."

> While all of it is enclosed in excellent tweed,
> with Mr. Baumer's personal attention to the
> shoulders and waist;
> All of it now roving, chatting amiably through
> space in a Plymouth 6
> With his soul (his own) at peace, soothed by
> Walter Lippman and sustained by Haig & Haig.
>
> *Collected Poems, 96–97.*

Fearing's poem is remarkable, it seems to me, for its early attentiveness to the way in which the limits of subjectivity were being reorganized by the intrusiveness *and* the creativity of the commodity in mid-twentieth century America. Not merely an appendage or an accoutrement of a stable, autonomous self, the commodities in Fearing's poem literally constitute the individual. But that individual is not thereby simply reified into a depthless object, as some critics of commodification would have it, but rather appears to us in Fearing's poem as a subject whose ineradicable emotional and moral life is as thoroughly colonized as his body is costumed. The poem therefore explicitly addresses the theoretical question that has since troubled virtually all critics of the culture of consumption—that is, it raises the question of the fate of individuality, subjectivity, and specificity in a world increasingly dominated by the production, consumption, and use of mass-produced objects, images, and representations. Simply put, the debate was, and still is, over whether the adoption of identical, mass-produced things, along with the adaptation of mass-produced emotions and thoughts through the use of popular narratives and musics, constitutes subjects as repetitive, conformist clones of each other, clones without significant intentionality; or whether that process is somehow mediated by prior historical constructions of a subjectivity that yet can affect its own negotiations with the world. What is finally at stake, of course, in this debate over the character and extent of the commodity's power to determine cultural life is the possibility of resistance and refusal, the fate of political opposition.

For audience researchers, there is no need to rehearse the full history of the mass culture debates. We are all familiar enough with the lineaments of the arguments mounted variously by the members and heirs of the Frankfurt School, by the liberal proponents of the "uses and gratifications" model of media use, and by the students of Althusserian structuralism. My concern here is with the place and fate of what I will call "the cultural studies consensus" in this debate over the consequences of mass production and consumption in an ever more integrated world economy. Histories and assessments of cultural studies have appeared virtually everywhere and with increasing frequency in recent years. Beginning perhaps with Stuart Hall's early and important account published in 1981, "Cultural Studies: Two Paradigms," this academic trend has swelled recently with the publication of Richard Johnson's "What Is Cultural Studies Anyway?" (1986/1987), Meaghan Morris's "Banality in Cultural Studies" (1988), Lawrence Grossberg's "Formations of Cultural Studies: An American in Birmingham" (1989), and Michael Denning's most recent "The End of Mass Culture" (1990). While each of these articles narrates a somewhat different history and maps a slightly divergent intellectual geography, all associate the nature of the cultural studies intervention in the mass culture debates with certain temporal and theoretical developments. Audience research, ethnography, and the theoretical contributions of Antonio Gramsci figure centrally in virtually all of the narratives.

Stuart Hall first acknowledged the debt to Gramsci in his effort to assess the relative contributions made by the culturalist and structuralist paradigms to the enterprise of cultural studies. Identifying the strengths of the culturalist approach with its interest in experience, subjectivity, and consciousness, he noted that culturalism has "insisted, correctly, on the affirmative moment of the development of conscious struggle and organization as a necessary element in the analysis of history, ideology and consciousness. (1981: 33)." With respect to Fearing's poetic subject, the culturalist might ask why, and to what effect, this man actively chose Plymouth 6, Walter Lippman, and Drs. Schultz, Lightner, Mannheim, and Goode as representations to figure his intentional self. Structuralism, on the other hand, according to Hall, has been useful for its emphasis on "determinate conditions" and therefore for its ability to conceptualize a cultural totality or whole that operates at least partially by producing an efficacious ideology that both positions and defines the subject. In Fearing's intrepid consumer, the structuralist would detect only the determined choices of the commodified, fully processed, "mass man" of the capitalist economy.

In a carefully reasoned argument that I cannot do justice to here, Hall characterizes the resulting distinction between the culturalist's intentional and conscious self and the more fully determined, structuralist subject as a "false polarization." He credits Gramsci with the theoretical breakthrough that might produce a more properly dialectical account of the historical process. Gramsci, he notes, "has provided us with a set of more refined terms through which to link the largely 'unconscious' and given cultural categories of 'common sense' with the formation of more active and organic ideologies, which have the capacity to intervene in the ground of common sense and popular traditions and, through such interventions, to organize masses of men and women (1981: 33)." Concluding his assessment of the field, Hall then identifies the core problem of cultural studies as that of "trying to think both the specificity of different practices and the forms of the articulated unity they constitute (p. 37)." What is finally at issue, he observes, is "the question of the relation between the logic of thinking and the 'logic' of historical process (p. 37)."

In the years between the publication of Hall's early metacritical assessment and Richard Johnson's subsequent 1986 evaluation, a large body of new work was undertaken as the methodological response to this theoretical dilemma. Much, though not all, of it was begun under the auspices of the Birmingham Centre for Contemporary Cultural Studies. Ethnographic in approach and highly localized in focus, the work of people like Paul Willis, David Morley, Charlotte Brunsdon, Angela McRobbie, Dorothy Hobson, James Lull, Ien Ang, and many others (and I need to include my own work here as well), attempted to address the question of how historical subjects actively engaged with the mass-produced representations available to them. At the same time, this work made an effort to ask whether media consumers were determined in their response to mass-produced significations by the character of their formal properties, or whether those consumers could make those representations into something more specific that they themselves could use.

Generally well-received and widely discussed within cultural studies—although not without criticism—this work has tended to produce at least a temporary consensus. This consensus prompted Johnson to write confidently that "*our* project is to abstract, describe and reconstitute *in concrete studies* the social forms through which human beings 'live,' become conscious, [and] sustain themselves subjectively" (1986/ 1987: 45; emphasis added). Further stressing the theoretical and political preeminence of "the concrete" throughout his account, Johnson pointed to ethnography as a privileged form of analysis because it could trace out the connections "between lived cultural ensembles and public forms (p. 72)." The best ethnographies, Johnson concluded, unite public and private forms, attending at once to structural determination and to private appropriation.

I have made Johnson sound more naive and optimistic here than is really fair, for despite his faith in ethnography, he also incisively noted that in 1986 cultural studies still lacked an adequate post-poststructuralist theory of subjectivity. What was missing from even the most sophisticated ethnographic work, he argued, was a satisfying theoretical account of the "*genesis* of subjective forms and the different ways in which human beings inhabit them (p. 63)." Referring here to the tendency to adapt highly abstract and ahistorical models of the subject from psychoanalysis in order to account for the readings of mass mediated forms produced by historical subjects, Johnson was actually calling for greater specificity and concreteness rather than for more theoretical abstraction. What he missed in cultural studies was a rich and complex understanding of the different, multiple, ever-changing configurations of subjectivity dialectically produced through the negotiation between historically produced individuals and material, social and discursive contexts. I will return to Johnson's question about the problem of subjectivity at the end of this chapter because I believe that ethnographic work in the cultural studies tradition has not yet grappled fully with the complexities of subject-formation in the commodified, postmodern world.

Whatever Johnson's reservations about cultural studies' still inadequate realization of the concrete in the early 1980s, it does at least seem clear that as a paradigm in its own right, cultural studies has become increasingly identified with the pursuit of lived specificity and with the search for an intentionality buried within the interstices of a machine-like system of production and consumption. Indeed, I am certain that most of us in the field could produce a compassionate and convincing cultural studies—type account of the struggles of Fearing's imaginary Plymouth consumer. This troubling prospect, in fact, prompted Meaghan Morris's acid comment in 1988 that "Sometimes, reading magazines like *New Socialist* or *Marxism Today* . . . , flipping through *Cultural Studies,* or scanning the pop-theory pile in the bookshop, I get the feeling that somewhere in some English publisher's vault there is a master-disk from which thousands of versions of the same article about pleasure, resistance, and the politics of consumption are being run off under different names with minor variations (1988: 15)." All of those articles, she observes, end with a restatement of the fundamental, enabling theses of cultural studies, those articulated earlier by Mica Nava in an important article, "Consumerism and Its Contradictions." Among those theses are the propositions that consumers are not cultural dopes; that consumption practices cannot be derived from or reduced to a mirror of production; that the act of consumption is more than an economic activity; and that consuming is "also about dreams and consolation, communication and confrontation, image and identity (Morris, 1988: 16)."

Morris's criticisms of audience-based cultural studies scholarship are significant and troubling. Not only does she demonstrate fairly effectively that such work can be seen as repetitive and formulaic, but she also manages to make her critique even more pointed by raising the more disturbing metacritical question of why such localized studies of the heroism of the ordinary and the creativity of the everyday have suddenly proliferated. Because I think that there *is* a danger in what Morris calls cultural studies' "voxpop style," which offers us, according to her, "the sanitized world of a deodorant commercial where there's always a way to redemption," I think it worth engaging with her critique in serious and extended fashion. Therefore, in the remainder of this chapter, I would like to address a few related questions prompted by Morris's indictment: what exactly *is* accomplished by the repetitive generation of locally specific articles demonstrating virtually identical theoretical propositions about the creative ability of varied postmodern subjects to struggle with and against historical conditions not of their own making? Has the historical moment passed in which theoretical propositions about the necessity to attend to the lived and the concrete might do their most progressive political work? Do we now need to go beyond ethnography and audience research in order to forge yet another paradigm capable of fending off the Pollyanna-like complacency vigorously nurtured by the recuperative powers of capitalist logic?

Although I will eventually argue that cultural studies cannot yet afford to give up on the pursuit of specificity, nor abandon ethnography entirely, I do not simply want to necessitate and enable one more repetition of the familiar propositions in the face of the familiar "discouraging Other," in this case Morris herself. Indeed, Morris observes astutely that this projection of the "mis-understanding Other," in the guise of "grumpy feminists and cranky leftists," has been used repeatedly in cultural studies to fend off irritating criticisms of the way in which the paradigm has chosen to speak on behalf of the popular. Thus I want to acknowledge the validity of certain of Morris's complaints in the hope of clarifying some of the risks *and* advantages of this localized approach to the possibilities of resistance in an increasingly gridlocked capitalist system.

To begin with, I believe Morris's analysis of the enunciative practices of ethnographic audience studies is largely correct. Despite the best of intentions to let popular voices speak, these studies (and I would include my own here) inadvertently construct "the people" as other to the critic precisely as a way of authorizing the critic's own pronouncements. As Morris observes, "[the people] are not simply the cultural student's object of study, and his native informants. The people are also the textually delegated, allegorical emblem of the critic's own activity. Their *ethnos* may be constructed as other, but it is the ethnographer's mask" (1988: 17). In endlessly dramatizing the people's "indomitable capacity to 'negotiate' readings, generate new interpretations, and remake the material of culture (p. 17)," the cultural studies scholar is also narcissistically offering herself and her own work up as a fuller realization of political opposition, constituted therein somewhat narrowly and very literally, as "oppositional reading."

Morris's charges here are really two, and both bear upon the uneasy position of the intellectual. I would like to take up each separately. To begin with, her critique of the enunciative practices of ethnography builds on a version of the familiar observation in theoretical anthropology about the limitations of the hermeneutic circle. That is to say, whatever information the ethnographer gathers in the field, he or she *must* construe that information precisely as data through an unavoidable interpretive process if he or she is not simply to repeat her informant's speech. The ethnographer's propositions, as Clifford Geertz has observed, are nothing more than interpretations, and, importantly, interpretations from *outside* the community being studied. It simply will not do, then, to stage the presence of the people within ethnographic writing as if such a presence circumvented all of the problems attendant upon the act of speaking as an intellectual and a scholar. What is embodied in ethnographic speech is, above all, a relation, and it is a relation that embodies a certain amount of transference on the part of the intellectual whose investment in the usually subordinated other is by no means simple.

I believe Morris is right to suggest, then, that what is needed to guard against the presentation of this hermeneutic circularity as something other than what it is, is greater theoretical self-consciousness about the way ethnographic relations are constructed and do their political work. And one of the first things we need to recognize about the ethnographic relation is that it is still largely initiated by an intellectual with specific cultural authority and power and that that relation is established precisely to generate knowledge whose first effects will be in the cultural or, more specifically, the academic realm. Whatever other effects the production of this knowledge generates, it will at least prompt discussion and evaluation of the scholar him- or herself, potentially leading to further credentialling and authorization. Self-interest, therefore, is a crucial part of the ethnographic relation.

Furthermore, although it is not the case that intellectuals have nothing in common with those they wish to study, they are often significantly different from them by virtue of their position as legitimated authorities. Thus, unless one wants to deny entirely that there is some validity to the acquisition of knowledge, or desires to abandon one's position as a professional writer and teacher, I see no way around the fact that the intellectual must always speak precisely as an intellectual, that is, as one implicitly regarded by some others as one with special claims to knowledge and authority. This is both the fate and the responsibility of the intellectual, and to deny either through ventriloquy, that is, by speaking as if through the mouths of one's informants, is to act, however unintentionally, in classic bad faith.

To say this, however, is to raise another troublesome question. If the ethnographer cannot avoid interpretation and therefore must speak always as an intellectual, why bother to engage with informants at all? Does not the absorption of another's speech into academic discourse fundamentally transform that speech into another register, thereby jeopardizing its status as evidence of an "other" reality? In such a case, does not academic language throw up a "*cordon sanitaire,*" in the words of Andrew Ross, around potentially competing and disruptive knowledges, making them safe for consumption? My perhaps unsatisfying answer is that it could, but it does not have to.

This process of containment can be at least partly thwarted, I think, by the adoption of two different writing practices. It is possible, for instance, through the use of explicitly literary characterization and careful narrative technique, to foreground the fact that a hierarchical *social* relation is the source of the offered interpretations. Thus one can attempt to characterize as fully as possible, through the use of personification and description, both one's informants and one's self as parties to an actual social relationship. At the same time, one can painstakingly transcribe and highlight the clash and conflict of different languages, the language of the informants and that employed by the ethnographer. These techniques can be helpful because they can fundamentally complicate the process of identification set into motion for any reader through the activity of reading itself. Readers situated differently in the social formation and constituted differentially by the possession of competing knowledges may well identify differently with different actors in the narrative, that is, with the ethnographer on the one hand, or with the informants on the other. By foregrounding conflict, the ethnographer might more effectively invite judgment, evaluation, even disagreement with his or her self-characterization and interpretations. As recent literary theory has suggested again and again, the narrator is not always in complete control, and this fact can be self-consciously foregrounded in a variety of ways.

I have been prompted to these observations by watching my own students engage with different ethnographic accounts of media audiences. The most interesting, exciting, and angry discussions are provoked by those studies that work harder to foreground the social and discursive clashes and conflicts at the center of ethnographic research. Political insights and epiphanies most often occur when students in the class identify differently with the actors in the narratives, when they *feel* the difference between the competing languages used to construct a single set of phenomena. Depending on the issues at stake, some tend to identify with the academic analyst, others with those individuals mobilized by the intellectual for his or her own purposes. Inevitably, this prompts reflection not only about the specific subjects and topics taken up in the particular studies in question, but more generally about the competing positions and statuses of difference knowledges—the everyday, common sense, and the legitimated formulations of academic discourse. When this occurs, it seems to me, localized ethnographies are doing extremely important and highly generalizable political work.

I want to foreground the place of ethnographic audience research in the pedagogical situation, therefore, in order to counter the familiar tendency among metacritics to privilege the larger, more abstract arenas in which the intellectual operates. Indeed, in criticizing cultural studies, Meaghan Morris makes no comment about how such work succeeds or does not in the classroom. While this is perhaps understandable from someone whose own principal identity has been constructed as a film critic and freelance writer, it seems inexcusable to me that few commentators on cultural studies or ethnography have addressed either as political agents in the classroom. To my knowledge, only Lawrence Grossberg and Henry Giroux and Roger Simon have addressed themselves to the full range of issues involved.

My own feeling is that we cannot afford to give up on highly specific and localized ethnographic studies of marginalized populations, precisely because they are extremely effective tools in the academic classroom. Because they can be used to foreground the conflict between academic and legitimated knowledges on the one hand, and those subordinate discourses used by others who are not generally permitted to speak in school on the other, ethnographies can be used to prompt discussion and questioning of the power distribution that prompts the distinction between the legitimate and the vulgar in the first place. They can be used, then, to highlight the ineradicable power of the intellectual as writer and teacher and to empower students to question more rigorously and self-consciously the very purpose and function of the process of schooling within which they are caught up. In a culture increasingly driven by the knowledge, authority, and power of the expert, this is not an insignificant political act.

But in focusing on the larger political possibilities opened up by the use of ethnography in the classroom, I do not want to discount the significance of the more specific and localized issues they address. It seems to me that Morris leaps too quickly to the admittedly similar theoretical conclusions demonstrated by so many cultural studies ethnographies of media consumers. In doing so, she thereby surreptitiously engages in the familiar devaluation of the local and the concrete in favor of the typical academic valorization of the abstract and the theoretical. To acknowledge that much of this work concludes by demonstrating a similar set of theoretical propositions does not necessarily mean that the material marshalled to demonstrate those points is *only* useful for the larger theoretical service it performs. It seems to me that localized ethnographies of media audiences, such as bikers, heavy metal fans, romance readers, Trekkers, soap opera watchers, and sci-fi addicts, address immediate and very real issues in the lives of our students. I do not think we can discount the electrifying effect an account of lesbians' interest in Madonna might have on a woman facing the difficulties of coming out in a college environment. Nor can we ignore the disconcerting sense of bewilderment produced in a young, white, male fan of heavy metal music upon reading of the gender and race politics encoded in the music. I am not suggesting that these self-confrontations necessarily produce political mobilization, only that they very often lead to both questioning and self-doubt, empowerment and self-validation. This, it seems to me, is what a politicized pedagogy is all about.

I want to suggest, therefore, that the moment has not yet passed in which localized ethnographic studies of the concrete and the everyday can do important political work. Indeed it seems to me that even narrowly focused ethnographies can serve as extremely useful mediators between the immediate, highly personalized concerns of our students and the more abstract theoretical questions they engage, such as those concerning the construction of race and gender, the consequences of commodification, and the effects of the legitimization of some knowledges in opposition to others. Thus, I think it urgent that we remember, in the process of making decisions about what groups and practices to study, that the largest and most predictable audience for the material we generate is not composed of our professional peers. Rather, it is made up of young men and women, aged roughly eighteen to twenty-two, ranged before us in the classroom, seeking not only professional, middle-class validation, but often guidance and reassurance about the appropriate emotional, moral, and political standpoint to take with respect to a confusing and oppressive world. I believe we should engage groups and issues in our research that will both prompt and capitalize on the identifications of our students, identifications that will be different and thus potentially at odds with one another and with our own. If we are lucky, disruption and discord will be the happy result.

The success of this process, however, will not turn simply on our ability to find subjects that will engage the interests of our students. It will depend even more heavily on the composition of our classrooms themselves. If those classrooms continue to bleach out, as they have in recent years with the declining admission of some minorities, if they continue to become ever more expensive with every passing semester, then we will increasingly face only our mirror-image as we talk. The possibilities for disruptive engagement, then, for constructive difference over the appropriate way to make sense of the world, will be significantly reduced. The result may be an even more perfect reproduction of the political status quo.

I raise this issue here in order to return to Meaghan Morris's other criticism of the structure of ethnographically based cultural studies, her observation that such work tends to hold up the creative and resistant reading of the ethnographer him- or herself as an adequate model for political activity and opposition. This is the result, it seems to me, of insufficient reflection upon the special conditions and limitations of intellectual activity. Although the activity of *reading* cultural discourses and practices "otherwise," or "against the grain," may look highly significant to an intellectual whose principal role is precisely *as* a professional reader, I think we mistake that activity as a sufficient cause of political change at our peril. It is not enough to demonstrate that alternate readings of commodified objects, narratives, and musics are possible. It is necessary also to provoke such readings and then to ensure that they connect with many other strategies in a social realm overdetermined by a vast number of interlocking practices producing enormous power differentials. What difference does it make if white teenagers purchase endless hip-hop and rap CDs as a way of empowering themselves and refusing bourgeois propriety if such cultural miscegenation leaves the fundamental infrastructure of racism largely intact? The activity of reading alternatively must connect with other actions.

Thus, it is not even enough for us to engage in oppositional readings within our classrooms, to content ourselves with the politics of reading. If we are to preserve, let alone enlarge, the possibilities for the exploration of local, concrete, and specific differences, then we must involve ourselves as well in the politics of admission and, even more broadly, in the politics of educational funding. What is at issue, here, is the business of articulation, that is, the task of hitching interest to interest, group to group, practice to practice. In its efforts to preserve and encourage opposition in a frighteningly windless system of increasingly global capitalism, cultural studies cannot afford to overestimate its efficacy or that of the people with whom it finds common cause. To do so is to risk inadvertently collaborating with the dominant system's need to cordon off opposition into safe and tightly contained arenas.

Having made a case for the continued political viability of ethnographies and audience research, I would like to return briefly to the subject of Fearing's consumer in order to say something about the adequacy of the ethnographic method to the task of understanding the extreme complexity of the postmodern subject. Despite my continued commitment to ethnography, I also believe we risk languishing at an impasse in audience research if we simply reproduce exactly the same sort of studies we have generated in the past. I am thinking here of studies like my own, studies that are genre-based despite their interest in audiences. Such studies are also predicated upon certain assumptions about the coherence of the reader, viewer, or consumer and, therefore, concomitantly upon certain assumptions about the identifiability and stability of audiences as aggregates of individuals. My concern is that such studies are still too closely tied to a residual literary formalism and to a kind of simple humanism to come to terms with the fluid and shifting processes of subject-formation in a postmodern world.

What I am getting at here is the fact that a significant portion of audience-based research begins with specific literary, cinematic, televisual, and musical forms, that is, with romances, horror movies, soap operas, television news, heavy metal, or punk. Sometimes, of course, this research is even more narrowly focused on individual forms like *Dallas* or *The Color Purple.* Such a research design is not surprising, given that our own professional lives have been organized by disciplinary divisions that began with the discrimination of forms and media. Thus we are trained to deal *either* with printed texts, with filmic and televisual forms, *or* with music. However, to import our confidence about the distinctiveness of media and genres to the study of their fans and consumers may well be to construct an unwarranted formalism of the audience, a formalism that renders invisible the complex ways in which media consumption is multiple, overlapping, contradictory, and stitched together at every turn with many other daily practices. To equate "women who read romances" with the category "romance readers" may well be to produce an algorithm that enables the easy formulation of a research project, but it may also fundamentally simplify the entangled set of activities whereby a woman is multiply, often contradictorily, constituted as a subject who is a mother, romance reader, feminist, *Thirty-something* fan, professional nurse, and volunteer at a battered women's shelter.

To construct her, then, as a "romance reader" may be to isolate only one small portion of her life and to mistake that part for the whole. Additionally, such construction may unfairly posit a stable subject, an individuality that exists prior to romance reading, a *woman* who reads romances. Hence, the familiar questions arise about the impact of romance reading upon a full and complete individual. This formulation is troublesome, however, because it reifies both text and reader, constructing them as self-contained, identifiable objects, objects whose connection is established via a relation of simple proximity. Consumption and use, then, are conceptualized on the basis of a phenomenal model adapted from Newtonian physics. Troped too crudely as a kind of corporeal or physical activity, whether as passive ingestion or creative labor upon raw materials, media use is thereby simplified as a linear process of reception and response.

But, as Fearing's poetic account suggests, and as many recent theorists of subject-construction have argued, such a model of consumption and media use may fail dismally at the task of capturing the intricate process whereby historical individuals are constructed *as subjects* by discourses embodied within practices and texts. Rather than impinging *on* individuals, discourses may in fact be dissolved *into* them, thereby functioning as the enabling condition of their existence at that moment, *as* a subject of a particular kind. The womanhood or femininity constructed through romance reading may well be at odds with the femininity constructed in the process of doing aerobics, watching *Roseanne,* or playing softball. But it is clearly not inconceivable that the same historical individual could engage in all of these and thereby be constructed quite differently at different moments in her life. The assumptions of coherence and continuity are just that—assumptions. As such, they need to be investigated as hypotheses rather than postulated as givens.

If the human subject is *not* unified organically or affected mechanistically by her interaction with her environment, how does she act and, more important for our purposes here, how does she change? These and other related questions, it seems to me, ought to generate new ethnographies, no longer animated by the desire to pin down the dynamic human subject in some lepidopteran attempt to still process and motion in order to facilitate categorization. Rather, a reworked ethnography ought to take the fluid *process* of articulation as its topic, that is, the process whereby the historical human subject is constructed through the linkage, clash, and confluence of many different discourses, practices, and activities. Such an ethnography would have to begin with the everyday, not with texts. It could start with historical individuals, but it would not assume that their unique individuality persisted throughout myriad activities unchanged. It would have to devise instruments designed to render comprehensible that osmotic process whereby individual and text, individual and practice dissolve into each other, thereby producing a distinct and recognizable *subjectivity.* Most important, it would try to understand how these multiple subjectivities interact within the individual, that is, how they affect and influence each other, thus prompting future intentions and actions. If we and our students could better understand the complexities of this fully colonized, postmodern subject, we might better be able to devise strategies for exploiting the dissonances, opening up the contradictions in the hope of promoting change.

Ethnography and Radical Contextualism in Audience Studies

16

Ien Ang

Our curiosity about the audience is never innocent. Specific interests and orientations, material and intellectual, generally shape the perspective from which we come to define our object of study, and the kinds of knowledge—its form and content, its scope and substance—we pursue. There is now clearly a sense of crisis in the study of media audiences; indeed, the conference that gave rise to this volume called ambiguously for theoretical "comprehensiveness," suggesting an awareness of a confusing lack thereof. The crisis is neither purely theoretical nor merely methodological (as misleadingly suggested in the oppositioning of quantitative and qualitative methods); rather, it is both deeply epistemological and thoroughly political. The current popularity of cultural studies approaches to the audience has not only produced considerable epistemological confusion over the status of the concept of "audience" as an analytical object, but has also reanimated the persistent critical preoccupation with the political standing of scholarship: What does it mean to do "audience research," and why do it in the first place?

In the past decade or more, the audience question has been especially acute in television studies. This is not only because the television audience has since the 1950s had the dubious privilege of being in researchers' spotlight, both within the industry and within the academy; the television audience has also become prototypical for the (real and imagined) "problem of the audience" that has risen to prominence in light of practical and theoretical concerns about the nexus of modernity, the media industries, and mass culture. More importantly, however, television's changing place in the late twentieth century has put our conventional understandings of the television audience under severe pressure. I would like to stress the notion of change here: we do live in a time of dramatic transformation of television's economic, institutional, technological, and textual arrangements. The eclipse of the national public service broadcasting systems in Western Europe, as well as the worldwide ascendancy of a multiplicity of transnational, commercially organized satellite channels, the proliferation of local and regional channels, and the ever more abundant availability of VCRs and other television-related technologies, have obviously thrown traditional models about television reception and consumption into disarray. This has been exacerbated by television's rising importance as a major actor in the enactment of global politics (as we saw, for example, in the Gulf Crisis), as well as by our growing theoretical awareness about television's specificity as a popular cultural form—its eclectic but repetitive narratives, its socially heterogeneous yet textually imposing modes of address, its stubborn always-thereness—which greatly challenge the validity of traditional, literary models of audiencehood, in which the discrete text/reader relationship forms the basic analytical focus. (If anything, viewing television today is often more like browsing than reading a book.) It seems to me that the crisis in audience studies should be understood in the context of this postmodern momentum of change.

It is often said, and often not without a sense of modernist nostalgia, that the television audience is becoming increasingly fragmented, individualized, dispersed, no longer addressable as a mass or as a single market, no longer comprehensible as a social entity collectively engaged and involved in a well-defined act of viewing. Indeed, television's proliferation has made it painfully clear that it does not make sense to speak about "television audience" as a neatly demarcated object of study. In my view, we should take this historical realization as an opportunity to finally mark out the productive end of the search for a "comprehensive theory of the audience," which has often been the implicit motif of the diverse paradigms of audience research within communication studies. Acknowledging the inevitably *partial* (in the sense of unfinished and incomplete) nature of our theorizing and research would arguably be a more enabling position from which to come to grips with the dynamic complexity and complex dynamics of media consumption practices.[1] In addition, a recognition of this sense of inexorable, *epistemological* partiality in the construction of knowledge would facilitate the foregrounding of the other, *political* meaning of being partial: the social and political importance of commitment, interestedness, and engagement in developing our understandings. I will return to the articulation of this double partiality in audience studies in a moment.

Recent culturalist audience studies are directly faced with the limits and limitations of comprehensiveness as an epistemological ideal. By "culturalist" audience studies I mean, very broadly, the kind of empirical and interpretive work that starts out from the recognition that media consumption is an ongoing set of popular cultural practices, whose significances and effectivities only take shape in the "complex and contradictory terrain, the multidimensional context, in which people live out their everyday lives" (Grossberg, 1988: 25). But how to turn this insight, this abstract hunch, into more concrete knowledge, more tangible understanding?

Most of us would agree that in order to do this we need to contextualize the media far more radically than we have done so far: we should stop conceptualizing television, radio, the press, and so on in isolation, as a series of separable independent variables having more or less clear-cut correlations with another set of dependent, audience variables. In the case of television, the consequences of this necessity of contextualization has been most resolutely problematized in David Morley and Roger Silverstone's research project carried out when they were affiliated with the Centre for Research into Innovation, Culture, and Technology (CRICT) at Brunel University in London.[2] It is not my intention to discuss this work substantially; instead, I will use it as a starting point to explore both the promises and the dilemmas, simultaneously epistemological and political, of what I would call "radical contextualism" in culturalist audience studies, and the significance of ethnography in this respect.

In their inclusive, almost totalizing vision, Morley and Silverstone state that television "has to be seen as embedded within a technical and consumer culture that is both domestic and national (and international), a culture that is at once private and public" (1990: 32). As a concrete starting point, Morley and Silverstone have decided to focus on two contextual concerns: television's place in the domestic context, and television's status as a technology. However, when these contextual concerns are pushed to their logical extremes, they inevitably lead to a fundamental shattering of the possibility of studying the television audience as a stable and meaningful psychological or sociological category.

First of all, the mundane fact that television is generally consumed at home (and not in a laboratory or a classroom) calls for the by no means new but still sobering observation that "the use of television cannot be separated from everything else that is going on around it" (1990: 35). That is, the activity so often simplistically described as "watching TV" only takes shape within the broader contextual horizon of a heterogeneous and indefinite range of domestic practices. As a result, the very notion of "watching TV" undergoes a dispersal: what the activity is—what it entails and what it means—cannot be predetermined, but depends on the influence of a plurality of interacting contexts. "Watching TV" is no more than a shorthand label for a wide variety of multidimensional behaviors and experiences implicated in the practice of television consumption. If this is the case, however, it becomes difficult to demarcate when we are and when we are not part of the television audience. In a sense, we are, as citizens, living in television-saturated modern societies, always inevitably incorporated in that category, even when we personally do not actually watch television very often. For example, even if we have never seen *Dallas* or *Murphy Brown* or have missed Saddam Hussein's television performance, we can hardly avoid being implicated in such television events through their general diffusion in the intricate networks of day-to-day social discourse (see also Jim Anderson's comments on what he calls the "emergent audience" in this volume.)

Considering television as a *technology*—rather than merely as a set of distinct messages or texts—only enhances the dispersal of "television audience" as a coherent category. The emphasis on television as technology enlarges the scope of what is generally known as the premise of the "active" audience. As a communications technology, television has what Morley and Silverstone call a double articulation: since it is both a set of hardware objects (i.e., the television set and connected technological items such as the VCR, the video camera, the computer, the remote control device, the satellite dish, the cable, and so on) and a vehicle for symbolic material, television creates an enormous open space for the ways in which it becomes integrated in the domestic flow of everyday life. This leads to a rather dizzying accumulation of the audience's meaning producing capacity. As Silverstone has put it:

> Television is potentially meaningful and therefore open to the constructive work of the consumer-viewer, both in terms of how it is used, or placed, in the household—in what rooms, where, associated with what other furniture or machines, the subject of what kinds of discourses inside and outside the home and in terms of how the meanings it makes available through the content of its programmes are in turn worked with by individuals and household groups who receive them (1990: 179).

Here, the scope of reception theory (which posits the indeterminacy of the meaning of the text outside of concrete viewer readings of it) is extended by applying the metaphor of textuality to the realm of technologies as well: Technologies, hardware, and material objects only take on meaning in and through their consumers' "readings" and uses of them. Television consumption, in short, is a meaning-producing cultural practice at two interdependent levels. Looking at television as a domestic technology implies for Morley and Silverstone a look at the television audience as "multiply embedded in a consumer culture in which technologies and messages are juxtaposed, both implicated in the creation of meaning, in the creative possibilities of everyday life" (1990: 51).

It is precisely the idea of profound embeddedness of television consumption (and of media consumption in general) in everyday life, and therefore its irreducible heterogeneity and dynamic complexity, that has been a central emphasis within culturalist audience studies, although the epistemological bearings of this emphasis, which amount to a form of *radical contextualism,* are not always thoroughly understood.

Of course it is true that the recognition of diversity in audience activity has been a major strand in the development of social-scientific audience research, ranging from uses and gratifications research to reception analysis, to observational work on television's social uses within the family. But many of these studies still seem to start out from a conceptualization of television itself as a given phenomenon with fixed features and intrinsic potentials, which can then be used or interpreted in different ways by different audience groups. From a radical contextualist perspective, however, television's meanings for audiences—textual, technological, psychological, social—cannot be decided upon outside of the multidimensional intersubjective networks in which the object is inserted and made to mean in concrete contextual settings.

For example, many research projects have been set up on the basis of the uninterrogated commonsense assumption that television is an "entertainment medium," implying that "entertainment" is not only an institutional or textual category but also a psychological need or preference, and that the two are more or less correlated in some functional fashion. If we take up the stance of radical contextualism, however, we must let go of such an ahistorical assumption of pregiven fixity of what television is, in the recognition that the meanings of television within the domestic realm only emerge within contextualized audience practices. That is, the precise "entertainment function" of television can only be determined ex post facto: Outside of specific articulations of television-audience relationships, we cannot meaningfully decide about "the entertainment value" of television. After all, the term "entertainment" itself can encompass a whole array of differential and shifting idiosyncratic meanings, depending upon the culturally specific ways in which social subjects experience and define as "entertaining" in any particular situation or historical setting.

To put it more generally, both "television" and "audience" are fundamentally indeterminate categories: it is impossible to list a priori which possible meanings and characteristics each category acquires in any specific situation in which people engage in television consumption. As a result of this contingency of meaning, the range of potential variety in audience practices and experiences becomes exponentially multiplied, indefinite, indeed, if not infinite. Which meanings are concretely actualized, however, remains undecided until we have caught the full, multicontextually determined situation in which historical instances of television consumption take place. From this perspective, what the audience researcher needs to do is to secure the "catch."

In his introduction to Morley's book, *Family Television,* Stuart Hall has pushed the notions of multiplicity, indeterminacy, and heterogeneity that underlie radical contextualism to their imaginative limits:

> We are all, in our heads, several different audiences at once, and can be constituted as such by different programmes. We have the capacity to deploy different levels and modes of attention, to mobilize different competencies in our viewing. At different times of the day, for different family members, different patterns of viewing have different "saliences." Here the monolithic conceptions of the viewer, the audience or of television itself have been displaced—one hopes forever—before the new emphasis on difference and variation. (1986: 10).

As is well known, this epistemological move toward radical contextualism in culturalist audience studies has been accompanied by a growing interest in *ethnography* as a mode of empirical inquiry. Ethnographically oriented research is considered the most suitable to unravel the minutiae of difference and variation as they manifest themselves in concrete, everyday instances of media consumption. What ethnographic work entails is a form of "methodological situationalism," underscoring the thoroughly situated, always context-bound ways in which people encounter, use, interpret, enjoy, think and talk about television and other media in everyday life. The understanding emerging from this kind of inquiry favors interpretive particularization over explanatory generalization, historical and local concreteness rather than formal abstraction, "thick" description of details rather than extensive but "thin" survey. But this ethnographic interest within audience studies is neither uncontroversial nor unproblematic. There is no need to go into the details of the controversy about ethnography here—many others have already done this—; suffice it to observe at this point that what is at stake in the problem of ethnography is not just its supposed lack of systematicity and generalizability (which is the conventional critique levelled against it), but also its potential political and theoretical relevance as a form of knowledge. In short, what's the point of the ethnographic rendering of media audiences? What is its politics?

That the drift toward "the ethnographic" is not merely a marginal academic matter but is also traceable in the belly of the beast itself, namely the commercial cultural and media industries, is exemplified by the crisis around ratings research, which is the most important and entrenched form of audience research circulating within the television industry.[3] This crisis gained momentum in the latter half of the eighties. Leaving aside the economic and institutional sides of the crisis, the controversy revolves around the alleged lack of "accuracy" of Nielsen's ratings figures, resulting in major discontent and antagonism in network and advertisers' circles. In response, a solution to the perceived imperfection of the current technology in use for measuring the television audience, the so-called people meter, is now being sought by the ratings firms in completely removing the subjective—and therefore allegedly unreliable—factor in present audience measurement procedures.[4] For example, the A. C. Nielsen Company is currently experimenting with a new technology, a so-called "passive people meter," which can identify the faces of those in the living room through electronic image recognition. Only faces directed towards the set will be counted as "watching." Clearly, this method approaches the utopian dream of perfect monitoring, by creating a simulacrum of unobtrusive naturalistic observation, of what's happening in the living room, so that there will no longer be any doubt about who is watching which channel, which program, which commercial, at any minute of the day.[5]

There is certainly an "ethnographic" flavor to this corporate initiative, insofar as being more empirically microscopic is envisaged as presenting an opportunity to improve measurement accuracy. More generally, marketing and advertising research circles are exhibiting increasing interest in qualitative and interpretive methods of gauging consumer behavior, in the conviction that more detailed and local knowledge is needed in order to make their strategies to attract, reach, and seduce the consumer more effective. In other words, even within market research the tenets of contextual variation and difference can be heard ever more frequently: There is talk of "recapturing (...) intimacy with the consumer" (Gold, 1988: 24) or getting in touch with "real persons" (Davis, 1986: 51).

But this industry flirtation with the particular and the qualitative that is also characteristic of the ethnographic moment in culturalist audience studies is intrinsically contradictory. Despite its increasing interest in more detailed information about consumers and audiences, market research must always stop short of fully embracing the theoretical consequences of the consistent radical contextualism that underpins the culturalist turn within academic communication theory and research. As I pointed out earlier, a radical contextualist perspective tends to lead to an unstoppable dispersal of the notion of "audience," to the point that it may become pointless to measure "it" (which is nevertheless an indispensable enterprise for an industry dependent for its functioning on determining the value of the audience commodity). For example, the recognition of the fact that the consumption and use of television is a multicontextually articulated, indeterminate, and overdetermined set of co-occurring, competing, mutually interfering activities at once, makes equating "watching" with "directing the face toward the screen" a rather nonsensical operationalization indeed, never mind practically difficult to determine in spaces where viewers are free to move about. It is hard to see how the quantity of the activity can ever be determined other than in an arbitrary, that is, discursively constructed, way, implied in the very method in use. "Size" of the audience is a discursive construction rather than an objective fact, accomplished by containing rather than recognizing irreducible difference and variation (see also Sepstrup, 1986).

Since market research is supposed to deliver informational products that can serve as the common symbolic currency for industry negotiations and decision-making, a too-detailed familiarity with the radically contextual ways in which people consume and use media would only be counterproductive. It would not fit with the requirements of prediction and control to be fulfilled by the research function within the industry. In other words, if market research selectively derives certain *methods and techniques* from ethnography, it certainly does not allow itself to adopt an ethnographic *mode of understanding*, in the sense of striving toward clarifying what it means, or *what it is like,* to live in a media-saturated world. It is toward the latter, I would argue, that we should proceed if the assumptions of radical contextualism are to make a critical difference in the way in which we comprehend and evaluate the quandaries of media audiencehood in contemporary society.

However, this vastly complicates our task as researchers. Since the premise of radical contextualism in principle involves the impossibility of determining any social or textual meaning outside of the complex situation in which it is produced, it is difficult to imagine where to begin and where to end the analysis. First of all, theoretically every situation is uniquely characterized by an indefinite multiplicity of contexts that cannot be known in advance. Furthermore, contexts are not mutually exclusive but interlocking and interacting, superimposed upon one another as well as indefinitely proliferating in time and space. A project that would strive to take into consideration the whole contextual horizon in which heterogeneous instances of media consumption acquire particular shape, significance and effectivity would be quite unwieldy and exhausting indeed, if not overambitiously megalomanic. This may be a reason why it seems easier to *talk about* ethnography than actually to do ethnographic work with audiences. This is also why the CRICT work is so significant, although not without its own problems and dilemmas. Let me sketch what I see as their main thrust.

As we have seen, Morley and Silverstone have singled out two contextual frameworks for television consumption, namely the domestic and the technological. But at the same time they (correctly) state that these contextual frameworks cannot be separated from "the wider context of social, political and economic realities" (1990: 32). The confusing consequence is that they seem to be somewhat unclear as to how to articulate the plethora of other contexts they theoretically envisage. Those of nation and gender are mentioned explicitly, but we could easily imagine a virtually endless, varied list of other contexts: race, class, ethnicity, regional location, generation, religion, economic conjuncture, political climate, family history, the weather, and so on and so on—into the main thrust of the project. If not held in check, awareness of the infinity of intercontextuality could lead to contextualization gone mad!

To put it differently, imagining the radical, that is eternally expanding contextuality of the particular meanings produced through media consumption would imply the taking up of an impossible position by the researcher, namely, the position of being "everywhere,"[6] ceaselessly trying to capture a relentlessly expanding field of contextually overdetermined, particular realities. Although such a position may be epistemologically adequate, it is in the end untenable ontologically, let alone pragmatically. No excursion into the real, no matter how ethnographic, can ever encompass such all-embracing knowing. As Jonathan Culler observes: "[C]ontext is boundless, so accounts of context never provide full determinations of meaning. Against any set of formulations, one can imagine further possibilities of context, including the expansion of context produced by the reinscription within a context of the description of it" (1983: 128).

How then to get out of this dead end? How can we come to terms with the inherently contradictory nature of the radical contextualist claim without succumbing to what Clifford Geertz (1988: 71) has called epistemological hypochondria? The answer, I would suggest, following Geertz, should not be sought in wanting to be epistemologically perfect, but in the uncertain trajectories of the politics of narrative and narration, of story and discourse. By admitting that the ethnographer cannot be "everywhere" but must always speak and write from "somewhere," we can leave the remnants of logico-scientific thinking (as embodied in the epistemology of radical contextualism) for what it is in favor of narrative modes of reasoning and representation, in which not only the contexts of media consumption, but also the contexts of ethnographic knowledge production itself are taken into account (see, e.g., Richardson, 1990).

It may be illuminating, in this regard, to turn briefly to some (meta)anthropological literature, where the status of ethnography has recently been discussed more extensively.[7] In practice, ethnographic studies of media consumption tend to take communities of audiences—such as family audiences, specific audience subcultures, or fan groups—as empirical starting point, treating them as sense-making cultural formations, just as anthropologists have for decades taken up the task of describing and interpreting other cultures as meaningful wholes. However, the very project of documenting a "culture" is being increasingly problematized within contemporary cultural anthropology. "Culture" as such can no longer, if it ever could, be considered as a transparent object of empirical enquiry, a finished entity that can be discovered and documented as such by the ethnographer. On the contrary, documenting a "culture" is a question of discursive construction that necessarily implicates the always partial (in both senses) point of view of the researcher, no matter how accurate or careful he or she is in data gathering and inference making. As James Clifford has remarked: 'Cultures' do not hold still for their portraits. Attempts to make them do so always involve simplification and exclusion, selection of a temporal focus, the construction of a self-other relationship, and the imposition or negotiation of a power relationship."[8] We do not need to succumb to the far-reaching but rather disabling poststructuralist postulate of the impossibility of description emanating from this insight (as Culler gestured toward) to nevertheless accept the assertion that all descriptions we make are by definition constitutive, and not merely evocative of the very object we describe.[9] Portraying a "culture" implies the discursive knocking up of a unitary picture out of bits and pieces of carefully selected and combined data, a picture that makes sense within the framework of a set of preconceived problematics and sensitizing concepts that the researcher employs as cognitive and linguistic tools to generate the descriptions in the first place.

However, while it might not have been too hard to hold such a picture romantically for a full and complete representation of a self-contained reality when the "culture" concerned is apparently some clearly limited and finite other culture, as in the case of classic anthropology's remote and primitive, small and exotic island in the middle of the vast ocean, inhabited by people whose daily practices were relatively untouched and uninfluenced by the inexorably transformative forces of capitalist modernity, it has become quite impossible in today's modern world system even to imagine a full and comprehensive portrait of any cultural formation, since contemporary culture has become a complex and thoroughly entangled maze of interrelated and interdependent social and cultural practices, ceaselessly proliferating in time and taking place in global space. In other words, there simply are no pristine, isolated, wholesome "cultures" any more that can be cut out from their surroundings in order to be pictured as such (see Marcus and Fischer, 1986, chapter 4). Today, all "cultures" are interconnected to a greater or lesser degree, and mobile people are simultaneously engaged in many cultural practices at once, constantly moving across multidimensional, transnational space. In Geertz's words, "the world has its compartments still, but the passages between them are much more numerous and much less well secured" (1988: 132).

This contemporary cultural condition—postcolonial, postindustrial, postmodern, postcommunist—forms the historical backdrop for the urgency of rethinking the significance of ethnography, away from its status as realist knowledge, and in the direction of its quality as a form of storytelling, as narrative. This does not mean that descriptions cease to be more or less true; criteria such as accurate data gathering and careful inference making remain applicable, even if their meaning and importance may become both more relative and more complicated, not just a question of technique but also perhaps one of ethics. It does mean that our deeply partial position as storytellers—a doubly partial position, as I noted earlier—should more than ever be seriously confronted and thought through in its consequences. Any cultural description is not only constructive (or, as some might say, "fictive"), but also of a provisional nature, creating the discursive objectification and sedimentation of "culture" through the singling out and highlighting of a series of discontinuous occurrences from an ongoing, never-ending flux, and therefore by definition always already falling short and falling behind. The point is not to see this as a regrettable shortcoming to be eradicated as much as possible, but as an inevitable state of affairs that circumscribes the implicatedness and responsibility of the researcher/writer as a producer of descriptions that, as soon as they enter the uneven, power-laden field of social discourse, play their political roles as particular ways of seeing and organizing an ever elusive reality.

This is what Geertz has called the "discourse problem" in anthropology (1988: 83). For Geertz, this problem is ultimately a problem of authorship: "The basic problem is neither the moral uncertainty involved in telling stories about how other people live nor the epistemological one involved in casting those stories in scholarly genres. . . . The problem is that now that such matters are coming to be discussed in the open, rather than covered with a professional mystique, the burden of authorship seems suddenly heavier" (1988: 138). This burden of authorship is all the heavier, I would suggest, as soon as we cease to conceptualize it as an individual predicament, but as a deeply social and political one. This implies two things. First, it is important not to reduce the anthropologist-as-author to a literary figure, engaged in writing ethnography as a self-indulgent, purely aesthetic practice. If ethnography is not science, it is not literature either.[10] Ethnographic discourse should retain its primarily hermeneutic ambition to provide representations that allow us to better *understand* other people's as well as our own lives. The choice for this or that literary style of writing, this or that form of storytelling, though essential considerations, should be explicitly related to this ambition.

Secondly, let us not forget that the burden of authorship conveys not only a problem of writing, but also one of reading; it is not only a question of producing texts, but also one of their reception. In short, the social context in which ethnographies are written, published, read, and used must be taken into consideration. Which stories to tell, in which form, to whom, where and when, and with what intention, are questions that academic scholars are not accustomed to asking themselves, but they are central to the politics of intellectual work. In this respect, I agree with Talal Asad's argument that a "politics of poetics" should not be pursued at the expense of a "politics of politics:"

> The crucial issue for anthropological practice is not whether ethnographies are fiction or fact—or how far realist forms of cultural representation can be replaced by others. What matters more are the kinds of political project cultural inscriptions are embedded in. Not experiments in ethnographic representation for their own sake, but modalities of political intervention should be our primary object of concern (1990: 260).

How then can culturalist audience studies benefit from this self-reflexive rethinking of ethnography within contemporary anthropology? First of all, we should recognize that just as representations of "cultures" are, in a manner of speaking, inventions of anthropologists (Wagner, 1981), so too are representations of "audiences" invented by audience researchers, in the sense that it is only in and through the descriptions conjured within the discourses produced by researchers that certain profiles of certain audiences take shape—profiles that do not exist outside or beyond those descriptions but are created by them. In this respect, academic audience researchers do not differ from market researchers: They are both in the business of creating audience profiles. But their politics, and therefore their rhetorical strategies and epistemological legitimizations—in short, the stories they tell—differ, given the disparate institutional conditions in which both groups have to operate.

Once again, this does not mean that people's involvements with media as audience members in everyday situations are not real or nonexistent; it only means that our representations of those involvements and their interrelationships in terms of "uses," "gratifications," "decodings," "readings," "effects," "negotiations," "interpretive communities," or "symbolic resistance" (to name but some of the most current concepts that have guided audience research) should be seen as ever so many discursive devices to confer a kind of order to the otherwise chaotic outlook of the empirical landscape of dispersed and heterogeneous audience practices and experiences.[11] The question, then, is what kind of representational order we should establish in our stories about media consumption. And in my view, culturalist audience studies, especially, should be in an excellent position to tell stories that avoid the objectification of "audience" for which market researchers inevitably strive in their attempts to make the chaos of media audiencehood manageable for the cultural industries.[12]

In a sense, radical contextualism is born of a creeping awareness of this chaos and a welcome attempt to do more justice to it in our representations of audience practices and experiences. It is, in the words of Janice Radway, one way of grappling with "the endlessly shifting, ever-evolving kaleidoscope of daily life and the way in which the media are integrated and implicated within it" (1988: 366). But as I have indicated before, the very desire for epistemological conquest implied in the will "to do justice" to endless contextualization could easily lead to a sense of paralysis, leading to the dictum "Don't do ethnography, just think about it." Of course, the opposite extreme, "Don't think about ethnography, just do it," is equally shortsighted (cf. Geertz, 1988: 139). For the moment, the middle ground can be held by doing the thinking with the radical contextualist horizon always in mind, but at the same time translating our limitations (i.e., our incapability to be everywhere at the same time) into an opportunity and a responsibility to make consciously *political* choices for which position to take, which contextual frameworks to take on board in our forays into the world of media audiences. Epistemological considerations alone are bound to be insufficient or even counterproductive as guiding principles for making those choices, as Morley and Silverstone's project suggests, because from an epistemological perspective all contexts relate to each other, even though one could theorize that not all contexts are alike or equally important. It is here that the "modalities of political intervention," to use Asad's phrase, gain their pragmatic relevance. It is within the framework of a particular *cultural politics* that we can meaningfully decide which contexts we wish to foreground as particularly relevant, and which other ones could, for the moment, within this particular political conjuncture, be left unexplored. Radical contextualism can then act as a stance not governed by a wish to build an ever more "comprehensive theory of the audience," which would by definition be an unfinishable task, but by an intellectual commitment to make the stories we end up telling about media consumption as compelling and persuasive as possible in the context of specific problematics that arise from particular branches of cultural politics. This is what Stuart Hall means when he argues that "potentially, discourse is endless: the infinite semiosis of meaning. But to say anything at all in particular, you do have to stop talking. ... The politics of infinite dispersal is the politics of no action at all" (1987: 45). Therefore, it is crucial to construct what Hall calls "arbitrary closures" in our storytelling practice (that is, epistemologically arbitrary), even though "every full stop is provisional" (p. 45). Or, as anthropologist Marilyn Strathern has succinctly put it, "I must know on whose behalf and to what end I write" (1987: 269).

In this respect, Strathern points to the success of contemporary feminist scholarship, a success that, in her view, "lies firmly in the relationship as it is represented between scholarship (genre) and the feminist movement (life)" (p. 268). And indeed, in much feminist scholarship the burden of authorship effectively transcends the tenets of liberal individualism that pervade conventional academic culture:

> Purposes may be diversely perceived; yet the scholarship is in the end represented as framed off by a special set of social interests. Feminists may argue with one another, in their many voices, because they also know themselves as an interest group. There is certainty about that context (p. 268).

This is not the place to enter into a debate about Strathern's confident assertion that feminism can provide a certainty of political context for academic work; after all, feminism itself is increasingly questioned in terms of its status as a general political roof for women's interests (see, e.g., Butler, 1990). Nevertheless, what matters here is the self-conception of feminism as an imagined community that manages to construct a commonality of interests, which enables feminist scholars to develop and entertain a sense of commonality of worldly purpose. For the academic and professional community of audience researchers, of course, determining the political context of their work is much more difficult, for they do not form, and cannot possibly form, a special interest group. They do not, in any sense, form an imagined community bound together by a unifying set of extra-academic social or political aims and purposes. But this is precisely the reason why it is all the more important for us to *construct* such aims and purposes, to define the modalities of political intervention that can energize our interest in knowing audiences, to actively create the "arbitrary closures" that can give audience studies a sense of direction and relevance in an ever more uncertain, complicated world. This is another way of saying, quite simply, that what we need more than ever is a renewed agenda for audience studies, one that is drawn up by considerations of the *worldly* purposes of our scholarship.

This brings me back, finally, to the conjuncture of change in our contemporary mediascape, which arguably poses the most pressing global context for audience studies in the years to come. It is clear that the initiatives of the transnational media industries are bringing about significant and confusing transformations in the multicontextual conditions of audience practices and experiences. At the same time, these large-scale structural developments have made the predicaments of postmodern audiencehood ever more complex, indeterminate, and difficult to assess, not least because of the ubiquitousness of these developments. There is no longer a position outside, as it were, from which we can have a total, transcending overview of all that is happening. Our minimal task, in such a world, is to explicate that world, to make sense of it by using our scholarly competencies to tell stories about the social and cultural implications of living in such a world. Such stories cannot be comprehensive, but they can at least make us comprehend; they should, in the listing of Geertz, "analyze, explain, disconcert, celebrate, edify, excuse, astonish, subvert (1988: 143–144)." How can we give substance to this claim? I can only give a partial—doubly partial—answer to this question, in the form some proposals that reflect my own concerns and interests. I offer them in the spirit of dialogue and discussion that were fruitfully engendered by the conference that led to this volume.

One political problematic that is rarely addressed in audience work concerns the problem of public policy in an age of so-called consumer sovereignty. In their search for viable antidotes to the hegemonic logic of commercialism, media policy makers—and I am thinking here especially of the European tradition of public service broadcasting—have, for better or worse, often resorted to a discourse of "quality" and "minority programming." But in doing this, public broadcasters still have not always managed to overcome the paternalistic or elitist attitudes toward the television audience that pervade classic public service broadcasting ideology. In my view, this is the result of broadcasters' real and symbolic *distance* from their audiences, a distance that tends to be intensified, not closed by the now common use of quantitative market research surveys in these circles. In this policy context, ethnographic understanding can be extremely useful; for example, it could potentially improve programming for ethnic minorities, now often suffering from lack of insight into the diverse and contradictory social experiences of its "target audience." In other words, only by an understanding of what it is like to live as non-European migrants in Europe can professional broadcasters hope to develop media provisions that these viewers find truly relevant. This is not to say that ethnography can save public service broadcasting as an institution; what I do suggest, however, is that ethnographic sensitivity for contextualized audience practices and experiences can enhance media production practices whose aim is more than the single-minded pursuit of profit (see Ang, 1990b, part three).

Of course, the politics of politics in audience studies does not always have to have such direct practical bearings. James Carey (1990) has astutely remarked that creating an understanding of media audiences is also creating an understanding of ourselves. However, even though we do indeed increasingly inhabit the same media-dominated world, entire fields of concrete practice and experience remain alien to us, precisely because we cannot be "everywhere"—neither literally nor symbolically. By ignoring this, we would risk succumbing to sweeping generalizations that could only slight the scope of difference and variation that still exist. I am referring here, of course, to the continuing concern over issues of cultural imperialism and globalization, issues that are likely to become more, not less pronounced in the coming decades. It is precisely ethnography that can help us to locate and understand the "gradual spectrum of mixed-up differences" (Geertz, 1988: 147) that comes with the progressive transnationalization of media audiencehood. I would suggest that the prime contextual factor to be highlighted here would be that of center-periphery relationships, especially important for North Americans and West Europeans who live and work in relative comfort in the centers of what Ulf Hannerz (1989) calls the "global ecumene." It is in telling stories about "a diversity in motion, one of coexistence as well as creative interaction between the transnational and the indigenous" (Hannerz, 1989: 72), that ethnography can, in Geertz's words, "enlarge the possibility of intelligible discourse between people quite different from one another in interest, outlook, wealth, and power, and yet contained in a world where, tumbled as they are into endless connection, it is increasingly difficult to get out of each other's way" (1988: 147).

To be sure, some audience work has already been done in this area, the most well known being Elihu Katz and Tamar Liebes's (1990) work on the cross-cultural reception of *Dallas.* But in general, this whole problematic has remained strangely unexplored. Given the persistent technological and economic determinisms in theorizing and research in this area, I want to highlight the necessity of cultural descriptions that can relativize what Morley has called the improper romanticism of consumer freedoms, on the one hand, and the paranoid fear of global control, on the other.[13]

Notes

1. I have explored this issue in greater detail in Ien Ang, 1990b.

2. For more information about this research project, see the chapters by Roger Silverstone and David Morley in this volume.

3. I have discussed the emergence and structural conditions of this crisis at length in Ang, 1990b (part two).

4. The people meter is an audience measurement technology that was introduced in the United States in 1987. It replaced the two earlier measurement methods, namely, the electronic setmeter and the paper-and-pencil diary. These two methods were generally perceived as having become obsolete as a result of their incapability of accurately measuring the increasingly fragmented consumption of cable channels, independent stations, and video recorders. But the people meter has from the beginning been seen as an imperfect

instrument, and as only a temporary solution to the quest for "better" measurement. Briefly, the problem with the people meter pertains to the disturbing observation that sample members do not seem to be very disciplined in reporting when they do and when they don't watch the tube, something they are supposed to do by pushing on a personal button on a keypad, which links the information to Nielsen's central computer.

5. For a further discussion of these developments, see Ang, 1990b (part two).

6. I borrow this spatial characterization of the epistemological quest of radical contextualism from Susan Bordo (1990).

7. Among the most well known publications in this respect are James Clifford and George E. Marcus (1986); George E. Marcus and Michael M. J. Fischer, (1986); and Geertz. (1988)

8. Clifford, "Introduction: Partial Truths," in Clifford and Marcus, 1990, p. 10.

9. Cf. Stephen Tyler, *The Unspeakable: Discourse, Dialogue, and Rhetoric in the Postmodern World* (Madison: University of Wisconsin Press, 1987).

10. This is what makes some recent discussions of ethnographic writing problematic. For example, John Van Maanen, in his *Tales of the Field* (1988), seems to have been seduced by the lure of literary effect in his favored "impressionist tales." As a result, he tends to slight the importance of theoretical categories in the construction of meaningful understandings in ethnographic discourse.

11. Relevant in this respect is Willard D. Rowland's study of the symbolic uses of effects research in his *The Politics of TV Violence* (Beverly Hills, Calif.: Sage, (1983). See also Ang, 1990b.

12. This is clear, for example, in the constant search for new strategies of "audience segmentation" within market research. At the same time, the very difficulty of producing satisfactory ways of segmenting the audience in clear-cut, mutually exclusive categories suggests that market researchers, too, are confronted with the ultimate intransigence of audience chaos. For the latter, see, for example, Arnold H. Diamond, "Chaos Science," *Marketing Research* 5(4) (1993): 9–12.

13. For a further discussion of this problematic, see Ien Ang, 1990a.

References

Ang, Ien (1990a). Culture and communication: Towards an ethnographic critique of media consumption in the transnational media system. *European Journal of Communication* 5(2–3): 239–260.

———. (1990b). *Desperately Seeking the Audience.* London and New York: Routledge.

Asad, Talal (1990). Ethnography, literature, and politics: Some readings and uses of Salman Rushdie's *The Satanic Verses. Cultural Anthropology,* vol. 5, no. 3.

Bordo, Susan (1990). Feminism, postmodernism, and gender-skepticism. In Linda Nicholson (ed.), *Feminism/Postmodernism,* pp. 133–156. New York and London: Routledge.

Butler, Judith (1990). *Gender Trouble: Feminism and the Subversion of Identity.* New York: Routledge.

Cary, James (1990). Paper presented to the conference, Toward a Comprehensive Theory of the Audience, Champaign-Urbana, Illinois.

Clifford, James, and George E. Marcus (eds.) (1986). *Writing Culture: The Poetics and Politics of Ethnography.* Berkeley: University of California Press.

Culler, Jonathan (1983). *On Deconstruction.* London: Routledge.

Davis, B. (1986). Single source seen as "new kid on block" in TV audience data. *Television/Radio Age*, 29 September.

Diamond, Arnold H. (1993). Chaos science. *Marketing Research* 5(4): 9–12.

Geertz, Clifford (1988). *Works and Lives: The Anthropologist as Author.* Chicago: University of Chicago Press.

Gold, L. N. (1988). The evolution of television advertising—Sales measurement: Past, present, and future. *Journal of Advertising Research*, vol. 28, no. 3.

Grossberg, Lawrence (1988). *It's a Sin: Essays on Postmodernism, Politics, and Culture.* Sydney: Power Publications.

Hall, Stuart (1986). Introduction. In David Morley (ed.), *Family Television: Cultural Power and Domestic Leisure.* London: Comedia.

———. (1987). Minimal selves. In *ICA Documents 6: Identity.* London: Institute of Contemporary Arts.

Hannerz, Ulf (1989). Notes on the global ecumene. *Public Culture* 1(2): 66–75.

Liebes, Tamar, and Elihu Katz (1990). *The Export of Meaning.* New York: Oxford University Press.

Maanen, John Van (1988). *Tales of the Field.* Chicago: University of Chicago Press.

Marcus, George E., and Michael M.J. Fischer (1986). *Anthropology as Cultural Critique.* Chicago: University of Chicago Press.

Morley, David, and Roger Silverstone (1990). Domestic communication: Technologies and meanings. *Media, Culture, and Society*, vol. 12, no.1.

Radway, Janice (1988). Reception study: Ethnography and the problems of dispersed audiences and nomadic audiences. *Cultural Studies*, vol. 2, no. 3.

Richardson, Laurel (1990). Narrative and sociology. *Journal of Contemporary Ethnography* 19(1): 116–135.

Rowland, Willard D. (1983). *The Politics of TV Violence.* Beverly Hills, Calif.: Sage.

Sepstrup, Preben (1986). The electronic dilemma of television advertising. *European Journal of Communication* 1(4): 383–405.

Silverston, Roger (1990). Television and everyday life: Towards an anthropology of the television audience. In Marjorie Ferguson (ed.), *Public Communication: The New Imperatives.* London: Sage.

Strathern, Marilyn (1987). Out of context: The persuasive fictions of anthropology. *Current Anthropology*, vol. 28, no. 3.

Wagner, Roy (1981). *The Invention of Culture.* Revised and expanded edition. Chicago: University of Chicago Press.

PART IV

LOCATING AUDIENCES

Hemispheres of Scholarship: Psychological and Other Approaches to Studying Media Audiences

17

Byron Reeves

In this chapter, I will elaborate on the following points:

1. Communication involves psychological and social processes. Communication includes the processing capabilities of individuals, as well as the social conveyance and sharing of personal reality. Individual and social processes are complementary rather than competitive, but they constrain each other.
2. A comprehensive view of audiences should explain the psychological abilities of individuals as well as the relationships people create and the symbols that define their culture. There is a need to address these domains separately, perhaps radically so, and a need to address their interaction.
3. A consideration of audiences as individuals includes the mental and physical capacities that people use to create and respond to information. Important mental abilities are those that allow people to focus, represent, organize, integrate, and retrieve information, as well as sensory abilities, emotions, and the capacity for physical action.
4. Theories about psychological abilities assume that people use information to accomplish goals and that people seek to maximize the efficiency of thinking. The processes that serve these goals are both controlled and automatic, and they are more similar than different across individuals.
5. Research that addresses psychological issues is quite varied, and it is sometimes difficult to defend theoretically. But in summary, psychological interests cannot be characterized as effects studies, administrative service, or transmission models. Psychological studies are attempts to explain *how* people communicate.
6. A psychological perspective on audiences has implications for how we should define media. Definitions of media will have psychological validity to the extent they are matched with concepts that describe abilities of individuals, rather than depending on categories of technology and message content that are based on professional or social considerations.

Throughout the chapter, I will mention studies and proposals for research that exemplify a psychological perspective on audiences. These examples are the home for most of my reading and work, and despite disagreements with others in my area, they define a place where I am comfortable.

The bulk of my outline, however, represents points I was encouraged to think about in preparing this essay. And they scare the hell out of me. These six points include assumptions about reductionism, functionalism, positivism, the advisability of interdisciplinary scholarship, and probably several other fighting words. It is probably wrong, I admit, but I do not think about these issues a lot. It is only when prodded to reconcile my narrow area with the broad list of contributions to this volume that I realize how difficult is the task of moving toward something that is truly "comprehensive."

Hemispheric Scholarship

The first two points in my outline pertain to the following questions: To what extent could any single perspective on communication offer a comprehensive theory of audiences? And if several perspectives are needed to represent the whole, how do we reconcile what appear to be competing proposals?

If I can generalize from my own experience, preparing comments pertinent to the goal of this volume is a dual task—we polish the details of our own perspective while trying to figure out what everyone else is up to. It may be possible, by analogy, to conceive of this task in the same way that psychologists treat dual processing in other domains. The analogy comes from the work on hemispheric specialization in the brain, and it is the basis for my concept of "hemispheric scholarship."

My particular use of this dichotomy does not refer to the more popular distinction between visual, right-brain abilities and linguistic, left-brain abilities. It is related instead to the more scientifically interesting dichotomy that describes the human ability to analyze simultaneously—details and the whole. Marcel Kinsbourne, a psychologist, summarized this dual talent in an article in *American Psychologist:*

> When confronted with a complex pattern, we see an undifferentiated mix of features. First one aspect or attribute occupies awareness, then another, and so forth, until the analysis is judged sufficient. Necessarily, when an attribute gains attention, it is temporarily *overstated* with respect to its ability to characterize the whole. When attention shifts to the next attribute, *it* is overstated, and the seeming importance of the one previously highlighted dwindles. Finally, the whole pattern is represented, not in its undifferentiated state, but in analytic detail, and preserving a proper balance between attributes.
>
> These analytic processes are *left hemispheric.* They involve—naturally, even inevitably, a shifting emphasis between ostensibly rival accounts, pending a final resolution of what turn out to be *different* perspectives, rather than incompatible ones.
>
> But what of the activities of the *right hemisphere?* The right hemisphere fits successive acts of information extraction, or insights, into a framework that is based on the initial overview of the problem, and *makes it possible to hold each insight in context.*

This passage captures my own hunch that different perspectives on the concept of audience, each representing valid "left hemisphere" analyses, must be pursued separately, and noncompetitively, *and* they must be characterized in the context of a sensible "right hemisphere" framework.

Left hemisphere scholarship is *against* context. Its object of study is the feature that defines what is central and separates it from what is peripheral. Psychological studies, a form of left hemisphere scholarship, attempt to say things about how people think that do not depend on politics, economics, social status, family situation, or social constructions of gender and age. And in the other direction, they do not depend on neural chemistry. An empty set does not remain when these influences are ignored. If they did, then psychology would be totally explainable in relation to other domains and should not constitute a separate area of study.

Right hemisphere scholarship *features* context. Its principal aim is to determine how each piece of the picture constrains a summary of the total. In communication, right hemisphere scholarship should emphasize the ways that social and psychological processes influence each other, not as determinants of how other processes must work, but via a process of constraining relationships. It seems quite reasonable to me to expect that the nature of intelligence will influence what we think about, and over time, even influence how we think.

What does this say about organizing scholarship that is comprehensive? First, the hemispheric distinction need not mean that there are an infinitely large number of useful "left hemisphere" perspectives, and that we should accept them all. Candidates for inclusion in the big picture could be wrong or irrelevant and unworthy of any consideration. But it also does not mean that among the approaches now available only one is right and that we should be quick about deciding which one it is. More than one perspective can be right, not because it is nice to include as many people in the field as possible, but because many different approaches are *needed* to characterize the whole.

The single most important reason why I believe this is often true in communication—why there are so many noncompetitive theories that can each be right—is that the concepts we choose *can* be about so many different things, even though they appear monolithic. I am convinced that the term "audience" could usefully refer to audiences, friends, communities, formal organizations, ethnic groups, nations—and individuals, like many other entities. Each domain will have its separate theories—and they will be competitive—but any comprehensive story about the whole will require that we combine theories that are about different things.

This immediately suggests two additional questions: What are the viable candidates for "left hemisphere scholarship?" And what does it mean to *combine* the *separate* theories? These are tough questions, especially the latter. I will briefly discuss the first question historically, and the second with an example.

I mentioned that there are many domains or levels of analysis that could define the concept "audience." Among all the ways to categorize these domains, I believe that a fundamental distinction is between (1) definitions that view audiences as a collection of *individual* psychological experiences; and (2) definitions that view audiences as individuals connected by relationships or common experience. For many reviewers in communication, this distinction describes historical epochs, and separates older, individualistic, and wrong definitions from newer, social, and correct ones. Raymond Bauer summarized this transition in the 1973 *Handbook of Communication:*

> Early treatment of the audience in the communication process literally referred to audience members as social "atoms," as isolated individuals who reacted to communication as though they had no social ties. Suffice it to say that in recent decades this view has been thoroughly replaced by one which places great emphasis on group membership.

In a footnote, he added the following:

> The reader may wonder how such a patently nonsensical idea gained currency. It can be explained largely in terms of the notion that the new industrial society had destroyed all the traditional patterns of associations that characterized the previous society. It was thought that group memberships, and patterns of interpersonal association, would have no role in modern society. This hypothetical view of things to come was imposed on the world in the face of substantial evidence to the contrary.

To many, the following may sound like bad news, but several people have not given up on "social atoms," although the current conceptions, I believe, have less to do with the effects of industrial society, and more to do with the maturing of psychology as a discipline and, specifically, with the revolution in cognitive science, which has shown that there *are* mental structures that can influence how people process media and that these structures may subsequently influence social constructions of reality. This does not mean, however, that social processes can be reduced to the domain of individuals. Instead, the premise that a psychological perspective on audiences is useful rests on the assumption that there are capacities of individuals, largely invariant across people, that will *influence* a "right hemisphere" big picture. How might this influence work? It is difficult enough to say anything about the psychology of communication, let alone to determine how those comments might be related to other theories. I will begin a short discussion of this question with an example.

My colleagues and I have recently begun to try to understand the psychological significance of a new technology, high-definition television (HDTV). Our work is proposed in the midst of policy deliberations at the FCC, technical and economic competition between Japan, Europe, and the United States, marketing concerns of electronics manufacturers, worries about who will be social winners and losers, and anticipated changes in programming at the networks and in Hollywood. But our interests here center only on a couple of small questions about people's psychological responses.

Briefly, our theory suggests that people's responses to moving pictures, and especially their involuntary and automatic responses, are influenced by a sense of "being there." To the extent that the physical cues of an audio/video image have high fidelity and take up a large portion of the visual field, they may be processed, at least at a primitive level, as if they were real rather than symbolic. The premise for this expectation comes from an acknowledgment that the human brain is not specialized to deal with twentieth-century media content. Rather, it is specialized to deal with primitive, natural cues—things like quick motion, visual surprises, sound/silence patterns, changes in visual field, and facial expressions—and these features can be part of a television presentation as easily as any other stimulus we encounter. And importantly, involuntary responses to these features may subsequently influence our more mindful evaluations of media content. Some students at Stanford call this the "jungle theory" of communication. I prefer the title of the Peter Sellers movie—*Being There.*

HDTV provides a good test of this theory because it has high-resolution pictures (and sound) that will typically be displayed on a big screen, with a wide aspect ratio. HDTV should increase the *intensity* of attention, because it will be more difficult, for example, to ignore quick motion and severe changes in camera angle. And it may mean shorter viewing sessions because people will tire more easily.

Also, the programs may be more emotionally arousing. For example, we are interested in how the increased resolution of facial features might enhance emotional response. This is based on good evidence that small changes in facial musculature, usually visible only during face-to-face contact, can influence our attributions about how others feel. Consequently, viewers may respond to pictures of people in ways that closely parallel responses to their actual presence. Also, the wide aspect ratio will increase the chances that action will appear in peripheral, rather than foveal vision, a known cause of arousal.

To many people, I am sure this sounds like pretty micro stuff. What I would like to argue, however, is that these comments, if supported, are not mere psychological curiosities. Rather, these findings could significantly constrain the social and cultural significance of this new technology. For example, at the organizational level, production companies, striving to create programs that knock the socks off of viewers, may alter the way they produce video. The FCC might decide (depending on who is president) to take protective action on behalf of children. And certainly, individuals will have different personal experiences to share as they, and scholars, attempt to define cultural changes attributable to a new symbolic form.

It seems quite reasonable to me that relationships between these domains of analysis might exist and that, taken together, they could form a plausible "right hemisphere" *story* about the significance of HDTV. An important question, however, is to what extent the different levels represented in this example—the psychology of individuals, the decisions of organizations and governments, and the quality of culture—represent unique domains that should have their own theories, or whether their combination itself constitutes a theory.

My belief is that any attempt to explain how something works—my definition of a theory—will be different for each domain. An explanation of how picture fidelity enhances arousal should sound quite different from an explanation of how media organizations make decisions, or how symbolic forms shape culture. Each domain represents a "left hemisphere" perspective that should be pursued separately. In fact, given the topical similarity of the theories, they should perhaps be radically separated, to ensure that they are not confused.

The different theories will, however, constrain each other. And any story about the relationship between the domains is essentially a story about the constraints. If current television pictures remain the broadcast standard, then producers will have no reason to change production techniques, and the relationship between picture fidelity and arousal could go unstudied. If we do adopt the new big pictures, then producers might change, and arousal is back on the agenda because there is now variance in picture fidelity to be studied.

But these contingencies, by themselves, do not constitute theories. It is impossible to have a *theory* about the relationship between two separate domains, without recasting the concepts so that each concept in the theory applies to the same things. And this recasting, by definition, dramatically alters the concepts.

This comment is one that I have addressed in another paper with Cliff Nass at Stanford, and it has gotten us both into a lot of trouble. It *seems* as though it should be possible to find a grand theory and that only a pessimist would discourage such an attempt. We still believe, however, that this is *not* possible. Consider the following two examples, each of which recasts two concepts so that they both apply to the same entities, though neither example constitutes a theory that crosses levels.

One strategy would be to observe different audiences for the products of different media organizations. We could determine whether the audiences become aroused or not, and then we could relate that variance to variance in organizational decisionmaking. This is an organization-level analysis, and a theory that explains this relationship is one about organizations. We might comment, for example, on differences in the arousal level between the audience for NBC products and the audience for ABC products. Alternatively, we could take the media products created by different organizational decisions and determine how they influence the arousal of individuals. Here, arousal is not an attribute of a collection of people, but an attribute of each person in the analysis. This is an individual-level analysis, and a relevant theory would explain differences in arousal levels between two or more individuals, or in the same person at two different times.

Other examples could be used that cast audience questions at the level of the family, local communities, nations, and culture. The problem with taking a single concept like audience and recasting it at different levels is that all definitions are potentially interesting, all could find a home in our field, and all *seem* as though they are about the same thing. But in terms of theory, they are radically different.

This is not to say, however, that the studies do not belong together. Even though *theories* will be unique to each domain, the "right hemisphere" *story* about how each domain constrains the other is obviously quite interesting and important. And the construction of stories about constraints may be a good occasion for people with different theories to meet.

Three other points can be made about relationships between theories in different domains. First, the *opportunity* to link theories from different domains is peculiar to the field of communication, and perhaps a couple other fields like political science, economics, and education. These fields cut across the traditional disciplines that occupy only a single point on a bio-psycho-social continuum. Consequently, this gives people in communication, at least those committed to "right hemisphere" scholarship, a positive role in providing a framework for studying individuals in a social context.

A second point concerns my belief that the competitive relationships between domains, especially the competition between individual and social processes, needs to be deemphasized. Rather than trying to figure out which process is correct, we should focus on how any one process works, and on determining which process will work when.

Toward this goal, we may be able to learn from the nature/nurture debate in biology and psychology. In spite of popular comments that still characterize this distinction competitively, most scientists have given up the adjudication of which is "right." Melvin Konner summarized this transition in his book *The Tangled Wing: Biological Constraints on the Human Spirit.* He spoke about biology and the environment, and I believe his comments apply more generally to individuals and social systems. He said the following:

> This fact of the matter is that no simple construct will ever subsume even what we know already, much less what we will soon know, about how [brain development is] partitioned between [the genes] and the environment. Now that the discussion of heredity versus environment has transcended the "versus," passing beyond the question, Which? ... to the mature question, How? we must prepare ourselves to face the fact that this last is not one question at all, but thousands. For each system, for each moment in development, we have on our hands a different balance, a different division of labor, a different integration of the functions of [the genes], and of the world. The roaring torrent of argument between the hereditarians and the environmentalists, narrow-minded bigots of different stripes, will undoubtedly continue; and, for the unsuspecting listener, it will obscure the complexity of the issues, and so in effect sabotage understanding. ... Things have come far enough, though, so that any analysis of the causes of human nature that tends to ignore *either* the genes *or* the environmental factors may safely be discarded.

I believe the same is true for the more general competition between psychological and social processes.

A final point is that the "How?" questions obviously emphasize explanation over prediction. Our aim should be to understand what people *can* do, and not exclusively what they *will* do. For example, asking *how* highly defined television pictures enhance arousal is quite different from asking *if* they will enhance arousal. The latter prediction, I believe, is nearly impossible. It depends on too many other things. Predicting that arousal follows exposure to HDTV would depend on who has access to new technology, whether they take the time to watch it, what content is available, what the viewing situation is like, and a hundred other things that are unrelated to arousal. Attempts to describe a system that is "closed" enough so that prediction *is* possible is an exercise that dooms research to concentrate on whether or not something will happen, rather than on how something happens, all other things being equal.

What Is a Psychological Perspective?

I would like to turn now to describing in more detail what might constitute a psychological perspective, and then to discuss, in my opinion, how that perspective should *not* be described. Perhaps most importantly in communication, designating a question as psychological serves to identify the "things" to which a theory refers—namely, individuals. This separates psychological theories from those that are about other things—groups, communities, nations, cultures.

Within psychology, however, this says very little. There are several subdisciplines that each claim a unique part of the story of individuals, and each of these areas has had some recent influence on audience research. Most histories of our work, however, agree that social psychology has been the most influential, especially for studies of politics, propaganda, and purchases. Second place might belong to developmental psychology, which has had a strong influence on studies of media and youth throughout the century.

But the definitions of psychological specialties have changed. Mid-century social psychology was a huge area representing everything that was not psychological or clinical. A 1958 reader in social psychology edited by Maccoby, Newcomb, and Hartley included sections on language, perception, memory, opinion change, interpersonal influence, reference groups, social stratification, socialization, group processes, and more. These titles describe specialties that are now substantially more differentiated, probably for the better, with separate journals, associations, and training. In fact, if anything has influenced the declining estate of social psychology, it was the insistence that *all* of these specialties had been subsumed in theory, a goal that assured a long list of situational and dispositional conditions for *any* finding. The most inclusive current label in psychology has shifted from social psychology to *cognitive science,* a more narrow, but intellectually more defensible amalgam, and it is from this perspective that I would like to venture a description of a promising psychological perspective on audiences.

The history and basic tenets of a cognitive science have been reviewed extensively elsewhere, and by real psychologists, of which I am not one. Nevertheless, I will briefly mention some of the themes of the perspective. I also wish to note that these themes are not at all novel prescriptions for research in communication. They characterize several current and exemplary research programs, and some of the best are represented in this volume.

Herbert Simon described cognitive science as simply the study of intelligence, including its development and application. Many disciplines have contributed—psychology, linguistics, artificial intelligence, philosophy, neurophysiology—but in communication research, the contributions from psychology, summarized as the information-processing perspective, have been most influential.

The cognitive perspective makes two assumptions about individuals:

1. First, people make discretionary as well as involuntary responses to their environment. This distinguishes the cognitive approach from the behavioristic assumption that human action is determined exclusively by physical input. Yet it acknowledges that the environment—media in our case—puts constraints on how intelligence is developed and applied.
2. The second assumption is that mental abilities are severely limited, which is either fortunate or not, depending on whether you are trying to be intelligent, or trying figure out how intelligence works. As a result of this limitation, mental efficiency is paramount. People try to accomplish multiple goals, and each requires that a large amount of information be processed in a short amount of time. The need for efficiency has several implications. Two primary ones are that (a) some processing proceeds automatically and therefore does not require significant mental resources; and (b) some mental structures, such as stereotypes, can be seen as reasonable and unexceptional thinking because they maximize mental efficiency, even if they have negative social consequences.

Most researchers in cognitive science also agree on the specific cognitive functions that are necessary for intelligence. The two most fundamental abilities are those of *representation,* and the manipulation or *processing* of information that is remembered. Processing usually includes the capacity to *focus,* to *store* in memory, to *retrieve* from memory, to make *inferences,* and to *implement* judgments. All of these abilities, I believe, are necessary for people to communicate, and they can help to describe and explain communication at any level.

These abilities also represent significant research programs in mass communication. I will mention a couple of examples. The issue of *representation* is about how people build internal images of their environments. Questions in psychology center on whether these representations are pictorial, or linguistic, or both. This has an obvious relationship to media studies, since different media could bias representation one way or another, because of the symbol systems they employ. And it also has implications for *translating* between pictures and words; for example, in constructing a verbal account of a news event, when the primary information source is pictures. The research of Gavriel Salomon, and especially his book, *The Interaction of Media, Cognition, and Learning,* exemplifies the study of representation.

It is also worth noting that several people outside of communication comment about media and the issue of representation. Kinsbourne, in the same review on hemispheric specialization from which I quoted earlier, has even suggested that the advent of lithography and television may have changed the way our brains work, by developing latent visual talents of the right hemisphere, home of pictorial processing. He even credits recent advancements in machine design to enhanced pictorial abilities caused by frequent attention to media, which at first reading led me to believe that McLuhan may finally have been a credible theorist. This proposal is also a good example of how culture may constrain psychology, a more rare comment, but a quite reasonable direction in which constraints may work.

The study of people's *focusing* abilities is concerned with how people *direct* their perceptual systems when they are faced with too much information. Even a simple television message contains more information than could possibly be processed, and consequently people must be able to *select* what is relevant and ignore what is not. This is done is relation to individual goals, with the acknowledgment that goals can be thwarted or changed by message features that compel attention regardless of goals.

There are several excellent research programs that examine how people attend to media. Most examine attention to television, and predominantly children's attention. Each program also has provided an excellent model for studying processing as it occurs over time, rather than limiting assessments to people's recollections that are solicited minutes, hours, or even days after the fact. Important contributions on focusing have been made by Dan Anderson, at the University of Massachusetts; Aletha Huston, John Wright, and Mabel Rice at Kansas University; and by Bob Krull and Jim Wright.

The other aspect of processing that is home for several research programs is that of information storage, or *memory.* Research programs about memory have studies on:

1. episodic versus semantic memory, that is, when are memories organized around temporal episodes or similar meaning
2. the short-term versus long-term endurance of memories
3. structural biases in recollection, that is, which memories are most available
4. changes in meaning that can occur as information is processed

The research has been conducted with almost every type of media content including advertising, news, and entertainment. There are many contributors in this area, including Ann Lang, Barrie Gunter, Kathy Kellerman, Esther Thorson, Joan Schleuder, Michael Shapiro, and Tom Srull.

Storage also considers the implications of memory organization for *retrieval* and use of information. There are several important reformulations of traditional topics in media effects that offer explanations for how media information is retrieved. Two important examples are about aggressive responses to television and agenda-setting. Berkowitz and Rogers have offered an explanation of aggressive behavior that depends on *priming*, a central concept in memory organization, and Shanto Iyengar has offered a similar explanation for agenda-setting. Theories about priming suggest that media activate related thoughts, and consequently behavior, through a network of mental relationships. The strength of these relationships is determined by semantic associations between concepts in memory and the frequency with which the links have been used.

Overall these efforts, and many others, have moved research on memory from studies that emphasized the transmission of *linguistic* information to studies that emphasize words *and* pictures, *forms* of memory other than short-term recall, and memory *content* other than topics deemed important by information sources. Most importantly, however, they consider how memory works, rather than whether or not memories occur.

The other important area of research in communication that is influenced by recent developments in psychology concerns emotions and arousal. While many of the questions in this area do not explicitly assess intelligence, and consequently are often excluded from reviews of cognitive research, they do acknowledge an important limit on intelligent processing. Recent research programs in communication include the following:

1. There are several attempts to explain how erotic films, through a process of arousal and excitation, influence individual sexual behavior, as well as beliefs and judgments about the seriousness of sexual crimes and how they influence interpersonal discussion of sex. This is exemplified by the work of Ed Donnerstein and Dan Linz at UC-Santa Barbara, and also the work of Jennings Bryant and Dolf Zillman.
2. Zillman and his colleagues have compiled an impressive number of studies that explore the psychological basis of enjoyment. This includes explanations of how humor, suspense, tragedy, and other content influence what we say about the quality of our media experiences.
3. John Newhagen, in a recent dissertation at Stanford, studied the relationship between emotion and *evaluation*. His experiments demonstrated that the inclusion of brief, but bloody segments in news stories, enhanced memory for pictures in the stories, and changed people's evaluations of the events portrayed. People were more willing to evaluate negatively the people in the news stories, even if there was no *obvious* connection between the people portrayed and the bloody scenes.
4. The issue of *unconscious* processing is part of a growing number of fascinating research programs about how emotions and thoughts can be influenced by stimuli that we cannot consciously identify. In psychology, there are several recent reports about perception without awareness, and we have worked on a demonstration that unconscious processing can influence evaluations of people on television. John Bargh has also written several good pieces about the general characteristics of automaticity in processing.

5. Finally, several people at Stanford have recently attempted to discover how emotions influence memory, both free recall and recognition of pictures and audio from television. We have been able to show a distinct memory advantage for negative information that is shown in public service announcements, political advertisements, and broadcast news stories. This advantage exists in spite of a strong conscious distaste for negative material.

All of these research programs (and others) share several characteristics that are worth noting, especially in relation to traditional assumptions about empirical research in mass communication. First, and most important, these research programs, when they are at their best, are guided by theories of *how* people think, even if on occasion their popularity can be accounted for by how the results are applied. Prediction, especially prediction that could generalize to the population at large, is deemphasized. Second, the explanations of how information is processed do not depend substantially on individual differences. Cognitive processes are most often conceptualized as invariant, although their invariance may be for limited time intervals, and it may be culturally bound. Third, there is a complex relationship, most often unstated, between neurophysiology, and cognitive science. I have never polled my colleagues, but I would venture that many believe that, in part, thinking could be explained in terms of electrochemical processes of the brain. This does not mean that any of us know much about the physiology of intelligence, but my guess is that this will change, although most easily for those just beginning their graduate study. Nevertheless many current theories, for example, in the area of memory, have been suggested by recent developments in neurophysiology. This is particularly true for the so-called "connectionist architectures" that have guided serious definitions for many mental concepts, including the popular s-word in psychology—schema.

A fourth assumption is that discoveries about how thinking works cannot depend solely on introspective abilities. It is difficult for people to characterize accurately how they think, even if motivated to do so. Consequently, the evidence for mental structures and processes comes from measures that are common in cognitive science, but quite different from the standard questionnaires of social psychology and communication.

It is also worth noting how these research programs *cannot* be characterized:

1. A first point is that this perspective represents a growing awareness that, *in part,* has *replaced* "effects" studies. There are many people whose research could be used as examples of psychological work in communication—many more than those I have named, more than there were twenty years ago, and more than double the number if interpersonal communication is included. If there has been any positive effect of the decline in regulatory interest in media, it was that the decline allowed the growing cognitive revolution to fill a wide gap left by studies that were ostensibly, if not actually, related to federal policy. Consequently, "effects studies," which is often used to mean psy-

chological studies, does not describe the entirety of current research about the psychology of communication.

2. Psychological studies are also not explicit efforts at administrative service to the media industry. They begin with theoretical questions central to an explanation of how people communicate, and not with the evaluative questions from the research agendas of professional communicators. They may *seem* as though they are more centrally related to the professions, but this may be due to an appearance of relevance perpetuated by academics and professional who mistakenly define practical questions exclusively as psychological ones.
3. It should be obvious that these efforts are not behavioristic. In the entire century, there have been few explicit attempts to apply strict behavioristic principles to communication. More likely, traditional effects studies are labeled behavioristic de facto, because it *seemed* as though they were committed to a theory that merely matched input with output. It is more likely, however, that past efforts were committed to no theory at all, and if they were, it was often more mentalistic than is acknowledged.
4. Finally, I do not believe that a cognitive perspective depends on a transmission model of communication. Many efforts, solidly within the field of mass communication, but not part of this review, *are* interested in the transportation of information from source to receiver, largely for the purpose of control, and largely in the service of politics and trade. But much of the cognitive work is uninterested in the relationship between message sent and message received. Individuals make sense of information by combining it with memories, adding and deleting information, and even substantially changing information. Any perspective that considers these altered representations as deviant, or as products of poor communication, is, in my opinion, wrong. These processes are to be celebrated rather than criticized; they are at the heart of communication. The fact that information changes as it is processed also bolsters the argument that any objective categorization of media or messages, independently of how information is represented by people, may not be important to the field of communication.

I myself am much more sympathetic to models of communication that do not emphasize transportation, although I would of course add a psychological twist to them. James Carey's description of a *ritual* perspective seems much more homey to me, even if his history of our field seems to me a place in a rival camp. Carey very much drew me in when he summarized the questions of the ritual perspective as follows:

> We create, express, and convey our knowledge of and attitudes toward reality through the construction of a variety of symbol systems: art, science, religion, journalism, common sense, mythology. How do we do this? What are the differences between these forms? What are the historical and comparative variations in them? How do changes in communication technology influence what we can concretely create and apprehend?

These questions all sound good. The last one precisely captures my interest in HDTV.

Admittedly, however, Carey emphasized that symbolic creations are *exclusively* a product of public, social interactions. My addition would be that these social processes are constrained by psychological abilities, and not in the fashion that Geertz described as "merely quaint, academic curiosities," but rather in a way that makes psychological abilities a necessary, if not a sufficient, part of communication, and hence a necessary part of any explanation of how it works.

Implications of a Psychological Perspective for Definitions of Audiences

Let me offer here a brief consideration of how a psychological perspective on audiences should influence definitions of media. Almost all definitions of media, regardless of the area of scholarship, the methods of study, or the domain of theory, are influenced by the popular names for technologies and journalistic entertainment products. Media are television, newspapers, and radio, and the content of media is television programs, news stories, and advertisements. It is my belief that these definitions, more than most other characteristics of our field, have determined and limited the success of our theories.

The standard definitions of media may disguise other descriptions of media that have more significance for theories about audiences, and indeed about any domain. Media are infinitely describable, and consequently many definitions of media *can* be accurate, in a sense, but they also can be totally unrelated to our theories. It is critical that variance between technologies and content be described with relevance to the particular theoretical questions being considered.

If research questions are about media industries or products, for example, then distinctions between film and television, ABC and NBC, or different television programs, seem reasonable. If the questions are about audiences, then these units may not help much beyond compiling a Nielsen rating. When media technologies and messages are examined with respect to psychological processing, then definitions of media must have psychological validity, and they should not be constrained by industry packages that have no relevance to how people process information. Memories, attention, and emotional responses likely occur in relation to units of media that do not exactly overlap traditional definitions. Consequently, psychological definitions of media should specify the relevant units—from visual surprises to facial expressions to story grammar—that match psychological concepts.

Conclusion

I would like to conclude with a subjective comment about the relative merits of the two hemispheres. My perception is that our field, at least by conviction, greatly values a right hemisphere perspective—at least it seems to me that this is where a lot of the action is. The person who gets to summarize, or characterize the whole, is a featured contributor. If the analogy of brain hemispheres holds, however, this may be mistaken admiration. The right hemisphere, in isolation, cannot offer a coherent approach to the resolution of problems any more than the left can. Kinsbourne reminds us that *only in combination do the hemispheres yield overview.* He concluded, and I agree, that a holistic or comprehensive approach that leaves features unspecified, is as alien to right-hemisphere function as it is inimical to rationality in general.

From Audiences to Consumers: The Household and the Consumption of Communication and Information Technologies

18

Roger Silverstone

In this chapter I present an account of the development of a research project into the household and the consumption of communication and information technologies. Previous research into the television audience has failed adequately to take into account the new communication environment of which television is but a part—an environment that is both technological and social. Here, I offer a model for the analysis of the television audience that does take this new environment into account. Drawing on qualitative empirical work among family households, I offer a model of the media consumption process that takes the social, economic, and technological aspects of the domestic sphere as central. In defining this domestic sphere as a *moral economy,* I suggest that the television audience has to be understood in terms of a set of practices, both routine and ritualized, that are firmly embedded in the various multiple dimensions of their domesticity.

An inquiry into the "audience" should be an inquiry not into a set of preconstituted individuals, but into a set of daily practices and discourses within which the complex act of watching television is placed alongside others and through which that complex act is itself constituted (Allor, 1988; Grossberg, 1988). It should be recognized also that this inquiry requires empirical as well as theoretical attention.

I will present an outline of the logic behind a study of the household and the consumption of communication and information technologies—a study twenty families in the southeast of England—that is both empirical and theoretical. The study involves, initially, three interrelated considerations. The first lies in the observation that historically the study of the television audience has been largely the study of *broadcast* audience, substantially without reference to other media. The dominance of television in the households of the Western world in the last forty years may explain some aspects of this, but it cannot possibly explain all of them. Yet, without doubt, television is no longer condemned to a single set (Gunter and Svennevig, 1987). Television is becoming increasingly one among many domestic technologies providing electronically communicated information and entertainment. The convergence of video, computer, and telephone-based services; the digitalization of communication and information delivery; the increasing possibilities for interactivity; the sheer amount of choice on offer (technologies, delivery systems, and program) are radically transforming (or at least threatening to transform) the social and cultural environment in which television is received and appropriated.

The second move involves consideration of the social and cultural environment itself and of the place of the audience within it. Again historically, within a behavioral as well as within a psychoanalytic tradition, the audience has been conceived as substantially decontextualized. Audiences have been constructed as socially isolated, vulnerable or otherwise to media influence or hegemonic domination through such processes as identification with, or interpellation from, mass media texts. Alternatively, "uses and gratifications" research, and more recently cultural studies, have constructed the audience as psychologically and socially active (albeit with different emphases, so that cultural studies emphasize the text, interdiscursivity, and a more "adequate" sociology (Curran, 1990), but rarely embedded in the complex relations of everyday life (see Livingstone, 1990). However, the sense of social isolation and decontextualization that pervaded much psychological research has been lessened by research within socially defined groups and in "natural" settings (Lindlof, 1987; Lull, 1988). The argument, particularly in relation to the latter, is that it is only as a result of constituting the audience both as social and as active that any sense of what I have referred to elsewhere as "audiencing" becomes possible (Silverstone, 1990).

Audiences are not simply or only watchers of television or listeners to radio: They are members of families, households, communities, and nations; they are gendered, aged, and members of social classes; they are skilled and unskilled, educated and uneducated; and they watch television while doing other things and in competition with other things, at times and in places, alone and with others, in ways that mark their activity as powerfully mediated by the social, economic, political, and technological systems and structures of everyday life.

The search for the audience, therefore, is not a search for a unitary psychological or social object. It is a search for a more or less consistent or inconsistent, more or less motivated or unmotivated, set of practices within a project that has its defining conditions in the domestic and the public spheres. The audience is neither a mirage, nor, *pace* Hartley (1987), is it invisible. It consists in, and is the product of, an infinity of more or less fragile and ephemeral interactions with an increasing variety of media and mediated texts, interactions that take place and become meaningful only within the private/public worlds of households, neighborhoods, or working environments.

The third consideration is, narrowly speaking, a methodological one. It requires a commitment to specific empirical inquiry. The requirement to generate an understanding of the contextual embedding of media use and to understand media use as embedded within the daily practices of everyday life suggested to us a predominantly ethnographic research strategy, designed to provide a detailed account of the domestic consumption and of the nature and significance of media and information consumption within the home. Our research (Silverstone, Hirsch, and Morley, 1990a, 1990b) aims to create a framework for the analysis of domestic media and technology use and to do so, by extension and intension of the "audience," within the various strategies and tactics of everyday life—strategies and tactics themselves constituted through the biographies and social and economic relations of households and families in their private and public worlds.

From Audiences to Consumers

The attempt to recontextualize the audience in the light of these three considerations has a number of dimensions and consequences. In the sections below, I first consider some of the elements of our position as they relate to the research we are conducting at Brunel University, then discuss some of the consequences of that position and sketch the outline of an analytic framework.

Families, Households, and the Domestic

We focus on the setting where most television and other media consumption takes place and where the meanings to be derived from the appropriation of texts and technologies are created: the private sphere. In doing so a number of substantive and terminological questions are raised. We can describe the private sphere in various ways. We can talk of the family, the household, the domestic. Each of these terms provides a different focus, but each is significant for an understanding of media consumption and of the television audience (Silverstone, 1994).

Families are social units, systemic, more or less clearly bounded through networks of kin, more or less coherent or secure in the patterns of the relationships through which they are defined, but the basis from within which individual identities are forged and sustained. Families are communicative environments in which decisions, including decisions about media use, are constantly being made. Families, too, have an ideological significance in the industrialized world. Perceived as one of the core institutions of modern society, they are often believed to be under threat, threatened by a whole slew of cultural and social changes, including technological ones, which are variously celebrated or condemned. It is the case that the nuclear family is the focus of considerable political and policy concern. It is also the case, of course, that the nuclear family is statistically on the decline.[1]

A family, fragile and fraught though it may be, in any individual case is nevertheless the social unit in which, for most of us, most media consumption takes place. The relationships that define it, the myths and values that sustain it, the conflicts and contradictions that threaten it, provide a basic social environment through which individuals struggle, on a daily basis, with the problems of everyday life. When media consumption takes place in the family, then, it takes place in a complex social setting in which different patterns of cohesion and dispersal are expressed in the various subsystems of conjugal, parental, or sibling relationships. These relationships are played out in variously expansive or cramped, highly differentiated or undifferentiated, domestic spaces. The patterns of media consumption are generated and sustained within these social and spatial relations. Individual identities, definitions of age and gender status, are expressed by, and plausibly constructed through, those patterns of consumption. The "audience" is constituted in and through them.

Research for the Brunel study takes place in families. However, it also takes place in households. The difference between the two is not simply a matter of social composition or of blood relations. Families live in households. Not all households contain families or consist only of families. These rather banal observations also have implications for an understanding of media consumption. Households are economic and cultural units that are constituted through sets of relations that involve individuals inside and outside the home, as well as activities that are equally dispersed. The household is the container for family and other relationships. It has a spatial expression, an economic function, and a cultural identity. We have characterized the household as a "moral economy" (Silverstone, Hirsch, and Morley 1990a), in others words, as an economic and cultural unit dynamically involved in a transactional system of the exchange of commodities and meanings, and as a complex economic and cultural unit in its own right (Pahl, 1990).

The consumption of media—of media technologies as well as media texts—takes place in, and is constrained and enabled by, the material and the symbolic dimensions of households and families. Decisions to buy, the mechanisms of appropriation, the objectifications and incorporations of appropriated commodities (be they physical or symbolic) into the private spaces and temporalities of the home, all are articulated through, and at the same time define, the integrity of the household as an economic and cultural unit. The boundaries around this unit can extend far beyond the physical expression of the home, but equally these boundaries can be permeable or impermeable, and the relations that a household has with the world beyond its front door are subject to constant negotiation and control. The boundaries around the household are therefore both material and symbolic, and communication and information technologies are crucially implicated in their definition.

Homes and households are constituted within a domestic sphere that can only be defined in terms of its differential relationship to the public sphere. The domestic, it is often argued, has been the subject of progressive encroachment by agencies of the state (Donzelot, 1979), and at the same time it has been seen as becoming increasingly isolated from, or irrelevant to, the public sphere. In an industrial society dominated by production and the exchange of commodities, in a capitalist society of bourgeois and privatized social relations, the domestic has, it is argued, become marginalized and insignificant (Zaretsky, 1976). At the same time the domestic is being seen as increasingly privatized, isolated, and removed from the mainstream of modern society, and only reachable through technical and heavily mediated forms of communication. Yet the domestic dies hard. The domestic is indeed subject to increasing regulative pressures and, certainly in an age of broadcasting, it can be seen to be becoming the site of a consistent attempt, through scheduling and ideological discrimination, to mobilize it into a subservient relationship to the temporal and patriarchal (and paedocratic) structures of an increasingly sclerosed public sphere. Yet publicly expressed identities and values are constructed in the domestic, and as the material pendulum swings toward consumption and fragmentation (as well as homogenization) as an articulating principle of postmodern or post-Fordist economies, the domestic is, plausibly, becoming a crucially important economic site and one increasingly the subject of academic concern and political struggle.

It almost goes without saying that the media are centrally involved in the relationship between the domestic and the public spheres and therefore in the definition of the domestic. The "audience" is both domestic and public. Within a media regime dominated by broadcasting it is clear where the balance of power lies. However, as new technologies and new services, as well as more of the old, emerge and become available to the domestic consumer within new regulatory regimes, the balance of power will perhaps settle slightly differently from hitherto: the "audience" will perhaps become increasingly less responsive to the loud-hailing rhetoric of restricted media channels, and audience members will increasingly, but of course never entirely, free to construct their own media environment. It is reasonable to suggest that it is in the domestic, the household, the family—it is in the home—where this new media environment will be worked and appropriated.

Technologies and Socio-Technological Systems

Thus, the contextualizing of audience activity within the family and the household has as one of its consequences the need to recognize the multiple layering, both of structure and of action, in relation to media consumption. This layering is, of course, enhanced in a domestic environment that incorporates a multiplicity of technologies and media. Television sets in different rooms; video recorders time-shifting or replaying hired material; computers no longer requiring an umbilical link to the television screen; telephones and telephone answering machines; cable, satellite, and advances in telecommunication services; stereos, Walkmans, clock radios, video and still cameras—all provide the basis for a domestic socio-technical system, systemic not necessarily in terms of the formal and technical links between machines, but in terms of the social relations that construct them and define their significance and patterns of use.[2]

Television may remain the household's "leading object,"[3] but no longer can it be studied, any more than it can be used, in isolation. The technological culture of the household provides a framework for domestic social and indeed political (Morley, 1986) relations, mediating between members of the households and offering objectifications of their identities and competencies as well as mediating between them and the outside world. The domestic socio-technical system consists of a bundle of skills, tastes, and competencies, expressed in styles and practices that construct and mark the cleavages of gender and age-based relations within and beyond the household.

We have argued (Silverstone, Hirsch, and Morley, 1989) that television and other media and information technologies are doubly articulated into the culture of the household, both insofar as they mediate public and private meanings, and also insofar as they are constructed as meaningful both as technical objects (they are bought for their aesthetic appeal, their features, their "label") and in the places they occupy in domestic space (see, for example, Leal, 1990); and they are significant as media (their messages are also consumed and rearticulated into private culture, or into subcultural or friendship networks outside the home).

There is in a sense, therefore, nothing "natural" in the placing of, or the practices associated with, media and information technologies in the domestic context. We have had to be taught how to use them and how to incorporate them into our own domesticity (Spigel, 1989), and they have, in the main, been domesticated. Yet all the design, marketing, and consumer advice in the world still leaves space for the appropriation of technologies and media into the domestic culture of the home. There is still a space for active—more or less creative—engagement with the commodities, both material and symbolic, produced by the formal economy and offered to the consumer (cf. Miller, 1987, and below).

Domestic Consumption

These commodities are, indeed, both material and symbolic. We consume objects. We consume texts. In making both our own, in incorporating them into our lives, in displaying and talking about them, we engage in a struggle over their meaning, and through that mostly unromantic and often inconclusive struggle (in our winning and losing) we in turn construct our own individual and social identities. Thus, the audience is a particular manifestation of a more general set of social practices, and while that does not absolve the researcher from inquiring into the specific nature of "audiencing," it equally well does not absolve him or her from the requirement to recognize and understand the nature of that specificity: to understand that our relations to television are a part of, and of a piece with, our relations to the dominating public world of commodities and meanings with which we are constantly engaged.

In a passage we have had occasion to cite more than once, the British anthropologist Daniel Miller characterizes consumption as a creative activity undertaken with objects and in settings that may not always be of our own choosing nor always providing much scope for transcendent appropriation. The possibilities for appropriation, for the transformative work associated with ownership and objectification, are nevertheless real for all that:

> All . . . objects [and most texts] are the direct product of commercial concerns and industrial processes. Taken together they appear to imply that in certain circumstances segments of the population are able to appropriate such industrial objects and utilize them in the creation of their own image. In other cases, people are forced to live in and through objects which are created through the images held of them by a different and dominant section of the population. The possibilities of recontextualization may vary for any given object according to its historical power or for one particular individual according to his or her changing social environment (Miller, 1987: 175).

Clearly, it requires empirical, as well as theoretical, inquiry to analyze what factors are relevant to the different households. The issue here is simply to note that there are both conceptual and empirical reasons for framing an understanding of the relationship between members of households (and of the household itself) and media messages within the same terms as Miller is advocating with regard to the consumption of material objects. Consumption of either involves an active engagement in the appropriation of mass-produced and distributed commodities. Consumption in both cases is not a matter simply of the fulfillment of utilitarian needs, but is a significant component in modern society, significant for the establishment of status and the construction of individual and social identities. As Bourdieu (1984) has noted in another context: "Taste classifies the classifier," and taste is expressed not just in the display of objects but in the consumption and conversion of messages and meanings both in the private spaces of the home and among the social networks served by gossip and talk.

The television audience has become the consumer of mediated messages and media and information technologies. Media and information technologies are profoundly implicated in the dynamics of consumption, through their double articulation. They are themselves consumed, of course, but they also enable consumption: Through their consumed and consuming messages, they bring news of consumption possibilities, and through them decisions to consume are communicated, goods ordered, objects and identities displayed. An understanding of media's status within the activities of consumption requires not just an understanding of the dynamics of a household's engagement with the media, but an awareness of the dialectic of dependence and freedom irresistibly inscribed into those dynamics.

We have come to see audiences, therefore, as domestic and as consumers. In both their guises we have come to understand them as firmly embedded in the social and cultural environments of both public and private spheres. Their involvement with media (that is, communication and information technologies) is both an expression and a constitution of the relationship between those spheres. But audiences are only intermittently audiences, and their "audiencing" activity is itself complicated and qualified by other dimensions of their social lives. The consumption of television, the consumption of media, consumption itself, cannot be isolated from the sociology, politics, and economics of everyday life.

The Domestic Consumption of Communication and Information Technologies

The Moral Economy of the Household

In attempting to conceptualize the household as a moral economy, we draw on a literature principally in anthropology (cf. Parry and Bloch, 1989) in which households are conceived as part of a transactional system of economic and social relations with the formal or more objective economy and society of the public sphere. The central idea here is that households are actively engaged with the products and meanings of this formal, commodity, and individual-based economy, and that in their appropriation of those commodities they incorporate and redefine them in their own terms, in accordance with their own values and interests.

The moral economy of the household is, in this sense, both an economy of meanings and a meaningful economy; and in both its dimensions it stands in a potentially or actually transformative relationship to the public, objective economy of the exchange of goods and meanings. The household is a moral *economy* because it is both an economic unit that is involved in the public economy, through the productive and consumption activities of its members, and at the same time it is a complex economic unit in its own terms (Pahl, 1990). The household is a *moral* economy because the economic activities of its members within the household and in the wider world of work, leisure, and consumption are defined and informed by a set of cognitions, evaluations, and aesthetics, which are themselves defined and informed by the histories, biographies, and politics of the household and its members. These are expressed in the specific and various cosmologies and rituals that define or fail to define the household's integrity as a social and cultural unit.[4]

In the continuous work of domestic social reproduction—and via the mesh of class position, ethnicity, and geography—the household engages in a process of value creation in its various daily practices—practices that are firmly grounded in, but are also constitutive of, its position in space and time and what Anthony Giddens characterizes as "ontological security"—a sense of confidence or trust in the world as it appears to be (Giddens, 1989: 278). Central to this project of the creation of ontological security, particularly in an advanced capitalist society, is the household's ability to display, both of itself and others, through the objectification of those practices (the fruits of its cultural labor), its status as a participant in the public economy.

Also central to this project of the creation of ontological security is the creation of the *home*,[5] which may or may not be a family home, and which is itself multstructured both spatially and temporally (Giddens, 1984: 119). And while mediated and nonmediated meanings, commodities, and objects are formed and transformed as they pass across the boundary that separates the private from the public spheres, it is the quality of the achievement of "homeness"—that which turns space into place, that which supports the temporal routines of daily life—which is the issue, which is particularly problematic in modern society, and in which communication and information technologies are centrally involved.

There are a number of things to be said about this conceptualization of the household as a moral economy. The first is that it is not intended as an evaluative term. The moral economy of the household is what is achieved by any and every household as a result of its efforts to sustain itself as a social and cultural unit. In characterizing households this way, we are not suggesting that there is a single "morality" that enables us to define sickness or health, normality or pathology, in relation to specific households or family life (nor is there any intention to hypostatize the family as the ideal or necessary social unit). The questions that are begged are not evaluative but descriptive ones, and the notion is not intended as a reification but as an heuristic attempt to provide a framework for understanding the similarities and differences between households as social, economic, and cultural units as they engage on an hourly basis with the economic and social relations of the public spheres. Equally, the characterization of the household as an economy does not imply that the household is necessarily organized as an economic unit according to the same rationality that characterizes the public sphere. On the contrary, it is precisely in its difference and in the quality of that relationship that the household (potentially or actually) constitutes itself, dialectically, as an economic unit of a particular order.

Finally, it can be said that (once again) communication and information technologies are crucially implicated in the activities of the moral economy, both as creators and resolvers of boundary and control problems for the household. The activities of the household, their relationship to communication and information technologies, are played out—that is, both determined and determining—within the symbolic and material spaces of the home and through the mesh of gender (and age) statuses. It is useful at this point briefly to explore some of the relevant issues here both in relation to the organization of domestic time and space, and in relation to the structuring of gender identities and relationships. In practice, of course, the two are intimately connected.

Communication and Information Technologies and Domestic Time and Space

The implication of communication and information technologies—media, computers, telephones, and their networks—in the space-time structure of the household can be approached on at least two different levels. The first is the phenomenological level at which individuals, families, or households include media and media-related activities into their own projects of fixing their identities in time and space: for example, the regularity of television or radio program; the connections with family or friends that the telephone facilitates (or conveniently distances); the social networks constructed around the exchange of computer games; the separation of domestic spaces through the display and incorporation of televisions or stereos in such a way that adults and children can occupy and "defend" their own personal spaces and times through their ownership and use. Within the moral economy of the household these technologies, in various fascinating ways, as our data are beginning to show, are centrally involved in the creation and maintenance of boundaries within the household, for example in the ways in which a family—amoeba-like—will come together or disperse around particular television programmes, or the ways in which adolescents may create a "wall of sound" with their stereos to insulate them from the dynamics of the rest of the house.

However, there is a second and interrelated level at which media are implicated in the space-time dimensions of the moral economy of the household. It is in their mediation of the public and the private spheres, in their mediation of the global and the local, above all in terms of defining the routes along which the particular balance between family and state—the particular character of the transactional system that links and separates the moral to and from the formal economy—is struck. Here the issues are, among others, regulative. Broadcast schedules, perhaps above all, are symptomatic of the attempt to regulate the household's temporality in accordance with the perceived "natural" but highly gendered domestic rhythms of the day, but also in accordance with the imposition of a national calendar and within a more or less self-conscious attempt to construct, through the audience, a nation of listener-viewer citizens (and families), sharing a common culture and experiencing the same set of national events.[6] Domestic spaces, too, have been restructured by, and on behalf of, the demands of broadcasting. Yet the movement is not all one-way. Although we have been taught how to incorporate the media into the moral economies of our households, we still do it in expressive and often idiosyncratic ways. And these particular expressions of domestic culture are not insignificant, for they mark, as we have argued, the site of a struggle for individual and social identity, a struggle which is at the same time a symptom of the family or household's need to resolve for itself some of the key contradictions of industrial and postindustrial societies—for example, in the relations between the public and the private or between the individual and the collective. New developments in media technology and services, of which the VCR is the forerunner, are likely of course to alter the terms under which this struggle takes place, though by how much and how quickly are still open questions.

Communication and Information Technologies and Gender Relations

The same two levels that define the space-time relations in the household are of course similarly relevant in relation to gender. The domestic has substantially been marginalized and colonized by a public sphere which itself is structured through male dominance and masculinity. The family-household has emerged as a result of a gendered division of labor and domesticity and has become a female preserve, in which, to put it crudely, her work creates space for his leisure. Domestic technologies, of course, gain their meaning—they too become gendered—in accordance, principally, with the domestic division of labor (Cowan, 1989; Faulkner and Arnold, 1985). But this is not uniform, and different technologies, both information and communication and other technologies, are gendered in variously uneven ways (Gershuny, 1982).

Within the bounds of individual households, of course, the gendering of technology is a function of the particular social relations that predominate and structure the patterns of activity in daily life. Newly purchased machines arrive into an already well-defined domestic culture of technology (Rogge and Jensen, 1988). They are placed, used, and displayed within domestic spaces that are equally already gendered. The domestic culture of technology consists in a set of social relations, skills and competencies, and values and aesthetics that are firmly inscribed (but not unchanging) within families and households. New technologies and new media texts arrive, of course, already more or less clearly marked in gender terms, through their marketed images, and in their design and their production values. But their incorporation into each household along these predefined lines or claims cannot be guaranteed, nor is it always significant or uncontested.

Televisions, telephones, videos, and computers are appropriated and incorporated, therefore, into a dynamic system of social and cultural relations. They provide also, in their double articulation, images and models of gender identities. We can ask, with Linda Rakow (1988), what role technologies play in constructing and maintaining gender relationships in the home (and vice versa), and we can ask how powerful they become in the assertion of individual gender identities. Telephones, for example, are of great significance, both functionally and phenomenologically for women (Moyal, 1989). Patterns of viewing (interrupted and uninterrupted, fragmented or continuous, guilty or wholehearted), control of the remote, the ability to program the VCR, the domination of the computer, the ownership of the stereo—all these technologies are incorporated into that domestic culture, and into the moral economy of the household in such a way as to create and reproduce, and also to challenge gender (and, of course, age-based) identities.

Thus, the gendering of the television audience should be understood not just in terms of a simple reproduction of the dominant structure of gender relations in the public sphere, nor even only in terms of the particular pattern of gender relations within the household, but as both cause and consequence of a set of technologically mediated practices in both private and public spheres. Central to our analysis, therefore, is an understanding of the multiplicity of media and information technology—focused activities and meanings available to a household that are necessarily interrelated, both through the organization of domestic space and time, and through the particular sets of social and cultural relations linking the moral economy of the household with the formal economy of the public sphere.

Conclusion: Methodology and Methods

The conceptual and empirical position that had informed our research has a number of consequences and raises a number of questions, two of which I would like now briefly to address. The first is the Brunel study's status and claims in relation to current debates around the role of ethnography in cultural studies. And the second is the relationship between ethnographic approaches to empirical research with the television audience and other methodological approaches.

Any claim for empirical social research must be based on the desire to understand, in a plausible and potentially convincing way, what it is that people actually do. However problematic such a position has come to be, Max Weber's formulation of the task of what he saw as social "science" still stands and informs all attempts to make sense of the world that are based in empirical investigation. As he put it: "We want to understand the reality of life that surrounds us, into which we have been placed, in its special nature—the context and the cultural significance of its individual phenomena in their present shape, on the one hand, and, on the other, the reasons why it has turned out as it has and not otherwise" (Weber, 1949: 72). The methodological agenda set in this characterization of the sociological project does not, of course, imply an exclusive concern with ethnography. But within recent cultural studies, and particularly in discussions of the television audience, as well as of the multiplicity of practices associated with consumption and of the hypostasization of "everyday life" as a site for critical reflection and analysis, ethnography is coming to be seen as something of a methodological panacea. Behind it lies a political-theoretical project that Meaghan Morris only partly accurately describes as one of "encouraging cultural democracy" (1988: 19). The lines of debate have been well drawn. On the one hand are claims for an active, culturally subversive, and creative consumer freed by the fragmentation of postmodern culture to construct and reconstruct a private culture or a subculture with the mass-produced products of late capitalism, on the other is a critique that identifies the exaggerated romanticism in such a position and the flawed methodologies (introspection, decontextualization, banalization) that characterize the investigation of these freedoms. These critiques too, of course, shade into absurdity as they argue for the impossibility of empirical investigation of any kind, let alone ethnography.

The task that we have set ourselves, we suppose, broadly speaking, is the Weberian one. It seeks to avoid, above all, the charge that Morris levels at cultural studies in general, that it "runs perilously close to [the] formulation: people in modern societies are complex and contradictory, therefore people using them produce complex and contradictory culture" (Morris, 1988: 22). And it seeks to do so not by a tautological appeal to "ethnosemiotics" (Fiske, 1990), but through a set of research strategies aimed at identifying the special and the specific in the social realities of the consumption of technologies and meanings, including television and television's texts. The charge of banality is, we think, resisted through our empirically grounded efforts at theorizing the dynamic relations between the strategies and tactics of the public and private spheres, respectively (de Certeau, 1984), at least to a point where claims can be made for the plausibility of a series of accounts (partial and always provisional, of course) that constitute *our* understandings of the phenomena under investigation. What in our view is missing in previous investigations of this relationship is the lack of both a structured analysis of everyday life (to match, as it were, the structural analysis of political and economic processes of the public sphere) and a conceptualization of the relationship between the two.

What is missing, despite the efforts of some recent work (de Certeau, 1984; Giddens, 1984; Miller, 1987), is an adequate, empirically grounded account of what constitutes the conditions for, as well as the consequences of, the various materially and symbolically mediated freedoms that individuals and groups have in their relationship to the commodities of contemporary culture. The central issue that we think our research once again raises (for it is only a reformulation of the classic inquiry into the power of the media) is how to confront methodologically and resolve substantively two superficially contradictory observations. The first rests in the recognition of the deeply embedded nature of media and media-dependent relations in the culture of the household, and the second rests on the recognition of the degrees of freedom asserted differentially within the moral economy of the household around and through those relations.

If these remarks can stand for a defense (or the beginnings of a defense) of empirical inquiry in general, there is perhaps more to be said on behalf of the ethnographer in particular. Challenged for its subjectivity and unreflexivity, challenged too for its inability to address the problems of power, for its weaknesses in the face of the problems posed by the shifts in the regulation and internationalization of the media, and challenged for its surrender to the particular and the ungeneralizability of its case-study material, ethnography (and ethnographic approaches) seems to stand quite naked and vulnerable. Yet the claims of context, of "thick description," of situated complexity, and the challenge precisely to ground an understanding of the relationship of political economy to social practice (Marcus and Fischer, 1986) that ethnography can and often does take on board—these claims are sustainable within a realist project that acknowledges, *pace* Geertz (1988), its limitations and seeks its authority both in its capacity to display the imperfectable world of the Other, and in the leverage it provides to enhance our understanding of our own. As Janice Radway cogently and sensitively puts it:

> I continue to turn to ethnography because as a practice it is predicated implicitly on the assumption that to move out into an alien world, the field as it were, is also necessarily to attempt to leave another one behind. It is ethnography's insistence that the social is always actively constructed by living subjects that I find potentially helpful. That insistence can authorize attempts to understand the complexities of a distant world's construction by those who inhabit it even as it can be turned back reflexively upon the construction of our own in order to expose the peculiarity of the articulations with which we produce it (Radway, 1988: 373).

Ethnography at home provides no special exceptions to this argument (but see Strathern, 1986). What it does raise, of course, is the problem of designing a study that is adequate to meet the particular problems generated by the questions that inform the project as a whole. We have discussed these issues elsewhere (Silverstone, Hirsch, and Morley, 1990b). Suffice it to say that the questions and problems raised by our study of the domestic consumption of communication and information technologies are not (and cannot be) exhausted by our or anyone else's version of ethnography. Ethnographic approaches close down as much as they open up. We are expecting to develop our own project in a spirit of methodological openness.

This article arises from an ongoing study of the household and the consumption of communication and information technologies, funded by the Economic and Social Research Council (U.K.) under its Program in Information and Communication Technologies (PICT). The study is being conducted by a team at Brunel University, including David Morley and Eric Hirsch. I am intensely grateful to both for their comments on this paper.

Notes

1. For a recent sociological defense of the study of the family as a dynamic social unit wider and more complex than the household, see Wilson and Pahl (1988).

2. On the concept of socio-technical system, though in a different context, see Hughes (1987).

3. The phrase "leading object" is one used by Henri Lefebvre (1971: 102—103) to describe the material and symbolic significance of the automobile in modern society: not without qualification I am implying that the television can be seen to have that status in the domestic context, reinforcing, in ways that would bear careful study, the significance of the car in the public sphere (both television and the automobile are communication technologies mediating the public and the private).

4. For a much more extensive discussion of the moral economy of the household, see Silverstone, Hirsch, and Morley (1990a), and Silverstone (1994).

5. "Home" is potentially the fourth of the terms that can be used to characterize the social, cultural, and economic unit we have already characterized in terms of family, household, and the domestic. It signifies, at least in the present discussion, a phenomenological reality perhaps underlying, but certainly informing and expressing, the project of domesticity.

6. The arrival of satellite television in Europe has extended this problem in a fascinating way. The [it]Independent[it] (19 July 1990) reports on the Chairman of the Broadcasting Standards Council's discussions with European colleagues on the rescheduling of the "watershed" hour dividing program suitable for children and adult viewing in the light of different practices across Europe. We have noted elsewhere how this watershed can be used to define the boundaries between childhood and adulthood in families. that is, the license to stay up beyond it is an indication of adult status.

References

Allor, Martin (1988). Relocating the site of the audience. *Critical Studies in Mass Communication* 5: 217–233.

Bourdieu, Pierre (1984). *Distinction: A Social Critique of the Judgement of Taste.* London: Routledge.

Certeau, Michel de (1984). *The Practice of Everyday Life.* Berkeley: University of California Press.

Cowan, Ruth Schwartz (1989). *More Work for Mother: The Ironies of Household Technology from the Open Hearth to the Microwave.* London: Free Association Books.

Curran, James (1990). The new revisionism in mass communication research: A reappraisal. *European Journal of Communication* 5(2–3): 135–164.

Donzelot, Jacques (1979). *The Policing of Families: Welfare Versus the State.* London: Hutchinson.

Faulkner, Wendy, and Erik Arnold (eds.) (1985). *Smothered by Invention: Technology in Women's Lives.* London: Pluto Press.

Fiske, John (1990). Ethnosemiotics: Some personal and theoretical reflections. *Cultural Studies* 4(1): 85–99.

Geertz, Clifford (1988). *Works and Lives: The Anthropologist as Author.* Cambridge: Polity Press.

Gershuny, J. I. (1982). Household tasks and the use of time. In Sandra Wollman et al. (eds.), *Living in South London*. London: Gower

Giddens, Anthony (1984). *The Constitution of Society*. Cambridge, Eng.: Polity Press.

———. (1989). "A reply to my critics." In David Held and John B. Thompson (eds.), *Social Theory of Modern Societies: Anthony Giddens and His Critics*. Cambridge, U.K.: Cambridge University Press.

Grossberg, Lawrence (1988). wandering audiences, nomadic critics. *Cultural Studies* 2(3): 377–392.

Gunter, Barry, and Michael Svennevig (1987). *Behind and in Front of the Screen: Television and the Family*. London: John Libby.

Hartley, John (1987). Invisible fictions. *Textual Practice* 1(2) (Summer): 121–138.

Hughes, Thomas (1987). The evolution of large technological systems. In W. Bijker, T. Hughes, and T. Pinch (eds.), *The Social Construction of Technological Systems*. Cambridge, Mass.: MIT Press.

Leal, Ondina Fachel (1990). Popular taste and erudite repertoire: The place and space of television in Brazil. *Cultural Studies* 4(1): 19–29.

Lefebvre, Henri (1971). *Everyday Life in the Modern World*. London: Allen Lane.

Lindlof, Tom (ed.) (1987). *Natural Audiences*. Norwood, N.J.: Ablex.

Livingstone, Sonia M. (1990). *Making Sense of Television: The Psychology of Audience Interpretation*. Oxford: Pergamon.

Lull, James (ed.) (1988). *World Families Watch Television*. London: Sage.

Marcus, George E., and Michael Fischer (1986). *Anthropology as Cultural Critique: An Experimental Moment in the Human Sciences*. Chicago: University of Chicago Press.

Miller, Daniel (1987). *Material Culture and Mass Consumption*. Oxford: Blackwell.

Morley, David (1986). *Family Television: Cultural Power and Domestic Leisure*. London: Comedia.

Morris, Meaghan (1988). Banality in Cultural Studies. *Block* 14: 15–26.

Moyal, Ann (1989). The feminine culture of the telephone: people, patterns, and policy. *Prometheus* 7(1): 5–31.

Pahl, Jan (1990). Household spending: Personal spending and the control of money in marriage. *Sociology* 24(1): 119–138.

Parry, Jonathan, and Maurice Bloch (eds.) (1989). *Money and the Morality of Exchange*. Cambridge, U.K.: Cambridge University Press.

Radway, Janice (1988). Reception study: Ethnography and the problems of dispersed audiences and nomadic subjects. *Cultural Studies* 2(3): 359–376.

Rakow, Linda (1988). Women and the telephone: The gendering of a communications technology. In Chris Kramarae (ed.), *Technology and Women's Voices*, pp. 207–28. London: Routledge.

Rogge, Jan-Uwe, and Klaus Jensen (1988). Everyday life and television in West Germany: An empathic-interpretive perspective on the family as a system. In James Lull (ed.), *World Families Watch Television*. Newbury Park, Calif.: Sage.

Silverstone, Roger (1990). Television and everyday life: Towards an anthropology of the television audience. In Marjorie Ferguson (ed.), *Public Communication: The New Imperatives*, pp. 173–189. London: Sage.

———. (1994) *Television, Technology, and Everyday Life: An Essay in the Sociology of Culture*. London: Routledge.

Silverstone, Roger, Eric Hirsch, and David Morley (1990a). Information and communication technologies and the moral economy of the household. *CRICT Discussion Paper,* Brunel University; also in Roger Silverstone and Eric Hirsch (eds.) (1992), *Consuming Technologies,* pp. 15–31. London: Routledge.

———. (1990b). Listening to a long conversation: An ethnographic approach to the study of information and communication technologies in the home. *CRICT Discussion Paper,* Brunel University; to appear in *Cultural Studies.*

Silverstone, Roger, David Morley, Andrea Dahlberg, and Sonia Livingstone (1989). Families, technologics and consumption: The household and information and communication technologies. *CRICT Discussion Paper,* Brunel University.

Spigel, Lynn (1989). The domestic economy of television viewing in postwar America. *Critical Studies in Mass Communication* 6(4): 337–354.

Strathern, Marilyn (1986). The limits of auto-anthropology. In Anthony Jackson (ed.), *Anthropology at Home,* pp. 16–37 London: Tavistock.

Weber, Max (1949). *The Methodology of the Social Sciences.* Translated and edited by Edward A. Shils and Henry A. Finch. New York: Free Press.

Wilson, Patricia, and Ray Pahl (1988). The changing sociological construct of the family. *The Sociological Review* 36(2): 233–266.

Zaretsky, Eli (1976). *Capitalism, the Family, and Personal Life.* London: Pluto Press.

Audiencing Violence: Watching Homeless Men Watch *Die Hard*

19

John Fiske and Robert Dawson

In this chapter we recount a small-scale ethnographic study of homeless men watching violent movies on television in a church shelter and use it to reflect upon some theoretical issues in audience research and upon the cultural politics involved in understanding "the problem of violence" in our culture.

We offer this study as an instance of "audiencing," by which we refer to the process in which audiences selectively produce meanings and pleasures from texts. "Audiencing" is an engagement in social relations, for the texts from which meanings are made and which constrain (but do not determine) them typically originate in a different position in the social order from that of most of their audiences. "Audiencing" involves negotiating this social difference and the different social interests inscribed within it. Homeless men were a productive audience to study, for they differed diametrically not only from Hollywood but also from the normative addressees of network television—members of middle-American families who earn and spend comfortable incomes. The men we worked with lacked homes, jobs, and families, and we hoped that the cultural practices they had developed to cope with their position of extreme marginalization would provide magnified examples of how the products of the dominant can be made, by audiencing, into the culture of the subordinate.

The homeless are a maximally deprived social formation; they are not a class, for at different times in their lives the homeless have been members of different classes. Homelessness is a social condition that people enter, and sometimes leave, through a number of social and personal causes—job loss, economic deprivation, family breakup, ill-health—but whatever the causes of their entry into this condition, the homeless constitute a social formation systematically deprived of the rewards, security, and empowerment that society offers its more "normal" members.

Homelessness is also a critical feature of contemporary U.S. society. Our culture circulates a range of meanings of homelessness that are pertinent to the social order that has created it. In this study we try to demonstrate that micro cultural instances such as audiencing a violent movie and macro socio-historical forces such as Reaganomics can best be understood conjuncturally, and that neither can be adequately explained without reference to the other. As such, we hope it contributes to the question of defining and studying an audience.

The Observations

The shelter in which these homeless men spend their nights is part of a Protestant church, yet is clearly separated from it. It was built onto the back of the church, is invisible from the front and the main entrance, and the men have to approach it through a back alley out of sight of the regular churchgoers. The shelter can be reached from the church itself but the door is kept locked and the homeless do not have access to the key.

The men are acutely aware of this "back door" access to their "home." Bill, who was one of the most forthright of them, made it abundantly clear that he found this troubling. He was pleased to have a place to eat and sleep yet, as he put it, "We can't offend the decent people who come to worship here." His sarcastic and slightly bitter tone of voice was evidence of his awareness of his subordination and his resentment at the way it was made so obvious and unavoidable. The supervisor himself is somewhat indignant that the comings and goings of the homeless men is through an alley. He talked of the shortcomings of "Christian brotherhood" with regard to the shelter, and thought that the men should be able to enter from the front and to feel more a part of the church itself. He is a Quaker who believes his denomination to be more humanely Christian than the Protestantism of this particular church. He works here only because his church does not have the money to run its own shelter. He cares deeply for the men and fights for their interests with the church administrator. Bill looks upon him as a friend and an equal, although he is puzzled by the supervisor's devout religious belief, which he interprets as evidence of naivety or ignorance of the "real" world.

Inside the shelter there is an open hallway with the supervisor's office on one side and two lounges for the men on the other. Between the lounges is a desk from which the night caretaker can oversee, to some extent, the men's behavior in them. The lounges are similar; the main difference is that smoking is permitted in one, but not the other. This seemed comical to some of the men, who joked about its absurdity and more seriously questioned why the church had such a desire to encourage them to quit smoking.

The lounges are where the men spend most of their time. Here they can watch television, play cards, read, nap, and sometimes fight and cry. The reading material is provided by the church, and is decidedly religious and/or middle-class: *People, Life, Time, Newsweek,* the local daily paper, and religious texts are scattered on a table in the back of the room. Some of the men complained openly about their lack of choice and the inappropriateness of the reading material. They recognized that their wishes meant little to the church, but refused to accept this put-down and smuggled in their own reading matter, almost exclusively tabloids and pornography. The tabloids were discouraged by the church and the pornography was strictly forbidden, so the men read them inside the covers of "respectable" magazines and newspapers. The repression of their own cultural tastes was one more deprivation suffered by these men, and they frequently grumbled about it. They were resentfully aware of the pressure to adapt to inappropriately middle-class sociocultural norms in order to qualify for charity. Their adaptation to them was, however, confined to surface behavior that would fool the authorities, so reading *Hustler* behind the covers of *Life* becomes an accurate metaphor for their ways of living in the space of the dominant Other.

The rules of the shelter, however benevolent or socially responsible their intention, are experienced by the men as part of society's process of minimizing their control over their own lives. They are aware of the inappropriateness of the social norms that are inscribed in the rules, and aware too of the social differences between them and the class from which those norms originate.

Nowhere is this more apparent than in the rule that the men must be out of the shelter between 8:00 A.M. and 5:00 P.M. This is the period of the "normal" working day when the "normal" man is out of the house, and these homeless men are expected to conform to these norms and to use these hours, if not to go to work, at least to look for it. The men, however, make no attempt to conform to such irrelevant norms and the social discipline enshrined in them. For most of them daytime is the time of maximum discomfort and the best way to pass it is asleep. Rather than looking for work, they spend their days looking for places to sleep. The night is when they can be warm, relatively comfortable, and can read, watch television, or interact with each other. They have no intention of wasting it by sleeping.

The rule was also seen, by Bill at least, as another means of achieving the invisibility that he felt was their lot: It ensured that the "decent" people visiting the church would not be disturbed by the sight of homeless men.

The lounges in which they spend their nights are furnished with a strange mixture of middle class castoffs and institutional furnishings. The "nicer" pieces of furniture are, predictably, in the nonsmoking lounge. The rooms were decorated by members of the church, and they look like someone's idea of what is pleasant and comfortable for others, certainly not for themselves. There is one couch in each of the lounges, and the rest of the furniture is upright chairs, most of which are hard and uncomfortable. There are three tables at the back of each lounge for playing cards or reading. Most of the card games played involved gambling, which of course is strictly against the rules. The stakes were cigarettes or money, neither of which could appear on the table, so the men used a system of notes to keep the accounts, which were settled under cover (like the reading of pornography) after the game.

While the men at the shelter have to live in an institutional setting, they do make a minimal space for themselves within it. This space may be seen as an equivalent of "home," for it is a space they feel they possess, however temporarily, a space where they can exercise some control over their immediate conditions of existence. Because of their lack of possessions and their transient lifestyle, this "home" is not a material place constructed from objects, but is a construction of lived practices, of ways of living within the place of the dominant Other. Breaking the rules (reading pornography or gambling) is a way of constructing some sort of personal space or home. This space can only be experienced as "theirs" when it contains elements of resistance to the social order as it is embodied in the regulation of life in the shelter, because the social order appears to them to work consistently to deprive them of what little autonomy they may have.

These practices of living within the place of the Other also inform their television viewing. One set of norms that the men reject are those inscribed in the schedule of television. The shelter did not receive cable, so the men were restricted to network programming and local independent stations. Broadcast television was, as a whole, rejected in favor of video tapes selected by the men from the public library. Bill considered it absurd to ask him why he did not watch regular television: "For what reason, there's nothing on there that interests me, I don't keep set hours, I'm not always around a TV and if I did watch a show and liked it then I'd miss it." His expressed lack of interest in network television may well reflect the irrelevance of its programs to his social situation, but it appears to be the regularity and repetition of its scheduling that he rejects most explicitly. This scheduling is an integral part of the normal structuring of time in our society–a structuring in which family routines, the regular alternation of work and leisure, and television scheduling are all active. To men without family or work, such "routines" are as irrelevant or alien as the programs within them.

There were, however, some exceptions to this. Some men in the nonsmoking lounge did watch scheduled programs during the night–but such programs conform less closely to the dominant social norms than those screened during the "normal" working day; they are programs for marginal people. Some men, too, would watch daytime soap operas in the local university union. In this study we were unable to follow up this viewing taste, though it obviously poses some very interesting questions.

The Taste for Violence

The consistent viewing preference of the smoking lounge men was for violent, R-rated movies, which they borrowed from the local library. During the time of the study, *Sudden Impact, Robocop,* and *Die Hard* were the most frequently chosen. We will describe one specific viewing of *Die Hard.*

In the movie Bruce Willis plays a cop from New York who comes to spend Christmas in Los Angeles with his estranged wife and child. His wife has left him because he would not allow her to develop her own career and she is now a senior executive in the Nakatomi Corporation. During the executives' Christmas party on the 32nd floor of their headquarters, terrorists take over the building and hold the executives to ransom. The plot of the movie involves the lone cop killing off the gangsters one by one and finally defeating them. In the process of restoring law and order, he wins his wife back.

The film starts and the men sporadically begin to watch the screen. Most have seen it before and know which bits to watch. The first scene to capture everyone's attention is the terrorists' invasion of the executives' party. The scene climaxes with the killing of Tagaki, the CEO of the Nakatomi Corporation, which elicits loud and enthusiastic cheers from the homeless men. He is shot coldly and unemotionally after refusing to give the terrorists the computer key to the corporation's vaults. The camera closes up on his impassive face, showing hardly a hint of fear, as he says, "I cannot give you the code, you'll just have to kill me." "Okay," says the terrorist, and he does. The screen explodes in red. Is the red horror or triumph? For the homeless, it appears to be the color of power and victory, suitably climaxed by the swelling crescendo of the music.

But their cheers are based as much on what has preceded the moment as the moment itself. The senior executives of the Nakatomi Corporation are herded together, panic-stricken, and silenced by the authoritative voice of Hans, the leader of the terrorists: "Due to the Nakatomi Corporation's legacy of greed around the globe they're about to be taught a lesson of the real use of power. And you will witness it." He then passes among them, looking for Tagaki, the CEO, reading his vita as he searches. Tagaki, we learn, emigrated from Japan in 1939, he was interned during the war, became a scholarship student at the University of California, earned a law degree from Stanford and an MBA from Harvard, and then became president of Nakatomi Trading and Chairman of Nakatomi Investment Group. The details of educational success leading to corporate power and wealth are an excessive display of what the homeless men lack, but which our ideology has taught them is the right that ought to be within their reach.

The men's reaction to the hero is more complex. In the early part of the movie, when he is an isolated male, armed only with a small handgun, his ingenuity and his physical prowess, in conflict with a well-equipped, well-organized team of terrorists, they appear to identify fairly closely with him. They side with the terrorists against the corporation, and the lone hero against the terrorists. As the movie progresses, and the hero becomes more closely aligned with official social power, embodied in the police force outside the building, their interest in him begins to wane. In fact, they switched off the tape before law and order were restored—and before the hero regained his wife.

The men were selective viewers of the movie—and what they chose to pay most attention to were representations of violence that was directed against the social order. The popularity of represented violence has long been a major concern in U.S. cultural politics. Underlying the concern is the assumption that representations of violence are, ipso facto, *bad.* The research agenda has consequently been directed to investigate how much, if any, of this "badness" is translated into social behavior.

Our theoretical framework differs considerably. It assumes a conjunctural connection between the popularity of represented violence, the committing of actual violence in society, and the contemporary conditions of U.S. capitalism that have produced and then exacerbated the gap between the privileged and the deprived. A conjunctural analysis will not look for direct causes and effects as a deterministic model will, though it will stress the primacy of the historical conditions within which the conjuncture of forces occurs.

The men's taste for antisocial violence arose at the end of a decade of Reaganomics, which minimized the role of the state in social life and maximized that of capital and the market. One of the main functions of the state in capitalism is to preserve the institutions and individual rights of a civil society independently of the interests of party government on the one hand or of the market economy on the other. The civil rights that have suffered under Reaganomics include the right of citizens to housing and to the means of sustenance. Changing these from the rights of citizens into charity given by the privileged to the deprived alters the meaning of homelessness: It makes significant changes in the social relations between the privileged and the deprived, and therefore in the social identities of the homeless. Equally it changes the meanings of being socially advantaged, it changes the social relations and experience of privilege.

The homeless men's sense of their relationship to the social order is a lived experience of Reaganomics: The way in which they watched *Die Hard*, the couch upon which they sat as they watched it, the shelter and the church that situated the couch are all specific, concrete instances of how capitalism in general and the Reagan/Bush inflections of it in particular inform and shape the practices of everyday life.

The qualitative changes in homelessness go hand-in-hand with quantitative ones. The total number of the homeless is almost impossible to assess, but in 1982 estimates varied from 250,000 to 1,000,000, and in 1988 from 500,000 to 3,000,000. The low figures tend to come from government sources, the high from welfare agencies such as the National Coalition for the Homeless. But all agree on the increase. Even H.U.D. admits that the use of shelters for the homeless tripled between 1984 and 1988.

The rise in the number of homeless has been accompanied by a decrease in government assistance. The funding for H.U.D. decreased from $35.7 billion in 1980 to $14.2 billion in 1987 to $7 billion in 1989. The number of housing units assisted by the federal government fell from 326,000 in 1978 to 95,000 in 1987.

Government tax policies under Reagan reduced the tax incentives for private investment in low-income housing and curtailed public housing authorities' ability to issue tax-exempt bonds. As a result, average rents doubled and low-income housing decreased by 30 percent. This loss of low-income housing accompanied a rise in the need for it. Poverty rose 36 percent between 1978 and 1983, AFDC monthly payments were reduced from a high of $520 in 1968 to $325 in 1985 (figures in 1985 dollars), and the economy as a whole lost approximately 16,000,000 jobs, most of them at the lower end of the wage scale.

Homelessness is structural. The majority of the homeless are not personally inadequate. The homeless population does contain more of the mentally and physically dysfunctional than the population at large, but this is because they are the most vulnerable to systematic deprivation.

When homelessness is a structural and systematic deprivation of the weak by a society which is also theirs, and whose material security they feel ought to be one of their rights, then the conditions are ripe for the taste for violence to develop. The violence may be actual or symbolic, it may be initiated by the social order to control the homeless, or by the homeless against the social order.

So riot police sweep the homeless from Tompkins Square Park in Manhattan and from Peoples Park in Berkeley. Police and demonstrators clash in Washington and Minneapolis. Mitch Snyder, a long-term activist on behalf of the homeless, commits suicide in despair over their condition. Activist homeless groups organize illegal squats in empty houses, and on the streets panhandlers confront passersby with more and more aggressive requests—or demands—for money.

The men we worked with showed few signs of physical violence; the most extreme was their intentional positioning of themselves so as to disrupt the walking patterns of pedestrians on the sidewalk or their stubborn occupation of park benches to prevent others from using them. These minor violations of the rights of "decent" people were deliberate and were a source of conscious satisfaction for the men. Society had deprived them of all means of asserting their rights and identities except their bodies, and they so used their bodies' abilities to occupy physical space as a way to assert their right to a position in the social space. The body is where violence is put into practice, so when social relations are reduced to the physicality of bodies, it is hardly surprising that one of the most accessible ways of engaging in them is by violence.

Such forms of violence, whether real or represented, are ways of engaging in social relations, not expressions of an uncontrolled, animal instinct for aggression. The origins of violence are many and complex, but what often organizes and activates them are the social conditions that produce the gap of privilege, particularly when deprivation is experienced by the deprived through the privileged's ideologies of individual freedom, equal opportunity, meritocracy, and masculinity. In saying this we do not wish to imply that the problem of violence is confined to the socially deprived. It is not. Domestic violence, for instance, is at least as common among the privileged. But we do believe social deprivation to be a key factor in many of the social conditions in which the taste for violence is encouraged. Israel et al. (1972), for example, found that male heavy viewers of television violence came disproportionately from lower-income groups with low levels of formal education. They were also disproportionately African American. Economic, educational, and racial axes of subordination came together and set up contradictory relationships with masculinity, which is an axis of domination.

When the homeless men chose to watch violent movies, the choice was made in an environment deeply inscribed by signs of social rejection and deprivation. Their taste for symbolic violence was significantly associated with their resentment toward and alienation from the society that systematically denied them access to the rewards that its dominant ideologies tell them are their rights.

The search for the origins of the taste for violence should not, then, be directed toward the aggressive instincts supposedly born into every member of the human species and developed especially in the males. The "basic instincts" or "lowest common denominator" theories cannot answer the question of why it is that the taste for violent images is stronger among subordinated and repressed social formations than it is among dominant ones.

The emphasis on social conditions rather than instinct is better able to account for the fact that not all violent images are popular. Many films and television shows with high levels of violence fail to be taken up and made into popular culture: the taste for violence is a discriminating one. It also enables us to address the fact that the disapproval of violent images originates largely from the socially advantaged, and the attempt to censor or control them is often part of the process of social domination.

Such a theory proposes, then, that certain images of violence are widely popular not because of a universal aggressive instinct, but because of widespread, almost universal, conditions of subordination. Young urban Aboriginals in Australia derive great pleasure from watching old Westerns on television; their pleasure peaks at the moment of the Indians' triumph, when they take the homestead or the wagon train (Hodge and Tripp, 1986). The supervisor told us of an almost identical practice of homeless Native Americans: They would watch Westerns on the shelter's VCR and switch off the tape at the moment of the Indians' victory, thus obliterating the restoration of white colonializing "law" in the second half of the narrative. Other Australian Aboriginals read Rambo's violence as representing the conflict between members of the Third World and the white officer class (Fiske, 1989). African Americans at the turn of the century read the violence in Buffalo Bill's touring Wild West Show as a representation of Indian genocide that paralleled their own history (Lipsitz, 1990). Working class male youths in London made Kung Fu movies from Hong Kong (Cohen and Robbins, 1979) into their popular culture just as readily as Australian Aboriginals made Rambo movies from Hollywood into theirs.

Certain representations of violence enable subordinated people to articulate symbolically their sense of opposition and hostility to the particular forms of domination that oppress them. These representations must contain not only violence, but also markers of ethnic, class, age, or national difference that are potrayed not as natural essences, but as structural agents of power and disempowerment.

The conventions by which television violence is represented support this view. The statistically typical hero is a white, middle-class or classless male, in his sexual and physical prime, who is pitted against villains who are statistically typified as non-white or alien, nonmiddle-class, younger or older than the hero—and ugly. The conflict between them is multidimensional. It is a conflict between the bodies of individual males, it is a conflict between law and order and social disruption, it is a conflict between good and evil, and simultaneously, it is a conflict between social centrality (or the dominant norms) and social subordination or marginality. Symbolic violence is a concrete performance of social inequality, and its popularity suggests that it can offer the subordinated both a representation of their own fighting ability and an articulation of their resentment toward the social order that oppresses them.

The violence in *Die Hard* is always associated with power. The terrorists frequently deny any political agenda—they do not wish to overthrow or change capitalism. What they want is the power that capitalism itself validates, the power of money, the $640 million in the corporation's vaults. The power is made material in the Nakatomi skyscraper itself, in its art objects and antiques, its elaborate computer systems, the numerous models of the corporation's capital projects around the world—all of which are spectacularly and systematically destroyed during the course of the movie. Spectacle invites popular participation, and indeed, is only spectacular by virtue of that participation. The spectacle here offers the deprived the opportunity to participate in vicarious revenge against those who have exploited the system that has deprived them.

The other dimension of power is power over people, power over immediate social relations—another social deprivation for these homeless men, who, as we have seen, are systematically disempowered by the rules and regulations of the shelter. The movie is full of struggles for power that always reverse the norm—the terrorists gain power over the corporation and defeat the police, the isolated hero gains power over the terrorists, the black subordinate cop gains power over the white chief of the LAPD—power is a site of struggle between those without social resources and those with; and those without are shown to have the ability to win tactical battles if not the final victory. In each of these social relations of power, the homeless men aligned themselves with the weaker party and gained great pleasure from their triumphs over the stronger, or from their endurance in the face of it. Violence is the way these power struggles are represented and indeed is the only way they can be enacted popularly, for if the deprived are to have their moments of victory, these moments can most spectacularly and effectively be represented through the only resources of which they cannot be deprived—their bodies, their physical strength and endurance, and their resourcefulness. The fights in the movie are spectacularly elongated, the body's endurance of pain and punishment and its ability to rise undefeated from them is displayed in fantastic detail. Toward the end of the movie the body of the hero is as physically battered, bloody, and weakened as the homeless are socially.

Both forms of power are articulated with and in masculinity. Corporate power is shown in the maleness of the boardroom (interrupted only by the hero's wife, who, significantly, is at this point corporate, separated, and thus defeminized). It is challenged by its mirror image—the organized, disciplined power of the terrorist group. Face-to-face power is articulated with and by the male body in action, particularly violent action, and the hero is popular with these viewers as long as he exhibits only this sort of power. As the movie progresses and he becomes more closely aligned with the police force, his popularity, with them at least, declines.

Another favored scene was that in which the terrorists easily repelled the police's attempt to retake the building. The climax of this scene occurred when an armored car attempting to batter its way in was disabled by a rocket launched by the terrorists. The hero, from his hidden vantage point, can see them preparing to fire another rocket, whose purpose is not to save the building from attack, but to complete the tactically unnecessary destruction of the machine and the men in it. In voice-over he begs them not to. They do, and as the camera lingers on the apparently pointless destruction and death, the homeless men cheer enthusiastically. The hero's plea works to position the viewer inside the norms of the dominant social order against the excessive violence of terrorists, but the homeless refused this positioning: They found the destruction both significant and pleasurable. Indeed, their pleasure probably stemmed as much from their refusal of the hero's socially correct way to understand the violence as from the spectacle of the violence itself. There is pleasure for the deprived in their ability to refuse and oppose the narrative position that is the equivalent of the social position of those whose privilege has produced their deprivation. The ability to read antisocial meanings against a prosocial text is the equivalent of reading pornography under the cover of a respectable newspaper.

Movies like *Die Hard* mattered intensely to the men. We observed more vitality, enthusiasm, and pleasure in their watching of representational violence than in any other aspect of their daily lives. It may well be that such imaginative experiences offer some of the only opportunities left to such men to validate their identity and difference from the social order that systematically deprives them of identity and self-esteem and reduces to an absolute minimum those conditions of their lives over which they can exert any control at all.

Problems of Censorship and Control

At the library from which the men borrow the videos, the librarian confirmed that the homeless consistently chose violent films to watch. He expressed extreme concern about the feared effects of violent films upon men such as these. He was explicit that "people like us" (white educated men) were able to watch these films without being affected, but he doubted if the homeless could.

His worries are typical of the socially advantaged and they lead easily into a desire to control or censor representations of violence in the fear that they may produce real violence among the socially deprived (who are implicitly characterized not as deprived, but as vulnerable, weak, incompetent, and thus in need of protection for their own sake). Such calls for censorship fail to take account of the social conditions of those among whom certain representations of violence are popular—they impose a top-down understanding and deny the validity of the meanings that such violence might have for the subordinate.

Censorship, therefore, becomes very problematic; whatever its motives, a top-down censorship always works with power and against a bottom-up sense of social difference. The calls to censor representations of violence, however ethical their rhetoric, can work to silence the culture of the oppressed and thus to contribute to their oppression. The taste for such violence originates in the structural effects of the same social order that privileges those who call most vehemently for its censorship. The assumption that social violence is caused by represented violence serves as a cultural alibi, for it enables the socially privileged to avoid the uncomfortable idea that their position of privilege itself is a significant feature of the social conditions that nourish both the taste for represented violence and the commission of actual violence.

The social conditions are indelibly gendered. Our ideology of masculinity equates the masculine with the ability to exert power over others publicly: Masculinity must not only be constantly performed, but its performance must also be visible and acclaimed. When a society restricts certain categories of men with this sense of their masculinity to social conditions that deprive them almost totally of the means to achieve it, then it creates the conditions under which the taste for violence, both real and represented, will flourish.

In their audiencing of *Die Hard*, however, the homeless men did not appear to engage with its gender politics. We observed no hostility toward the hero's estranged wife, and they were not at all interested in watching him regain his marital control over her, for they switched off the tape before he achieved it. But at other times, some of the men did express misogynist sentiments, particularly by blaming their unemployment on women taking the jobs that were "rightfully" theirs. The librarian had told us of one homeless man who had viewed the shower scene from *Psycho* over and over for hours. We observed no such obsessive behavior, and, given the librarian's position on violence and homeless men, we suspect that the story has become exaggerated in the telling. But, nonetheless, there are grounds for being deeply worried that, in some deprived men, the combination of the belief that women are the cause of their deprivation, a taste for pornography, and an obsessive pleasure in sexualized violent images such as those in *Psycho*'s shower scene is part of a culture that all too often normalizes violence against women. We do, however, wish to raise the question of the extent to which a taste for violent images may be either causal or symptomatic in such a culture, and thus to question how strategically important an objective for social action are the images themselves. The dangers in overemphasizing their causality are that the effort to control them may divert energy from more effective objectives and that if such control is achieved its effect may be demoralizingly less than that which had been desired and predicted.

In the debate about violence, censorship is a red herring. Censorship is an effect-oriented "solution" that reduces images to stimuli that mean the same to everyone and therefore produce singular effects. The social dimension of texts, however, cannot be read so simply from the texts themselves, but requires us to study *which* of their potential meanings are taken up socially by *which* audiences in *which* conditions. Popular images are rarely homogeneous, and the contradictions in them are often activated by audiences in ways that may surprise critics and political theorists. Some women fans of horror movies, for example, dissociate themselves from the "weak" women who are the typical victims of the genre and refer to them as "GVs" (gratuitous victims) who are stupid enough to walk past the graveyard after dark! Their pleasures in this genre lie in contrasting the actions of "GVs" with those of the "strong" woman who bravely fights against the over-masculine monster (Bernardi, 1990). For them, images of male violence against women are images not of the victimization of women but of the terrain over which women have to fight. Censoring the violence would also censor representations of resistance to the patriarchal order within which the violence originates. Censorship has rarely solved the problems it has addressed. Except in the most extreme cases, contradiction and argument are generally better ways of opposing gendered violence in the domain of representation: "Labeling" images so that they are not unwittingly viewed and controlling their distribution to minors may also be viable measures in certain cases, but prohibition rarely is. Prohibition is more appropriate to, and more effective in, the material domain of social behavior. Actual violence against women requires unequivocal legal action, but images of violence are not the same as actual violence.

Images of violence do not constitute a homogeneous category: not only are they contradictory both in themselves and in the meanings that may be produced from them, but also, even at the level of manifest representation, they require differentiation. Representations of male violence against women are categorically different from those of male violence against the social order. In patriarchy women are a politically deprived group. If certain representations of violence can be shown to contribute to this deprivation there may be an argument for regulating either the images or their distribution. But no such argument can be mounted for images of the violence of the subordinated against the social order that subordinates them. No "speech" of the subordinate against the dominant should ever be censored. But dominant speech, for example, sexist or racist, which contributes to subordination, needs to be thought of quite differently.

The issue of censorship is further complicated by the fact that the politics of popular taste are often deeply conflicted. In particular, in our current social conditions, class, race, and gender politics often contradict one another—progressiveness in one is often accompanied by repressiveness in the others. Representations of violence may well offer potentially progressive meanings in class or racial politics but repressive ones in those of gender. Controlling such representations may be progressive in gender politics and work toward *diminishing* power difference between the sexes but in class or racial politics its effect may be quite different: In these cases it may work to maintain the power difference between the privileged and the deprived—and it is this difference that, we argue, nourishes the taste for violence both in society and on the screen, and therefore encourages what it seeks to prevent.

Censorship is a red herring because it ignores the conditions under which a taste for violence may flourish, and which violent images may help to maintain. But the red herring itself has had real effects on media research. It produced a generation of studies that confined themselves to effects and ignored larger causes. Politically, this is hardly surprising. The government agencies that funded most of the research and the social lobby groups that called for it had no interest in putting the U.S. social system itself on the research agenda. They had no interest in investigating the correlation between the circulation of violent images in a society and its socioeconomic policies and political system. They did not wish to interrogate the evidence that the society that produced the most violent images (i.e., the United States) was the one that had the fewest mechanisms for equalizing wealth and for caring for the welfare of its weakest and most disadvantaged citizens. The research agenda they funded excluded the gap between the haves and have-nots from "the problem of violence" and implied, therefore, that in the event of a mismatch between the values of the social order and the values or behavior of some people within it, reforming efforts should be directed solely to the behavior and not at all to the social order. Such a research agenda situates the "problem of violence" within the behavior of media producers, the images they produce, and the audiences to which they are distributed. Regulation is recommended, then, when behaviors of producers and audiences around images fail to accord with the values and interests of the social formations that are privileged by that same social order that has ruled itself out of the possible causes of the problem and therefore out of the possible objects of reform.

Methodological and Theoretical Implications

This "problem of violence" has occupied much of television audience research since the early 1960s. This study differentiates itself clearly from most of this work in a number of ways, of which the most important is what we might label, simplistically, the relationship between representation and reality.

Early "effects" research defined television (representation) as a stimulus and the audience-as-miniaturized-society (reality) as the site of response to it. Predictably, reducing representation to a stimulus and audiencing to a response meant that such studies failed to translate from the clean, controlled world of the experiment into the muddied, confused conditions under which social beings live. Its exclusion of sociohistorical conditions from either representation or reality disqualified it from any effective intervention into the inescapably sociohistorical realm of policy and regulation to which it commonly addressed its findings. Equally, its unidirectional model of the relationship (from stimulus to response, from representation to reality) led it to the simplistic assumption that images affect society and not vice versa. Seeing the culture industries as the originary producers of culture led to the assumption that control over the images produced by them could be extended to control over the social behavior produced in turn by those images. Fortunately, neither cultural production nor social behavior are as controllable as that.

During the same period, more critical and theoretical studies, such as those of the Frankfurt School, were equally reductionist and unidirectional. For them, the system of representation was an agent of commodification, and its effect upon the audience was either to turn it into a commodity itself to be exchanged between producers and advertisers or to turn it into a buyer of commodities and thus into a producer of capitalism. The model is remarkably similar to that of effects research, with capitalism added—capitalism produces commodified culture, which produces commodified people.

The cultivation theory developed by George Gerbner may not be quite as reductionist, but it is still effect-centered. It does at least recognize that the system of representation has a content that is worthy of analysis and that the content (defined as manifest, systematic, predictable patterns) bears some relation to the more abstract structures of values and social differences in society at large. It recognized, too, that change is possible and indeed has tried to track that change, but it has no theory of how change might occur. By casting the audience as those who were acculturated by the system it still modeled them as the recipients of effects. Like other effect-oriented models, it never saw the audience as an agent in the social circulation of meanings in which television plays such an important part, and nor, particularly, did it see the audience as a potential agent of social change.

Uses and gratifications theory recast the audience from acted upon to active, and television from active to being acted upon. The question then became not what television does to people, but what people do to television. The theory has been rightly criticized for being too individualistic and for defining circularity the needs met by the media in terms of the gratifications that people report from their media use. It differs also from our study in its assumption that people's needs are actually gratified, however partially, by their media use. Our study suggests that media use articulates and clarifies social needs but in no way gratifies them. The homeless men's sense of unjust social deprivation was validated and confirmed by their watching of *Die Hard,* certainly not assuaged by it.

The turn to Gramsci and Volosinov, initiated by Stuart Hall and his colleagues at the Birmingham Centre for Contemporary Cultural Studies blew unidirectional and effect-oriented models out the window. For them culture was seen not as something imposed upon a helpless populace, but a site of struggle: The struggle over meaning was not just a reflection of the social struggle but a part of it. If it is generally true that no social system can survive without a system of meanings and values to hold it in place, then such a truth is particularly pertinent to democratic free-market societies with their need to maintain the social order by means of consent rather than coercion. The organization of consent around the interests of the power bloc involves constant negotiation between different and conflicting social interests, and culture is central in this process.

In societies as complex and elaborated as that of the United States consent is never finally achieved, but is constantly subject to contestation, and at certain moments, in certain social domains, its precariousness means that social change becomes possible. Homelessness may well be such a domain and such a period. If so, a conjunctural analysis should be able to reveal how a variety of social forces come together in a unique and historically specific configuration to destabilize the point of consent to a point where a significant change becomes possible.

At such conjunctures the relationship between the various social forces or agencies is never deterministic or unidirectional: One can never be said to produce direct effects upon another. Conjunctural associations are always multidirectional and interactional. Our homeless men cheering the death of the CEO constitute a minute but material and significant activity in such a conjunctural association. We have tried to indicate some of the other agencies that are conjuncturally related: They include Reaganomics and its reconfiguration of social relations, particularly along the axes of class and gender; charity and volunteerism and their connections with organized religion; the welfare function of the state and its ability to produce politically laden meanings of civil rights; and the media industries and their production of commodities that can only meet their own economic needs by also meeting the cultural needs of people who live under and contribute to the social conditions within which their popularity and profitability have to be achieved.

Hollywood's production of a crop of violent movies in the late 1980s and early 1990s is conjuncturally related to the widening of the gap between the privileged and the deprived that Reaganism produced. The popularity of some of these movies is due in part to the potential they offer to people experiencing this acute difference of privilege to articulate and validate their social position in relation to it. The violence in the streets, the violent images in the movies, and the violent pleasures in the minds of audiences are conjuncturally related; they are all part of that social circulation of meanings and values that we call culture. *Die Hard* offered certain images of violence from which the homeless men produced meanings and pleasures that were pertinent to their social situation: the industrial producers of the movie and the semiotic producers of meanings from it (i.e., its audiences) are dependent upon each other—both are agents and both are acted upon: Audiencing a movie is both productive and constrained, as is producing one.

The aim of our study was to understand how audiencing fitted into the immediate social conditions of those engaged in it, and to suggest how these immediate social conditions might be significantly related to the macrostructural forces of national politics and economics. The ways in which meanings of homelessness are generated and kept in circulation among, for instance, Reaganomic policy, religious and welfare institutions, the media, the givers of charity, and the homeless themselves occasionally take a material form that makes them available for empirical analysis. The cultural analyst needs to select sites where this circulation does become visible, but the analysis of any such site is significant only when it is set into a theorized relationship with other sites, however "invisible" these relations might be. Empirical analysis and theoretical analysis need to support (and challenge) each other.

An audience is a cultural site, but it is significant only in its social relations; it cannot be abstracted from them, dressed in a white coat and put in a laboratory. The advantage of "laboratory" study is its ability to control key variables. In a more naturalistic study, the impossibility of controlling the conditions of investigation must not be seen as a problem to be overcome (or apologized for) but as a benefit to be taken advantage of. Our study is, in its small way, a contribution to the social circulation of meanings of homelessness, and we wanted the homeless men to contribute actively to it, and not to be merely objects within it. We did not want to control them.

Our relationship to them was, therefore, integral to the study and not just a precondition for it. It took a long time and much discussion to convince the supervisor and the church authorities that the study was a valid one, and that it would be conducted sympathetically and sensitively. Much more problematic, both theoretically and ethically, was the attitude of the men. Many were suspicious and saw the study as another way in which "They" were attempting to interfere in their lives; others simply wanted nothing to do with it, figuring, quite accurately, that there was nothing in it for them. Robert Dawson made all the initial approaches, and visited the shelter at least twice a week for three months. We intended that when the men were more comfortable with his presence John Fiske should join him. In practice, this point was never reached, so all the observational work was by Dawson (see Dawson, 1990). Even then, it was only one man, Bill, who was willing to talk at all freely, but within limits that he set. Some other men would make some comments when they chose to, and some kept as distant as they could. After the viewing of *Die Hard*, for instance, Dawson tried to get the men to discuss their reactions to the movie. They saw no point in this and preferred to watch the next tape. We might have got more data if Dawson had been more insistent, but the status of that data would have been questionable. As it is, our information comes from those who chose to give it to us, and what we have is what they chose to give. This is quite reasonable, for what we wanted was some access to their point of view. The fact that this point was one from which to view us and our study as well as their homelessness was both inevitable and proper, as was their desire to control and limit our access to their position.

They would not allow themselves to be tape-recorded, and Dawson soon found that any note-taking aroused their suspicion. Our data, then, are drawn from notes made from memory after each visit, which is why there are so few direct quotations; the ones that there are have been filtered through the observer's memory.

Conversations with ethnographers produce data that exist only because of the ethnographers' presence; they are not "natural" but have been generated by and for the process of study. This does not necessarily invalidate them, particularly if the ethnographer has played as nondirective a role as possible in the conversation, but it does require us to check our interpretation of them against other types of data, particularly those that are found or observed rather than generated.

The placement of the shelter and its access, for instance, or the rules and the furnishings within it were in no way affected by the presence of the observer. We like to think, too, that Dawson's presence did not significantly affect the men's behavior in watching television, or their undercover reading or gambling. His observational method was to be as unobtrusive and unassertive as possible—to talk with those who wanted to, to share cigarettes, to watch television, and just to pass the time there, much as the men did.

The interpretation of each type of data needs to be informed by that of the other. The meanings of social rejection that we are trying to interpret inhere in the structural relationship between different orders of data, not in any one set of data on its own. Bill's verbalized resentment of how "decent" people viewed the homeless (data generated by the presence of the observer) is related to the hidden access to the shelter and to its furnishings (found data) which are also related to the way Bill and his fellows watched *Die Hard* (observed behavior).

Interpretation is a structural and theoretical process. The relationship among the specific data supplied above is only made significant by its insertion into a macrotheoretical understanding of the social order and its class relations. Indeed, the choice of these data as worthy of study is a theoretically and politically informed choice. So, too, is the process of transcribing the data into words. Our description of the homeless as "deprived" rather than "unfortunate" or "inadequate" is part of the same political and theoretical framework as our description of the furnishings and decoration of the lounges in terms of the social differences between their point of origin and their current use. If no description can be objective because discourse is not, then the discursivity of the description must be recognized as part of the interpretive process, and not as part of the object to be interpreted. In practice, there can be no categorical distinction between observation, description, and interpretation. Choosing which data to observe and which to ignore (and however "thick" the description, more must be omitted than included) and choosing the words or conceptual categories by which to identify and transcribe the data are already acts of interpretation, already theory in practice.

The relationship between theory and data is a major problem for the ethnographer. Even in its most "scientific" forms, ethnography can never produce the purely objective, factual data that empiricism claims to; no ethnographer will assume that his or her observations are repeatable by others. Ethnographic data can never have the property of objective reality. On the other hand, there can be little point in empirical (rather than empiricist) investigation of any sort if theory is granted a totally predetermining role. There is real value in empirical data that put flesh on a theoretical skeleton and show how a macrotheory can be revealed in a particular analysis, but this is not all that such data can offer: Ethnographers must always expect to be surprised, must always expect, indeed desire, their theoretical preconceptions to be challenged and to require modification. Ethnographic data frequently set up productive tensions within a theoretical framework: They are not "facts" with a nontheoretical existence and meaning of their own.

This tension between theory and data has its equivalences in the tension between macro and micro social experiences and explanations or, to put it another way, between structures and practices. Culture is, in Raymond William's words "a whole way of life" (1989) that is experienced by people at multiple levels from the macro to the micro, the analysis and interpretation of which must therefore encompass an equivalent perspectival range. The instance of these homeless men watching a specific scene from *Die Hard* is not an isolated or self-sufficient moment in their culture. The television screen and its viewers interact in a symbolic environment that precedes both of them; the shelter's lounge is already densely packed with social meanings through which macrostructural factors inform micropractices. The newspapers, the furniture, and the regulations all bear macrosocial meanings of class difference, of power and privilege that prestructure the micropractices of reading pornography or of refusing to sleep. The specifics of living in this shelter show how some of the structural contradictions between Christianity and capitalism are experienced in practice, where donating cast-off furniture is materially benevolent but symbolically repressive, where discouraging smoking may promote the physical health of the men while oppressing them socially. Such acts of charity, such well-intentioned regulations, always naturalize control as the right of the class with power, however benevolently they wish to make it appear. Gambling against the rules of the shelter, not using the daytime to look for work, and refusing to watch the hero restore law and order at the end of *Die Hard* are cultural practices that are structured within a particular historical social order. So too, the desire to censor representations of violence on television and the refusal to recognize any legitimacy in some of the specific cultural pleasures and meanings that they offer is structurally equivalent to the back alley entrance to the shelter: Both are manifestations of the systematic disempowerment and delegitimation of the subordinate.

Ethnography is not a discipline but a methodology that can be used in sociology, anthropology, linguistics, or cultural studies. Our use of it is within cultural studies, so our purpose was to investigate in specific instances the generation of meanings of social identity, of social relations, and of social experience, and the role that texts and other signifying objects and practices played in that process. We were not concerned to produce an ethnography of the homeless (which would serve an equally valid, but more sociological, purpose). Cultural ethnography of the sort we undertook makes no claims to being a social science and thus differs from the well-established effects and survey research traditions. Rather, it takes its place alongside textual and ideological approaches, for the data it produces require analysis in the same way as do texts and ideological practices. What ethnography does is to extend the object of analysis to include socially specific practices that textual or ideological approaches often overlook.

The instances of culture in process that form the stuff of cultural ethnography are not statistically representative, so they do not claim to be generalizable in the way that quantitative and laboratory studies do. But context-specific practices of culture are not simply isolated fragments, they are, rather, *systemic.* By this we mean that they are concrete instances of a system in practice, and through their systematicity we can generalize out, not to specific groups of people, nor to their social acts, but to the structural forces that shape the social order. The methodological model here is drawn from linguistics. A language utterance is systematic, for in its concrete specificity the linguist can trace the abstract, structuring principles of the language system. Linguists do not ask of the utterances under study how statistically representative they are of other utterances, nor do they expect to be able to use them to predict other utterances. Their concern is to explore the concrete uses of an abstract system and the interrelationships between the two. In particular, this school of linguistics, with its roots in Volosinov on the one hand and sociolinguists such as Sapir and Labov on the other, explores how it is that linguistic systems and social systems interface in the moment of utterance.

There is another difference between this systemic model and the statistically representative one, and that lies in its evaluation of norms. Quantitative social science is a normative epistemology: It finds greatest significance in what is statistically most normal, and its politics tend to be conservative. A linguistic model, on the other hand, allows for greater significance to be granted to the specific, the marginal, and the abnormal. To understand the full capabilities of a language system, we have to be able to understand its extreme or deviant uses. Linguists also believe that linguistic change is more likely to originate in deviant or marginal uses than in more normal ones.

So, too, the tension that motivates social change may well be seen more clearly on the margins of a social system than at its center; it may well reveal itself more urgently in deviant rather than normal social experience. Another motor of social change can be found in those social differences that are maintained by the subordinate in their own interests, rather than in those that are produced and controlled from above. In both linguistics and cultural studies ethnography can help us to understand the practices of the subordinate and marginalized by which means they exploit the flexibility of the dominant system (linguistic, cultural, or social) and turn it, at times, to their own advantage.

The "bottom-up" view of culture afforded by ethnographic observation has not yet contributed much to the debate on violence. In general, this debate has been framed by a conservative agenda and conducted from a top-down perspective. We hope that this study will contribute by arguing against that framework and that perspective, and by suggesting that cultural ethnography can offer research methods that are capable of producing more critical and progressive insights into the problem.

References

Bernardi, D. (1990). Masks of pleasure: Female forms of the contemporary horror film. Unpublished paper, University of Wisconsin at Madison.

Cohen, P., and D. Robbins (1979). *Knuckle Sandwich.* Harmondsworth: Penguin.

Dawson, R. (1990). Culture and deprivation: Ethnography and everyday life. Paper presented at the International Communication Association Conference, Dublin, July.

Fiske, J. (1989). *Understanding Popular Culture.* Boston: Unwin Hyman.

Hodge, R., and D. Tripp (1986). *Children and Television.* Cambridge, Eng.: Polity.
Israel, Harold, W. R. Simmons and Associates, and John P. Robinson (1972). Demographic characteristics of viewers of television violence and news programs. In George A. Comstock and Eli A. Rubinstein (eds.), *Television and Social Behavior,* vol. 4, *Television in Day-to-Day Life: Patterns of Use,* pp. 87—128. Washington, D.C.: U.S. Government Printing Office.
Lipsitz, G. (1990). *Time Passages: Collective Memory and American Popular Culture.* Minneapolis: University of Minnesota.
Williams, R. (1989). *Resources of Hope.* London: Verso.

The Geography of Television: Ethnography, Communications, and Community

20

David Morley

> *For most people there are only two places in the world—where they live and their TV set.*
>
> Don DeLillo (1985)

Soja (1989) argues that up until now, time and history have occupied a privileged position in critical theory while, as Foucault puts it, "Space was treated as the dead, the fixed, the undialectical, the immobile. Time, on the contrary, was richness, fecundity, life, dialectic" (quoted in Soja, 1989: 4). Thus capitalism itself has been treated as an historical but only incidentally geographical process, the geography of which, when seen at all, has been recognized only as an external constraint or as an almost incidental outcome. Geography, for Marx himself, was little more than an "unnecessary complication." At the same time, as Soja notes, modern geography itself was "reduced primarily to the accumulation, classification and theoretically innocent representation of factual material, describing the 'real differentiation' of the earth's surface—to the study of outcomes, the end products of dynamic processes best understood by others" (1989: 4).

Soja's own project involves the recognition of the fundamental distinction between space per se—space as a given, natural backdrop to human affairs—and the created space of social organization and production—the "second nature" that is the proper object of a materialist interpretation of spatiality. As Harvey observes:

> Marx . . . Weber and Durkheim all . . . prioritize time and history over space and geography and, where they treat the latter at all, tend to view them unproblematically, as the stable context or site for historical action . . . . The way in which spatial relations and geographical configurations are produced in the first place passes . . . unremarked, ignored (Harvey, 1985: 141—142).

Moreover, as he also argues, "It is invidious to regard places, communities, cities, regions or even nations as 'things in themselves' at a time when the global flexibility of capitalism is greater than ever . . . . Yet a global strategy of resistance and transformations has to begin with the realities of place and community" (quoted in Robins, 1989: 145).

Soja's declared aim is to spatialize the (conventional) historical narrative, to reveal "how relations of power and discipline are inscribed into the apparently innocent spaciality of social life" and thus transcend the "fixed dead . . . Cartesian cartography of spatial science" (1989: 6—7), which sees only "natural forms," susceptible to little beyond measurement and phenomenal description.

Foucault observes that "the great obsession of the 19th Century was, as we know, history . . . (but) the present epoch will perhaps be above all the epoch of space" (1986: 22). Jameson (1984) argues for the spatial specificity of the cultural logic of (postmodern) "Late Capitalism." Some years ago, John Berger argued that "Prophesy now involves a geographical rather than historical projection; it is space, not time, that hides consequences from us" (quoted in Soja, 1988: 22). It is in this context that we should heed Foucault's injunction: "A whole history remains to be written of *spaces*—which would at the same time be the history of *powers* . . . from the great strategies of geopolitics to the little tactics of the habitat" (1980: 149). I have, with Kevin Robins elsewhere (Morley and Robins, 1989; 1990), begun an exploration of the issues at stake once we try to think of communications processes within the terms of a postmodern geography, and once we begin to consider the role of communications in the ongoing construction and reconstruction of social spaces and social relations. At a meta-level Robins (1989) has argued that in the present period we are involved in fundamental processes of political and economic restructuring and transformation that presage (if not already reflecting) a shift beyond the Fordist system of accumulation and social regulation. Robins's central point is that at the heart of these historical developments is a process of radical spatial restructuring and reconfiguration, which is "at once a transformation of the spatial matrix of accumulation and of the subjective experience of, and orientation to, space and spatiality. Its analysis . . . demands a social theory that is informed by the geographical imagination" (1989: 145).

The central point, for my present purposes, concerns the fact that the image industries, as Robins notes, are implicated in these sociospatial processes in significant and distinctive ways. Thus, as Robins argues, "issues around the politics of communication converge with the politics of space and place: questions of communication are also about the nature and scope of community" (1989: 146). The further point, for the argument of this paper, is that theoretical work that has begun to take on these questions—for instance, in the context of debates around satellite television and cultural identity [1]—has done so at a very abstracted level, principally in the context of international geopolitics. However, the force of Foucault's remarks quoted earlier is, of course, to remind us that the "geographical imagination" and its refocusing of the relation of communications and geography, needs to be applied, as he puts it, to the "little tactics of the habitat" every bit as much as to the "great strategies of geopolitics" (1980: 149). If one of the central functions of communications systems is to articulate different spaces (the public and the private, the national and the international) and, necessarily, in so doing, to transgress boundaries (whether the boundary around the domestic household, or that around the nation), then our analytical framework must be capable of being applied at both the micro- and macrolevels.

It is in this context that I address in this chapter the question of the place of ethnographic studies of media consumption in the analysis of the simultaneous dynamic of globalization and localization in contemporary culture. The key issue here is that of the status of small-scale studies of microprocess(es) in the analysis of these macroissues. My argument is that it is precisely through such detailed "local" studies that we will most effectively grasp the significance of the processes of globalization and internationalization that have been widely identified as central to contemporary culture. Though the current expression of this argument tends to be couched in terms of "postmodern" theory, it is as well to remember that the ground for it was laid many years ago (and in terms of a technologically determinist argument that has never quite been shaken off) by Marshall McLuhan, who, as is well known, argued that the effect of television and computer technology was to erase space-time differences and to herald a new audiovisual age of global Gemeinschaft. Thus, McLuhan and Fiore argued: "Electric circuitry has overthrown the regime of 'time' and 'space' and pours upon us incessantly and continually the concerns of all other men. . . .Ours is a brand new world of "allatonceness." "'Time' has ceased, 'space' has vanished. We now live in a global village . . . " (McLuhan and Fiore, 1967: 16). My basic (indeed simplistic) argument is that before we jump to too many conclusions about the nature of this global village life it may be worth our while to visit some of the tents and see what the people living in them are up to. We could even ask them what they thought about all this themselves, and we might (possibly) discover interesting variations in how things are done in different places. The first part of my argument will be concerned with issues concerning "texts" and their "readings," which have been the subject of some debate in recent media analysis. The second part will principally be concerned with the role of communications systems in constituting (and transforming) the elementary division between the public and private spheres, and with the role of communications in articulating these domains across their shifting boundaries (involving issues of separation, articulation, transgression, and transformation).

Ethnographers, Questions, and Answers

In recent years there has been considerable debate concerning the methodological issues surrounding the practice of ethnographic and other types of qualitative audience research (see, inter alia, Hartley, 1987; Ang, 1991; Clifford and Marcus, 1986). At the most fundamental level, Feuer (1988) queries the very point of undertaking empirical work with audiences at all. Feuer's argument is that much of this audience research is involved in a logical regress, in which the attribution of meaning is endlessly deferred—from text to analysis and to audience.

Feuer argues that, in displacing the text onto the audience, the reception theorist constantly risks falling back into an "empiricism of the subject" by granting a privileged status to the interpretations of the audience over those of the critic, and only succeeding in producing a new "text" to be interpreted—the "text" of the audience's discourse. Feuer argues that once one recognises the textual status of the audience's response, the problem of interpretation is exacerbated (as the researcher then has to attempt to read the unconscious of the audience without benefit of the therapeutic situation, and thus may well tend to privilege the conscious or easily articulated response). Feuer concludes that studies of this type are not necessarily gaining any greater access to the spectator's unconscious responses to texts than the more speculative attempts by film theorists to "imagine" the possible implications of spectator positioning by the text.

Certainly, much of the audience work discussed here is inevitably subject to the problems of reflexivity that Feuer raises. My own previous research, for example, offers the reader only a "reading" of the text supplied by the respondents—those texts themselves being the respondents' accounts of their own viewing behavior. However, in relation to the problems of the "status" of any knowledge that might be produced as a result of this process of "reading of readings" I would still argue that these techniques of empirical research remain a fundamentally more appropriate way to attempt to understand what audiences do when they watch television than for the analyst to simply stay home and imagine the possible implications of how other people might watch television.[2]

Qualitative media researchers face the difficulty, finally, of telling stories about the stories that their responders have chosen to tell them. These problems are both irreducible and familiar. As Geertz remarked long ago: "What we call our data are really are own construction of other people's constructions of what they and their compatriots are up to. . . . Right down at the factual base, the hard rock, in so far as there is any, of the whole enterprise, we are already explicating: and worse, explicating explications" (Geertz 1973: 9). However, as Geertz also notes, rather than to give up and go home, on realizing this, the ethnographer's alternative is to try to pick his or her way through the piled-up structures of inference and implication that constitute the discourse of everyday exchange. In this context it is of course also necessary to take note of the serious debates (Clifford and Marcus, 1986; Marcus and Fischer, 1986) that have developed in recent years about the epistemological and moral/political issues of empirical audience research. Hartley (1987) and Ang (1989) have addressed the difficulties arising generally from the constructivist nature of any research project and have warned against the dangers of failing to see that our data are inevitably products of the research process. The insistent question is that of the politics of audience ethnographies. Ang (1989) rightly insists that doing research is itself a discursive practice, which can only ever hope to produce historically and culturally specific knowledges, through equally specific discursive encounters between researcher and informants. Research is thus, from this point of view, always a matter of interpreting, indeed constructing, reality from a particular position, rather than a positivist enterprise seeking a "correct" scientific perspective that will finally allow us to achieve the utopian dream of a world completely known in the form of indisputable facts.

It is on these issues that recent debates concerning postmodern (or poststructuralist) anthropology and ethnography have centered, especially in the United States. The central issue has been the relationship between the observer and the observed and the basis of the ethnographer's authority to convey the cultural experience of others. Fiske refers to "the imperialist ethnographer who descended as a white man into the jungle and bore away back to the white man's world meanings of native life that were unavailable to those who lived it" (1990: 90). Among other commentators, Marcus and Fischer (1986) have talked of a crisis of representation, and Said (1978) has cogently argued for a more reflexive analysis of the process of "Orientalization"—the process of imaginative geography that produces a fictionalized Other as the exotic object of knowledge. In these debates, the object of criticism is a form of naive empiricism or ethnographic realism that would remain insensitive to issues of reflexivity instead presuming both a transparency of representation and an immediacy of the problematic category of "experience" (see Althusser, 1972).

For critics like Clifford, and Marcus there can be no "place of overview (mountain top) from which to map human ways of life, no Archimedian point from which to represent the whole world. Mountains are in constant motion . . . we ground things, now, on a moving earth" (1986: 22). This, then, also requires media researchers to specify who writes, about whom, and from what position of knowledge and power. In response, Geertz (1988) has referred to what he calls the "pervasive nervousness" and "moral hypochondria" engendered by poststructuralist and postmodern writing about ethnography. These "Jesuits of the future" or "diehard apostles of the hermeneutics of suspicion," Geertz argues, start from a quite proper suspicion of the Malinowskian ideal of "immersionist" ethnography and of the naive invocation of the ethnographer's sincerity and authenticity—Being There—as the founding authority of the ethnographic account. The point for Geertz, however, is that if the traditional anthropological attitude to these questions ("Don't think about ethnography, just do it") is the problem, then, equally, to fall into a paralyzing (if vertiginously thrilling) trance of epistemological navel-gazing ("Don't do ethnography, just think about it") is no kind of answer for anyone with a commitment to empirical work. Even Clifford (1986: 7) himself has expressed the hope that this "political and epistemological self-consciousness need not lead to ethnographic self-absorbtion, or to the conclusion that it is impossible to know anything certain about other people."

For Geertz there is an important limit to what can be conceded to the poststructuralist argument. To recognize the subjective component of ethnography is no more than common sense; the burden of authorship is inescapable. In Geertz's words, "to argue . . . that the writing of ethnography involves telling stories" could only have seemed contentious on the premise of "a confusion . . . of the imagined and the imaginary, the fictional with the false . . . making things out with making them up. . . ." (1988: 140). The value of ethnographic methods lies precisely in their ability to help us "make things out" in the context of their occurrence—in helping us to understand television viewing and other media consumption practices as they are embedded in the context of everyday life.

The Death of the Text?

In recent years, one further cause of concern in media studies has been the viability of any concept of texts as independent of their "activations" and "readings" in particular contexts. The most significant recent work in this respect has been that of Bennett and Woollacott (1986), Fiske (1987), Grossberg (1987), and Browne (1984), all of whom have queried in various ways the viability of the concept of an independent text. Recently Charlotte Brunsdon (1989) has responded to these developments with an argument in defense of the status of the text. She argues that the need to specify context and mode of viewing in any textual discussion, and even the awareness that these factors may be more determining of the experience of the text than any specific textual feature, does not, in and of itself, either eliminate the text as meaningful category, or render all texts "the same." The fact that the text is only and always "realized" in historically and contextually situated practices of reading does not demand that these categories collapse into each other. As I have argued elsewhere (Morley, 1989), in recognizing the complex nature of the domestic setting in which television is viewed, one does not necessarily abandon concern with texts and what they communicate. Rather, what is necessary is to examine the modes and varieties of viewing and attention that are paid to different types of programs at different times of the day by different viewers. Though it is against a baseline of expectation of fragmented and distracted television viewing that the variations in viewing behavior must be traced, one does not necessarily conclude that intensive and attentive viewing never occur. Similarly, as Brunsdon argues, the fact that texts may in certain circumstances be interpreted differently from how they were intended should hardly mean that we cease to concern ourselves with the structuring of texts in particular ways by particular types of institutional processes (see also Brunsdon, 1990), and there is no warrant for "dissolving" the text into its "readings" or "uses" on the spurious moral/political grounds of readers' rights (or visions of readers' "liberation" from the tyranny of the text).

As we all know, in the bad old days, television audiences were considered as (bad) passive consumers to whom things happened as television's "miraculous powers" affected them. According to choice, these (always other) people were seen as being turned into zombies, transfixed by bourgeois ideology or filled with consumerist desires. Happily, so the story goes, it was then discovered that this was an inaccurate picture because, in fact, these people were out there, in front of the set, being "active" in all kinds of ways—making critical/oppositional readings of dominant cultural forms, perceiving ideological messages selectively/subversively, and so on. So it seems we need not worry—as a result of the work of people like Katz and Liebes (1990) on "differential decoding," it is often now argued, the passively consuming audience is a thing of the past, according to the new "Cultural

Studies" orthodoxy.[3] Meaghan Morris (1990) acidly sums up what she takes to be this cosy "orthodoxy." As she notes, many versions of this theory have now been offered—from John Fiske's (1987) notion of a "readers' liberation movement" through Mica Nava's (1987) analyses of the "contradictions of consumerism" to Iain Chambers's (1986) accounts of counterhegemonic forces in popular culture, all extolling the creative energies of the much maligned consumers of popular culture. As far as Morris is concerned, the "Ur-thesis" of this kind of cultural studies runs perilously close to the banal observation that, as she puts it, "people in modern mechanized societies are complex and contradictory; mass cultural texts are complex and contradictory; therefore people using them produce complex and contradictory culture" (Morris, 1990).

Though I would agree that the work of the authors Morris criticizes is problematic, from my own point of view it is the lack of a sufficiently sociological or materialist basis to Fiske's or Chambers's work that is the key problem. Certainly if, as Morris suggests, all one could say was that "it's always complex and contradictory," it would hardly seem fair to the trees to bother to do so. The point, however, is in my view, an *empirical* one: The question is one of understanding (and here I continue to believe that Bourdieu (1984) has much to offer in this respect) just *how* "complex" or "contradictory" it is, for *which* type of consumers, in *which* social positions, in relation to *which* types of texts or objects. The "distinctions" are all, in this respect, and if Fiske and Chambers can be faulted for failing to help us see the sociological groundings of these "distinctions," Morris seems not to realize that they are what we need to look for. Everything might be simply "complex and contradictory" at one level of abstraction—but the banality of that observation is, to my mind, ultimately a function of the level of (over) abstraction of Morris' argument and of the lack in her own analysis of an explicitly sociological perspective. As Nice (1978) argued long ago in this connection, an implicit sociology is often an erroneous sociology, the more insidious for being unrecognized.

Understanding Viewing: The "How" Questions

Central among the points of explicitly sociological interest that are at stake here, I would argue, is the need to develop an understanding of the practice of television viewing in its "natural" (contemporaneously, mainly domestic) setting. Lindlof and Traudt (1983) have rightly argued that an awful lot of media research has focused on questions of *why*, to the exclusion of *what* and *how*, attempting to analyze the causes and consequences of television viewing without an adequate understanding of what it is and how it gets done. Certainly in my own work I have become increasingly concerned with the problem of how to integrate the analysis of questions of ideology and interpretation (what Hall [1988b] has described as the "vertical" dimensions of communications) with the analysis of the uses and functions of television in everyday practice (the "horizontal" dimension of communications in Hall's terms). So far as I can see, this is, among other things, the only way to avoid the danger of analyses of the interpretation (or "decoding") of texts (in relation to questions of ideology) without reference to the salience (or otherwise) of these texts within the context of the everyday life—worlds of people concerned (see Morley, 1986, for an autocritique of the overprivileging of interpretation over salience). The issue, then, is how to integrate the analyses of ideology and ritual, cultural power and social process.

Certainly, any analysis that offers us *only* an understanding of the microprocess of television consumption in this or that domestic context, without reference to the broader cultural (and indeed ideological) questions at stake, is ultimately going to be of only limited value. If we simply pile up an endless set of descriptions of the process of consumption, however fine-grained our analyses of these processes, we confront the "so what" problems. It is not a question, finally, of understanding simply television's ideological (or representational) role nor simply its ritual (or organizing) function. It is a question rather of how to understand all of these issues in relation to each other.

Many years ago Kaarl Nordenstreng (1972) conducted a very interesting piece of research on news viewing in Finland that is instructive in relation to these issues. He estimated that 80 percent of adult Finns watched the news at least once a day and found that many of them claimed that the news was among their favorite programs. However, when interviewed shortly after viewing, the majority of viewers were found to have retained little or none of the factual information contained in the news broadcasts that they had watched. From this, Nordenstreng concluded that "for most Finns, viewing the news is a *mere* ritual . . . a way of carving up the daily rhythm . . . a custom serving to maintain a feeling of security" in which the *content* of the news is a matter of indifference to them. (Nordenstreng, 1972). Clearly, there are problems with this conclusion, insofar as to equate media influence with the retention of factual information is to ignore a whole other level of potential (and possibly more significant) influence—in terms of agenda setting and conceptualizing of issues, for example. However, I would want to argue that the recognition that *one* dimension of news watching is the regulation of domestic time in a ritual manner is an important advance. It is clearly not *all* that "watching the news" amounts to, but it is a very important part of the story.

As an anecdotal indication of the domestic significance of television news watching as ritual activity, consider the following account:

> When I am writing I knock off to cook a very easy lunch and then work till about five. Then Leslie knocks off too, and always at six o'clock we sit down with a drink of gin and Cinzano and watch the news with dear Sue Lawley and lovely Nicholas Witchell. We always have a date with them and they don't know us from Adam! She makes you feel so right about everything, whatever the news is.[4]

In this connection Silverstone (1988) has argued that our watching of television involves us in a rite of passage, away from and back to the mundane, in an often equally taken-for-granted, but nonetheless significant immersion into the "other worldliness" of the screen:

> Our nightly news-watching is a ritual, both in its mechanical repetitiveness [cf. Nordenstreng—DM], but . . . more importantly in its presentation . . . of the familiar *and* the strange, the reassuring *and* the threatening. In Britain, no major news bulletin will either begin without a transcendent title sequence [London at the centre of the planet Earth, Big Ben at the centre of the metropolis—DM] nor end without a "sweetener"—a "human interest story" to bring viewers back to the everyday. Indeed the final shot is almost always of the two newsreaders, tidying up their papers and soundlessly chatting to each other, thereby announcing the return to normality (Silverstone, 1988: 26; my emphasis).

The challenge lies in trying to construct a model of television consumption that is sensitive to both the "vertical" dimension of power and ideology and the "horizontal" dimension of television's insertion in and articulation with the context and practices of everyday domestic life. We need to develop a double focus on television viewing, so that, for instance, we can understand viewing as simultaneously both a ritual whose function is to structure domestic life, and to provide a symbolic mode of participation in the national community and a process operating within the realm of ideology. To debate whether we should regard television as either one or the other is equally to miss the point. News watching is not to be understood as either "mere ritual" or a simple process of transmission of ideological categories, but precisely as operating along both dimensions at once. Indeed, the notion of "mere ritual" is itself problematic for, as Silverstone (1981) and others have argued, an understanding of the rituals of television is an essential component of any understanding of its place in everyday life and, as such, a crucial aspect of ideology. Our objective, therefore, ought to be the production of analyses of the specific relationships of particular audiences to particular types of media content, which are located within the broader framework of an analysis of media consumption as domestic ritual. These analyses must, of course, be sensitive to empirical variation.

A Retreat into Domestic Space?

In recent years some critics have argued that researchers such as Lull (1990), Silverstone, and myself, in our concern for the domestic context of television viewing, were busy conducting an ill-considered (if not hasty) "retreat" into the private realm of the domestic and away from the important "public" issues of power, politics, and policy that constitute the "proper" subjects of the study of communication. I shall argue that this critique is misguided, on a number of counts. It is not only that the average sitting room (in my experience) is the site of some very important political conflicts—it is also, among other things, one of the principal sites of the politics of gender and age. It is also that, in my view, the sitting room is exactly where we need to start from if we finally want to understand the constitutive dynamics of abstractions like "the community" or "the nation." This is especially true if we are concerned with the role of communications in the continuous formation, sustenance, re-creation, and transformation of these entities. The central point concerns television's role in articulating, for example, the "familiar" or domestic and the national spheres and in sustaining both the image and the reality of the "national family."

It is a commonplace that in the advanced capitalist countries of the West, we increasingly live in a "privatized world"—a result of the long-term historical process that Donzelot (1979) has described as the "withdrawal to interior space." Certainly, in Britain, the recent past has seen the development of a home-based consumer culture whose characteristics include a substantial increase in expenditure on consumer durables (including home-based entertainment equipment and "software") for the home, a significant increase in the amount of leisure time spent in the home, and a corresponding decrease in overall rates of participation in out-of-the-home leisure and public life, for all except an affluent minority. Currently, industry commentators are forecasting a further, internal process of fragmentation, inside the home–with the move to media as "personal delivery systems" for the increasingly individuated members of the "multi-cellular family."[5] Among the issues at stake here, we need to give careful consideration to the role of various forms of communication media—on the one hand as sustaining a "mediated" relationship between the increasingly privatized realm of the domestic and the space of public life (in both its national and international forms), and, on the other hand, as increasingly problematic in their transgressive potential in relation to the private/public boundary. To the extent that home and family are considered to be a private shelter from the pressures of public life, television and other ICTs are problematic, insofar as these media function to disrupt this separation of spheres. It was exactly on this premise, for instance, that the Annan Committee in the U.K., in its deliberations on broadcasting policy, based its argument for the need for some form of censorship and control of broadcasting:

> The audience for a programme may total millions: but people watch and listen in the family circle, in their homes, so that violations of the taboos of language and behaviour, which exist in every society, are witnessed by the whole family . . . in each other's presence. These violations are (therefore) more deeply embarrassing and upsetting than if they occurred in the privacy of a book, or in a club, cinema or theatre.[6]

The private or domestic realm is not, of course, some pre-given or natural realm, into which the state intervenes post hoc: the very constitution of the citizen's rights in the private realm is itself a legal process. The private is a juridically constructed space into which the state and its agencies can intervene (for instance, in the realm of child care, where such interventions are legitimated by reference to the state's proper concern for the welfare of its future citizens) and whose very privacy is itself constituted and ultimately guaranteed by the state.

As Paterson argues (1980 and 1987), it is on the basis of considerations of this kind that broadcasting scheduling policy is ultimately based—by reference to the state's concern to "police" the socialization (and thus the moral welfare) of children in the family (and especially that of children in the "chaotic" families of subordinate groups—*vide* the "video nasties" debate in the U.K.). The relation of the public and private spheres and the implication of television and other ICTs in articulating their connection is a complex one. However, it is also, commonsensically, a largely taken for granted one, in which the central place of these media in the microgeography of the home (and indeed, the sitting room) is so naturalized as to easily escape scrutiny. In this respect the historical work of writers such as Moores (1988), Spigel (1990), and Boddy (1986) is vital in reminding us of the deeply problematic and contested nature of the historical process through which first radio, and then television (and now a whole range of ICTs) first made their entry into the home.

Having begun by referring to the centrality of the home in contemporary Western culture, we must also consider, again from a historical point of view, the media's own role in increasing the attractiveness of the home as a site of leisure (Moores, 1988) and in promulgating what Frith (1983) has described as the "pleasures of the hearth" (cf. Lodziak, 1987).

The Socialization of the Private Sphere

Scannell (1988) has analyzed the role of broadcast communications technologies in the "socialization of the private sphere" and the significance of broadcasting's role in the domestication of standard national time. Scannell's key point concerns the role of communications technologies (especially in the form of national broadcasting systems) in organizing (both at a calendrical and at a quotidian level) the participation of the population in the public spheres of national life (whether through the occasional viewing of a royal wedding or the regular domestic ritual of "watching the news" as a structuring activity in the daily cycle of life in the home). As Scannell notes, modern mass democratic politics has its forum in the radically new kind of public sphere that broadcasting constitutes.

Cardiff and Scannell (1987), in their historical analysis of the development of British broadcasting, focus on broadcasting's crucial role in forging a link between the dispersed and disparate listeners and the symbolic heartland of national life, and its role in promoting a sense of communal identity within its audience at both regional and national levels. Historically, the BBC, for example, can be seen to have been centrally concerned to supply "its isolated listeners with a sense of the community they had lost, translated from a local to a national and even a global level" (Cardiff and Scannell, 1987). As Cardiff and Scannell note, the audience has always been seen as composed of family units—as "a vast cluster of families rather than in terms of social classes or different taste publics." Brunsdon and Morley (1978) argue that the central image of much contemporary current affairs and "magazine" programming is precisely the family—the nation as composed of families. In this type of broadcasting the nuclear family is the unspoken premise of much program discourse: Not only is the programming addressed to a "family audience," but also this domestic focus accounts both for the content ("human interest stories") and the mode of presentation (the emphasis on the everyday aspects of public issues). What is assumed to unite the audience is the experience of everyday life, as a nation of families. Broadcasting does much more than simply to make available experiences (the Cup Final, the Proms, etc.) that were previously only available to those who could be physically present. Beyond this, the "magic carpet" of broadcasting technologies plays a fundamental role in promoting national unity at a symbolic level, linking individuals and their families to the "center" of national life, offering the audience an image of itself and of the nation as a knowable community–a wider, public world beyond the routines of a narrow existence, to which these technologies give symbolic access.

In a similar vein, in his analysis of the development of radio light entertainment, Frith observes that the radio did more than simply to make public events accessible by bringing them into the home—more importantly, "what was on offer was access to a community . . . what was (and is) enjoyable is the sense that you too can become significant, by turning on a switch" (1983: 121–122). And thus, while domestic listening (or viewing) might be "a very peculiar form of public participation" it offers above all else a sense of participation in a (domesticated) national community. From this perspective it can also be argued that the usual focus of concern within media studies—questions of broadcasting's representational role—might usefully be complemented by a parallel concern with broadcasting's role in the social organization of participation in national life—critically, the synchronization of social experience in time.

King (1980) argues that the development of both physical and symbolic technologies of communication has played a vital role in the standardization of time in industrial societies—bringing metropolitan time into what were, previously, the differential rhythms of local and domestic modes of temporal organization. E. P. Thompson (1967) has analyzed the importance of the standardization of time as a part of the process of synchronization of labor activity in the development of capitalism. In a similar way, Giddens, following Mumford, argues that "the clock rather than the steam engine should be regarded as the prototype of the era of mechanical production" (Giddens, 1979: 210). Broader questions are also inevitably involved—questions of the imposition of a standard or national time and of the relationship between time and modes of communication. Thus as King (1980: 198) argues in his analysis of nineteenth-century British society, with the diffusion of clocks and watches, urbanization, and the development of railways, there emerged a totally new orientation to, and organization of time, with local time being suppressed in favor of London time. It is not simply that the analysis of communication needs to be situated in the context of an understanding of the spatiotemporal organisation of society; it is also that modes of communication both physical (e.g., the coming of the railways) and symbolic (e.g., the coming of broadcasting) themselves transform the modes of temporal organization.[7]

In a similar vein, Moores's historical analysis of the development of radio points to the way in which broadcasting was responsible for bringing the precise measurement of time into the home—via what he calls the "domestication of standard national time" (Moores, 1988: 38). Here we see the role of broadcasting spanning the private and public spheres at its most elementary (and perhaps most ontologically significant?) level—where nationwide time ("Big Ben") can be relayed directly into the private sphere, thus providing all those who listen with the temporal authentication of their existence as members of a synchronized national community (Anderson, 1983). In this sense broadcasting has to be seen both to work with preexisting social divisions of time and, through its scheduling, to offer, if not impose, its own segmentations as a model for viewers' and listeners' private routines.

Filson Young (quoted in Moores, 1988: 38) comments on the significance of the "broadcasting of time" as both one of the most commonplace and regular features of the daily program and also one of the strangest of the new things that broadcasting invented. If one thinks for a moment of the insistence of the time checks on many radio stations (in between the statutory announcements of the quarter and half hours) and of the ritual of the "news on the hour" in the context of many radio listeners' habit of having the radio on all day, we begin to see that for many listeners one of the principal things that radio is, is the "speaking" clock that synchronizes their private experience and activities with those of the larger (local, national, international) communities.

As Bausinger (1984) notes, a variety of communications technologies (including the morning newspaper) can be seen to function in similar ways as articulating or "linking" mechanisms between the rituals of the domestic or "private" sphere and the construction of the "memberships" of national (and other "public") communities. The central point concerns the ontological significance, for the viewing audience, of modes of viewing, the motivation of which, as Rath notes, is not so much " . . . 'I see', but 'I also will have seen.' . . . a formation of the collectivity around a shared visual perception . . . [where] . . . the spectator can feel part of this imaginary totality" (Rath, 1988: 37). Here we approach another dimension of the articulation of public/private spheres, this time between the nation and the family (or individual) viewing in the sitting room. In this connection Hartley has argued that "television is one of the prime sites upon which a given nation is constructed for its members," (1987: 124), drawing on Ellis's definition of broadcast television as "the private life of the nation-state" (1982: 5) and Anderson's concept of the nation as an "imagined community," the construct of particular discourses (Anderson, 1983). The point lies in the central role of broadcast media schedules in regulating a simultaneity of experience for their dispersed audiences (cf. Hartley, 1987, on the function of newspapers as the basis of "mass ritual,") and thus in providing them with a temporary authentication of their existence as members of a synchronized national community, despite their physical dispersion. If television is one of the key sites at which a sense of national (or other) community is constructed, and if television is a domestic medium, then it may well be that one of the places to find the nation is in front of its television sets, in its sitting rooms.

Television, Community, Nation: Diasporic Ceremonies?

Martin-Barbero (1988) has identified the key role of the communications media in "converting the masses into a people and the people into a nation," noting that in many countries it was above all the development of the mass media that "provided the people of different regions and provinces with a first daily experience of the nation." To this extent the "nation," as a lived experience, is only made possible by broadcasting technologies—among whose achievements is the "transmutation of the political idea of the nation into lived experience, into sentiment, and into quotidian" (Martin-Barbero, 1988: 455–456). From this perspective, one of the key functions of broadcasting is the creation of a bridge between the public and the private, the sacred and the profane, the extraordinary and the mundane. Thus, as Silverstone argues:

> In Durkheimian terms, television provides a forum and a locus for the mobilization of collective energy and enthusiasms, for example, in the presentation of national events, from coronations to great sporting fixtures, and it also marks a consistently defined but significant boundary in our culture, between the domestic and taken-for-granted world and that of the unreachable and otherwise inaccessible world of . . . show business, *Dallas* and the moon landings (Silverstone, 1988: 25).

In a similar vein Chaney (1986) analyzes the role of broadcasting in enabling the public to participate in the collective life of the nation. As Chaney points out, a "nation" is a very abstract collectivity, insofar as it is too big to be experienced directly by the individual. To that extent, the "we-feeling" of community has to be constantly engendered by opportunities for identification, as the sense of a "nation" is manufactured. Chaney is particularly concerned with the role of mass media in relaying civic rituals (coronations, royal weddings, etc.). As he notes, if such rituals are "dramatization" of the nation as a symbolic community, then the infinite reproducibility of media performance makes the "audience" for them possible on a scale previously unimaginable (Chaney, 1986: 121). Recalling Silverstone's definition of television's role in establishing "the space of intimate distance" (1986: 23), Chaney analyzes the "quasi-democracy of intimate access"[8] created by the presence of the TV camera, "representing" the public in the most intimate moments of symbolic ritual. At the heart of the process is an ambivalence, in which the public figures are simultaneously humanized through vicarious observation (and, I would add, the camera often gives the audience at home a closer view than that those physically present), but also distanced by the dramatic conventions of media presentation.

Chaney is concerned with the spectacular character of ceremonial occasions, arguing finally (in a curious kind of reverse echo of Ellis, 1982) that "spectacular forms of mass communication are the public life of mass culture" (1986: 132). Contrary to the established view that "ritual" is less significant in secularized societies than it was in earlier times, Chaney argues that because of the scale and nature of these societies (where the citizenry simply cannot be personally acquainted and a sense of collective identity must be continually invented), ritual becomes *more* salient as a mode of dramatizing (indeed constituting) "community." Thus Chaney notes that "collective ceremonies have patently not disappeared from the calendar of institutional identity and reproduction, indeed they have been made more accessible and less arcane through their dramatization as media performances" (1986: 132). This is, in part, a question of "access"—thus Chaney notes the significance of the radio broadcasting of George VI's coronation in 1937—in involving a huge proportion of the national public who "spent the day listening in and thus partaking in the central events" (quoted in Cheney, 1986: 129). However, it is not only a question of access—thus in an earlier article Chaney notes that, in the end, the media's role transforms these events, so that the national festivals become media occasions, rather than occasions to which the media have access.

It is this "interfacing" of the public and the private that concerns us here. On the one hand, the audience for such national events is usually atomized, either attending individually or in small groups such as the family or peer group. On the other hand, each group sits in front of a television set emitting the same representation of this "central" event; the "public" is thus experienced in the private (domestic realm)—it is "domesticated." But at the same time the "private" itself is thus transformed or "socialized." The space (and experience) created is *neither* "public" *nor* private in their traditional senses.

In unraveling these connections, the work of Dayan and Katz (1983) on the representation of the royal wedding of 1981 on British television may be of some help. Drawing on Austin's (1976), theory of "performative" speech acts, Dayan and Katz are concerned to analyze television's role in constructing (literally "performing") media events such as the royal wedding. In this connection, they argue, television should not be seen as "representing" the event so much as constructing the experience of it for the majority of the population. Television is not so much transmitting, commenting, or reporting on the event, as they are actively involved in "performing" it—in bringing it into existence. In this sense, they argue, "broadcasters double as monument makers or priests . . . unlike news, media events are not descriptive of a state of affairs, but symbolically instrumental in bringing that state of affairs about" (1983: 183). To this extent, on these occasions, television goes beyond the role of witness, and rather is concerned to create the conditions of public participation in the event on a mass scale,[9] equalizing access (you no longer need a special invitation to be physically present) and offering to all that (peculiar) "experience of *not* being there" (1983: 189; emphasis in original).

The question that Dayan and Katz pose is that of what happens to such ceremonies when instead of attending in person, they are delivered to us at home. Here they touch on a crucial transformation in the fundamental logic of public events—insofar as the dominant mode of "publicness" is, they argue, changing from a theatrical one (based on physical presence) to a new mode, founded in television and based on the separation of performers and audiences and on the rhetoric of narrative (whereby public occasions acquire the formal characteristics of fiction texts) rather than the virtue of contact.

As Dayan and Katz note, physically distanced as they are from the ceremonial forms and also isolated from one another, television audiences do nor form "masses" or "crowds" except in the abstract statistical sense (cf. Ang, 1991). The question they pose is that of whether we can still speak of a public event when it is celebrated at home—and whether we can speak of a collective celebration when the collectivity is scattered. As they note, under these conditions:

> The very hugeness of the audiences has paradoxically transposed the celebration into an intimate register. Ceremonial space has been reconstituted, but in the home. Attendance takes place in small groups congregated around the television set, concentrating on the symbolic center, keenly aware that myriads of other groups are doing likewise, in a similar manner, and at the same time (1983: 194).

The analogy that Dayan and Katz offer is that of the Jewish Passover seder ritual—a collective ceremony without a central "cultic temple," which translates the public celebration into a "multiplicity of simultaneous similarly programmed, home-bound micro-events," (1983: 195). Thus, they imply, the television audience, as a dispersed community, can usefully be seen as regularly united (both by its occasional viewing of special events or by its regular viewing of the news or favorite soap operas) by precisely this kind of "diasporic ceremony."

From the "Government in the Sitting Room" to Global Totemic Festivals

General De Gaulle's concept of television as the "face of government in the sitting room" can, of course, be argued to apply only to broadcasting under quite particular conditions, specifically where broadcasting is allowed very little autonomy from direct government control. However, if we take our lead from the work of Chaney and Dayan and Katz, we can begin to see not only the crucial role of television in articulating governmental (or "public") with domestic space, but we can also pose the more fundamental question—as to extent to which it still makes sense to speak of broadcast media as "reporting" on political developments. The problem is that to pose the question this way is to presume that there exists some separate realm of "politics" on which the television subsequently reports. In an age when international sporting events are routinely arranged to suit the convenience of broadcasting schedules and acts of war are timed not so much with reference to military requirements as to maximizing PR advantage, this may seem obvious.

The fundamental issue is of some long standing. As long ago as 1974, Trevor Pateman argued a similar point in relation to electoral politics. His point was that television can only "cover" an election when a campaign has an existence independent of the presence of television, and that nowadays, these campaigns no longer have any such existence, being principally designed and planned—in terms of "photo opportunities," "sound bites," and the like—with reference to their televisualization. Thus Pateman argued that we do not have television coverage of elections, we have television elections. Pateman's point can be extended well beyond the specific field of "elections" to cover "politics" in a much more general sense. For the majority of the population, politics is principally a "media event," and their participation in this realm is a heavily mediated one. We are back, once again, to the politics of "being there."

This is increasingly a complex issue. *The Guardian's* South African correspondent, David Beresford, offered a telling account of his attempt to report Nelson Mandela's speech in Cape Town on his release from prison in early 1990—where "being there" physically unfortunately meant being unable to see or hear Mr. Mandela as well as the viewers at home could do. Beresford describes this as an experience of "being there and not being there," where being the "man on the spot" in fact has the seemingly perverse effect of making one unable to witness the images available to the rest of the global village (Beresford, 1990). In a similar vein, Dayan and Katz refer to the seemingly puzzling (but increasingly common) behavior of those physically present at public events who, if they can, also take with them a portable television set, so they can see "what is happening." Physical contiguity does not necessarily equate with effective participation—and of course, vice versa.

From this angle we could usefully reconsider all the debates that arose concerning the TV spectaculars of the 1980s–from "Band Aid/Live Aid" onward. Many commentators were critical of the ways in which these events expressed a "mythology" of international (if not universal) community. However, in a very important sense this was no "mythical" achievement. If a sense of community was created, this may have something to do with the fact that all over the world millions of people were (in reality) watching these simultaneous broadcasts–and to that extent, in Dayan and Katz's terms, participating quite effectively in a "diasporic ceremony" that was anything but illusory.

The traditional *equation* of community with geographical boundary and physical place may well be something that we simply have to ditch if we are to understand contemporary culture and communications. This is *not* to say that these terms will have *no* effective relation–simply that it is increasingly misleading to reduce the former to either of the latter. As long ago as 1933, the art historian and psychologist Rolf Arnheim foresaw the social consequences of television as a means of distribution, meaning that

> it renders the object on display independent of its point of origin, makes it unnecessary for spectators to flock together in front of an "original" . . . it takes the place of other means of distribution. . . .Thus TV turns out to be related to the motor car and the aeroplane–as a means of transport for the mind (quoted in Rath, 1985: 199).

Thus, as Rath argues, new communications technologies like television have to be understood as bringing with them new forms of sociocultural reality—such as, for instance, "communities" built up around broadcasts and made up of "the invisible electronic network between isolated homes and dwellings" (Rath, 1985: 200). In this situation frontiers of a national, regional, or a cultural kind no longer count. What counts much more is the boundary (or satellite "footprint") of the territory of communications or transmission. Thus, Rath argues, we confront a new situation of "TV geography," where "The space of transmission . . . a new geographic entity . . . cuts across the geographies of power, of social life and of Knowledge, which define the space of nationality and culture. . . .What is at stake is not "influence" upon a reality, but the effect of constituting reality," (p. 203). In this context, viewers experience themselves as "belonging to a new kind of electronically constituted society . . . [where] . . . the experience of watching TV may therefore be described not so much by the words "I see" as by the words–"I am among those who will have seen" . . . [a] . . . sense of shared collectivity established by shared visual perception" (Rath, 1988: 89).

It is also a question, as Robert Stam (1983) argues, of understanding the specific form of the pleasure offered to the viewer by television, and in particular by television news, in its most general sense. Stam is concerned with what he calls the "metaphysics of presence" of television and the ways in which television news promotes the regime of the fictive "we" as a community. Stam's argument is that epistemophilia can only offer a partial account of the motivation of news viewing (cf. Groombridge, 1972, and Morley, 1989). Beyond this, argues Stam, we must attend to the ways in which the pleasure offered are narcissistic and are "designed to enhance the self-image of His of Her Majesty the Spectator," (1983: 27). The principal point, he argues, is that television transforms us into "armchair imperialists" and "audio-visual masters of the world." In this respect, Stam argues, while "live" television is only a small part of broadcast television, it sets the tone for much of what television offers. As he puts it, television

> allows us to share the literal time of persons who are elsewhere. It grants us . . . instantaneous ubiquity. The telespectator of a lunar landing becomes a vicarious astronaut. . . .The viewer of a live transmission, in fact, can in some respects see better than those immediately present on the scene (p. 24).

Communications and (Postmodern) Geography

In recent years writers such as Carey (1989), drawing on, among other sources, the work of Innis (1951), have rightly drawn our attention to the historical role of communications systems (both physical and symbolic) in transforming our senses of space and time. Thus at one point, for example, Carey speaks of the "United States (as) the product of literacy, cheap paper, rapid and inexpensive transportation and the mechanical reproduction of words–the capacity, in short, to transport not only people but a complex culture and civilization from one place to another . . . between places that are radically dissimilar in geography . . . and . . . climate . . . the eclipsing of time and space" (Carey, 1989: 2–3).

Carey is concerned, among other things, with the role of communications in the construction of empire and the administration of power and with deconstructing such taken for-granted "facts of life" as the existence of standard national time.[10] Thus, Carey notes, the economic influence not only of the coming of the railways, but more dramatically perhaps, of the coming of the telegraph–which "permitted for the first time, the effective separation of communication and transport . . . allowing messages to be separated from the physical movement of objects," (1989: 203), thus freeing communication from the constraints of geography, and to that extent "making geography irrelevant" and "diminishing space as a differentiating criterion in human affairs" (p. 222).

In a similar vein, Meyrowitz (1985) has offered a fascinating (if overblown) analysis of the impact of electronic media on social behavior in its transforming the "situational geography of human life." Meyrowitz's concern is with the way in which electronic media have undermined the traditional relationship between physical setting and social situation, to the extent that we are "no longer 'in' places in quite the same way" (1987: 333) as these media "make us . . . audiences to performances that happen in other places and give us access to audiences that are not physically present," (1985: 7). Meyrowitz's central argument is that these new media redefine notions of social "position" and "place," divorcing experience from physical location. Thus, to return to our previous example of global broadcasting, Meyrowitz argues that "Live Aid was an event that took place nowhere but on TV," (1987: 329), the ultimate example of the freeing of communications experience from the "restraints of social and physical passage" (1985: 117).

Meyrowitz argues that the electronic media have transformed the relative significance of live and mediated encounters, bringing "information and experience from everyplace to everyplace" as "state funerals, wars . . . and space flights are dramas that can be played on the stage of almost anyone's living room" (1985: 118), and in Horton and Wohl's terms (1956), viewers develop forms of "parasocial interaction" with media figures and stars they have never met. In this way, these media, according to Meyrowitz, create new "communities" across the spaces of transmission, bringing together otherwise disparate groups around the "common experience" of television in a cultural homogenization. Thus, argues Meyrowitz, "the millions who watched the assassination of JFK . . . were in a "place" that is no place at all . . . the millions of Americans who watch TV every evening . . . are in a "location" that is not defined by walls, streets or neighborhoods but by evanescent "experience" . . . more and more, people are living in a national (or international) information system rather than in a local town or city" (1985: 145–146).

Thus Kirby (1989) notes that Meyrowitz's central argument is that the electronic media are destroying our sense of locality, so that "places are increasingly like one another and . . . the singularity . . . and importance of . . . locality is diminished" (1989: 323). This may be to overstate the case, as Meyrowitz admits in his reply to Kirby, but, minimally, the function of these electronic media is certainly likely to "relativize" our sense of place—so that "locality is no longer seen as the center stage of life's drama," (Kirby, 1989: 330). That center stage is then, according to Meyrowitz, taken by national television, in the home, bringing us news of the "generalized elsewhere" of other places and "non-local people" and their similar "simultaneous experiences"–thus undermining any sense of the primacy of "locality," as the "unifying rhetorical space of daily TV extends into the living rooms of everyone" (Berland, 1988: 147).

These new media, of course, are multivalent in their potentialities. Thus if television, radio, and the telephone turn once private places into more public ones by making them more accessible to the outside world, conversely, portable television and Walkman sets have the potential to "privatize" what were formerly spaces of shared public experience. Part of the point is that, for instance, access to nonlocal people (via the telephone) is often faster and simpler than access to physical neighbors. The "community" is thus "liberated from spatial locality," and many intimate ties are supported by the telephone rather than by face-to-face interaction (cf. the American telephone advertisement, "Long distance is the next best thing to being there"). Thus, it seems, we should no longer conceive of community so much in terms of local clustering of relationships, but rather in terms of types of social relationship, whether local or distant–a "psychological neighborhood," or a "personal community" as a network of (often nonlocal) ties (Wellman, 1979, quoted in Meyrowitz, 1987). Thus "community" is transformed: Living physically near to others is no longer necessarily to be tied into mutually dependent communication systems; conversely, living far from others is no longer, necessarily, to be communicationally distant. Locality is not simply subsumed in a national or global sphere; rather, it is increasingly bypassed in both directions: Experience is both unified beyond localities and fragmented within them (Castells, 1983).

Against Homogenization

The vision offered (even celebrated) by Meyrowitz of this emergent "placelessness" (Berland, 1988: 147) can be criticized on a number of counts. It offers little recognition of the dimension of power in which what emerges across this electronic "placeless" network is what Mattelart et al. identify as the "time of the exceptional and the spectacular," the product of an international entertainment culture (Mattelart et al., 1987: 97), a heavily standardized televisual language that will tend to displace and disqualify all others. Equally, as Ferguson (1987) argues, the "techno-orthodoxist" world view, which proclaims that satellite and other new ICTs have effectively reduced time-space differences to insignificance, can be seen to be badly overstated. Principally this is because the argument has little empirical grounding and operates at a level of overabstraction that does not permit us to answer questions about *how* these media shift our everyday understandings of time and space, nor *which* media forms influence *which* people in *which* ways in their conceptualization of duration and distance (Bryce, 1987).

As Ferguson notes, despite the grand claim of the techno-orthodoxist "homogenizers" it remains true that "just as they have differential access to new and old communication media, so do different cultures, social groups, and national sources of power perceive, categorize and prioritize temporal and spatial boundaries differently" (Ferguson, 1987: 153) (cf. Morely and Silverstone, 1990). Moreover, as she argues, rather than perceiving a uniform effect in which, from a crudely technological determinist perspective, new ICTs impose new sensibilities on people across the globe, it may be more realistic to conceive of them as overlaying the new upon the old, so that rather than the new media's promoting a "boundless media land of common understandings," a variety of senses of "temporal elasticity and indeterminacy" may be the more likely result, where "formerly finite absolutes take on a notably relativistic character . . . and old certainties . . . [are undermined, to some extent by] new ambiguities" (Ferguson, 1987: 159). This seems a more realistic (cf. Miller, 1990) and a richer perspective from which to analyze the interaction of local definitions and larger communication systems. Thus, to the extent that imported television programs penetrate local meeting systems, rather than "homogenizing" diverse cultures, their principal effect may rather a variable one, insofar as they introduce a relativizing perspective, as an "uncertainty principle" that may work to undermine established and dominant frameworks of meaning in a variety of ways.[11]

As Ferguson notes, what is needed in this respect is "qualitative research into *how* electronic communications magnify [or otherwise–DM] time-space imperatives and which forms produce *which* kind of intended and unintended consequences" (1987: 171).

Conclusion

If the claims of some commentators as to the determining effects of new communication systems on our contemporary sensibilities are overblown (whether in their original McLuhanite forms, the revisionist techno-orthodoxy, or their more recent postmodern manifestations), neither would I want to see any further "romanticization" of the "creativity" of media consumers. If the television is not simply the "government in the sitting room" and if the homogenization of space and time in contemporary culture has not yet abolished all differences, still we must attend to the need to construct a properly postmodern geography of the relations between communications and power and of the contemporary transformation of the public and private spheres.[12]

Research at Brunel University (see Morley and Silverstone, 1990; Silverstone, Morley, et al., 1989) has begun to explore some of the issues concerning the modes of domestic consumption and use of television and other ICTs, in both their material and symbolic dimensions, as these pertain to the articulation of the public/private boundary. Elsewhere Kevin Robins and I have begun to sketch out the issues involved in the analysis of the complex interplay of the global-local dialectic in contemporary culture (Morley and Robins, 1989 and 1990). In a similar vein, Miller (1990) has begun to develop an analysis of the production of local cultures utilizing global resources. These are no more than beginnings. By way of indication of some of the issues involved in developing this work further, we can usefully refer to the work of Gillespie (1989), who offers an insightful analysis of the role played by the video recorder in the negotiation of ethnic identities among Asians in Britain, who utilize the video to arrange regular showings of Indian films and similar material unavailable on broadcast television in Britain–a process that can also be found among ethnic minorities (Turks, Moroccans, etc.) in other European countries. In this way, new communications technologies are mobilized in the (re)creation and maintenance of traditions, of cultural and ethnic identities that transcend any easy equation of geography, place, and culture, creating symbolic networks throughout the various communities in diaspora. The point here is that such groups have appeared in the research frame on the understanding that theirs is a particularly problematic position–as "immigrants." In this respect Hall (1988) usefully reminds us of the increasing centrality of the "migrant" experience throughout contemporary culture. If we are to understand how any (post) modern sense of identity, community, or nation (at any level) is produced, then we shall need to confront, among other things, the domestic setting of its production, via the consumption and use of broadcasting and other ICTs. To this extent, a "retreat" into the domestic sphere may precisely be the detour we need to make, if we are effectively to understand these "larger," more obviously "political" questions.

Notes

1. cf. Richard Collins "Natural Culture: a contradiction in terms?" paper to International Television Studies Conference, London, 1988; Philip Schlesinger "On National Identity," *Social Science Information,* vol 26, no 2, 1987.
2. cf. E. Seiter et al "Towards an ethnography of soap opera viewers" in E. Seiter et al (eds) *Remote Control,* London: Routledge, 1990, especially for their critique of Modleski (1986).
3. See James Curran "The New Revisionism in Mass Communications Research," *European Journal of Communications,* Vol 5, nos 2—3, June 1990, for a trenchant critique of the "new orthodoxy" in audience studies, delivered with all the wisdom of hindsight.
4. "A Life in the Day of Celia Fremlin," *Sunday Times,* 1988.
5. As one trade commentator notes: "Whereas in 1980, TV was a family mechanism, it now provides a more personal service for each of the various members of the household." (London: *Marketing Review,* June 1987, p 15).
6. Annan Committee, *Report on the Future of Broadcasting,* London: HMSO, 1977, p 246.
7. cf. James Carey (1989) on the role of the railways and the telegraph in the construction of "standard national time" in the USA in the late 19th century.
8. cf. Dayan and Katz (1983) "Television is that which abolishes distance" (op cit, p 88).

9. cf. W. Benjamin (1977) on the effects of reproducibility on "aura."

10. Introduced in the USA on 18—11—1887, according to Carey (1989, p 223).

11. cf. D. Hebdige (1988) and K. Worpole (1983) on the effect of "foreign" cultural artefacts in undermining the established hierarchies of national taste cultures.

12. See Morley (1991) for a differently focused analysis of some of these issues. The present article attempts to develop the arguments with more specific reference to a micro and macro geographical framework.

References

Althusser, Louis (1972). Ideological state apparatuses. In *Lenin and Philosophy.* London: New Left Books.

Anderson, Benedict (1983). *Imagined Communities.* London: Verso Books.

Ang, Ien (1989). Wanted: Audiences. In E. Seiter et al. (eds.), *Remote Control.* London: Routledge.

———. (1991). *Desperately Seeking the Audience.* London: Routledge.

Austin, J. (1976). *How to Do Things with Words.* Oxford: Oxford University Press.

Bausinger, Herman (1984). Media, technology, and everyday life. *Media, Culture, and Society,* vol. 6, no. 4.

Benjamin, Walter (1977). The work of art in the age of mechanical reproduction. In J. Curran et al., (eds.), *Mass Communications and Society.* London: Edward Arnold.

Bennett, Tony, and Janet Woollacott (1986). *Bond and Beyond.* London: Macmillan.

Beresford, David (1990). Article in *The Guardian,* 17 April.

Berland, Jody (1988). Placing television. *New Formations,* no. 4.

Boddy, William (1986). The shining centre of the home. In Philip Drummond and Richard Paterson (eds.), *Television in Transition.* London: British Film Institute.

Bourdieu, Pierre (1984). *Distinction.* London: Routledge.

Browne, Nick (1984). The political economy of the television supertext. *Quarterly Review of Film Studies,* vol. 9 (Summer 1984).

Brunsdon, Charlotte (1989). Text and audience. In Ellen Seiter et al. (eds.), *Remote Control* London: Routledge.

———. (1990). Problems with quality. *Screen,* vol. 31, no. 1 (Spring 1990).

Brunsdon, Charlotte, and David Morley (1978). *Everyday Television: "Nationwide."* London: British Film Institute.

Bryce, Jenifer (1987). Family time and television use." In Tom Lindolf (ed.), *Natural Audiences.* Norwood, N.J.: Ablex.

Cardiff, David, and Paddy Scannell (1987). Broadcasting and national unity. In J. Curran et al. (eds.), *Impacts and Influences.* London: Methuen.

Carey, James (1989). *Communication as Culture.* London: Unwin Hyman.

Castells, Manuell (1983). Crisis planning and the quality of life. *Society and Space,* vol. 1, no. 1.

Chambers, Iain (1986). *Popular Culture.* London: Methuen.

Chaney, David (1986). The symbolic form of ritual in mass communications. In P. Golding (ed.), *Communicating Politics,* Leicester, U.K.: Leicester University Press.

Clifford, James, and George Marcus (1986). *Writing Culture.* Berkeley: University of California Press.

Dayan, Daniel, and Elihu Katz (1983). Performing media events. In J. Curran et al. (eds.), *Impacts and Influences.* London: Methuen.

DeLillo, Don (1985). *White Noise.* London: Picador.

Donzelot, Jacques (1979). *The Politics of Families.* London: Hutchinson.

Ellis, John (1982). *Visible Fictions.* London: Routledge.

Ferguson, Marjorie (1987). Electronic media and the redefinitions of time and space." In Marjorie Ferguson (ed.), *Public Communications.* London: Sage.

Feuer, Jane (1988). Dynasty. Paper presented to International Television Studies Conference, London.

Fiske, John (1987). *Television Culture.* London: Methuen.

———. (1990). Ethnosemiotics. *Cultural Studies,* vol. 4, no. 1.

Foucault, Michel (1980). The eye of power. In Colin Gordon (ed.), *Power/Knowledge,* pp. 148–165. New York: Pantheon.

———. (1986). Of other spaces. *Diacritics,* vol. 16.

Frith, Simon (1983). The pleasures of the hearth. In James Donald (ed.), *Formations of Pleasure.* London: Routledge.

Geertz, Clifford (1973). Thick description. In *The Interpretation of Cultures.* New York: Basic Books.

———. 1988. *Works and Lives.* Cambridge, Eng.: Polity Press.

Giddens, Anthony (1979). Central Problems in Social Theory. Berkeley: University of California Press.

Gillespie, Marie. (1989). Technology and tradition. *Cultural Studies,* vol. 3, no. 2 (May).

Golding, Peter (1989). Political communication and citizenship. In Marjorie Ferguson (ed.), *Public Communication.* London: Sage.

Groombridge, Brian (1972). *Television and the People.* Harmondsworth: Penguin.

Grossberg, Larry (1987). The in-difference of TV. *Screen,* vol. 28, no. 2.

Hall, Stuart (1988a). New ethnicities. In *Black Film, British Cinema.* London: Institute of Contemporary Arts.

———. (1988b). Introductory address to International Television Studies Conference, London.

Hartley, John (1987). Invisible fictions. *Textual Practice,* vol. 1, no. 2.

Harvey, David (1985). The getopolitics of capitalism. In D. Gregory and J. Urry (eds.), *Social Relations and Spatial Structures.* London: Macmillan.

Harvey, David (1989). *The Condition of Post-Modernity.* Oxford: Basil Blackwell.

Hebdige, Dick (1988). Cartography of taste. In *Hiding in the Light.* London: Comedia/Routledge.

Horton, Donald, and Richard Wohl (1956). Mass communications and para-social interaction. *Psychiatry,* vol. 19.

Innis, Harold (1951). *The Bias of Communications.* Toronto: University of Toronto Press.

Jameson, F. (1984). Postmodernism: The cultural logic of late capitalism. *New Left Review,* vol. 146.

Katz, Elihu, and Tamar Liebes (1990). The export of meaning: Cross cultural readings of "*Dallas.*" Oxford: Oxford University Press.

King, Anthony (1980). *Buildings and Society.* London: Routledge.

Kirby, Andrew (1989). A sense of place? *Critical Studies in Mass Communications,* vol. 6, no. 3.

Lewis, Justin, et al. (1986). *Art: Who Needs It? The Audience for Community Arts.* London: Comedia.

Lindlof, Tom, and Paul Traudt (1983). Mediated communications in families. in Mary Mander (ed.), *Communications in Transition.* New York: Praeger.

Lodziak, Conrad (1987). *The Power of Television.* London: Frances Pintar.

Lull, James (1990). *Inside Family Viewing.* London: Comedia/Routledge.
McLuhan, Marshall, and Quentin Fiore (1967). *War and Peace in the Global Village.* New York: Bantam Books.
Marcus, George, and Michael Fischer (1986). *Anthropology as Cultural Critique.* Chicago: University Press.
Martin-Barbero, Jesus (1988). Communication from culture. *Media, Culture, and Society,* vol. 10.
Mattelart, Armand, et al. (1987). *International Image Markets.* London: Comedia.
Meyrowitz, Joshua (1985). *No Sense of Place.* Oxford: Oxford University Press.
———. (1987). Reply to Kirby. *Critical Studies in Mass Communication.*
Miller, Daniel (1990). "The Young and the Restless" in Trinidad. Paper presented to Workshop on Domestic Consumption, Brunel University, CRICT.
Modleski, Tania (1982). Loving with a Vengeance: Mass-Produced Fantasies for Women. Hamden, Conn.: Archon Books.
Moores, Shaun (1933). The box on the dresser: Memories of early radio. *Media Culture and Society,* vol. 10.
Morley, David (1986). *Family Television.* London: Comedia.
———. (1989). Changing paradigms in audience studies. in Ellen Seiter et al. (eds.), *Remote Control.* London: Routledge.
———. (1991). Where the global meets the local. *Screen,* vol. 32, no. 1.
Morley, David, and Kevin Robins (1989). Spaces of Identity. *Screen,* vol. 30, no. 4.
———. (1990). No place like Heimat. *New Formations,* no. 12.
Morley, David, and Roger Silverstone (1990). Domestic communications. *Media, Culture, and Society,* vol. 12.
Morris, Meaghan (1990). Banality in cultural studies. In Patricia Mellencamp (ed.), *Logics of Television.* Bloomington, Ind.: Indiana University Press.
Murdock, Graham (1990). Television and citizenship. In Alan Tomlinson (ed.), *Consumption, Identity, and Style,* London: Comedia/Routledge.
Nava, Mica (1987). Consumerism and its contradictions. *Cultural Studies,* vol. 1, no. 2 (May).
Nice, Richard (1978). Pierre Bourdieu: A "vulgar materialist" in the sociology of culture. *Screen Education,* no. 28.
Nordenstreng, Kaarl (1972). Policy for news transmission. In Dennis McQuail (ed.), *Sociology of Mass Communications.* Harmondsworth: Penguin.
Pateman, Trevor (1974). *Television and the General Election.* London: British Film Institute, TV monograph No. 3.
Paterson, Richard (1980). Planning the family. *Screen Education,* no. 35.
———. (1987). Family perspectives on broadcasting policy. Paper Presented to British Film Institute Summer School, August.
Rath, Claus-Dieter (1988). Live/Life: Television as a generator of events in everyday life. In Philip Drummond and Richard Paterson *Television and Its Audience.* London: British Film Institute.
Robins, Kevin (1989). Reimagined communities. *Cultural Studies,* vol. 3, no. 2.
Said, Edward (1978). *Orientalism.* Harmondsworth: Penguin.
Scannell, Paddy (1988). Radio Times. In Philip Drummond and Richard Paterson (eds.), *Television and Its audience,* London: British Film Institute.
Schiller, Herb (1988). The erosion of national sovereignty. In Michael Taber (ed.), *The Myth of the Information Revolution.* London: Sage.

Seiter, Ellen, et al. (eds.) (1989). *Remote Control.* London: Routledge.
Silverstone, Roger (1981). *The Message of Television.* London: Heinemann.
———. (1988). Television, myth, and culture. In James Carey (ed.), *Media Myths and Narratives.* London: Sage.
Silverstone, Roger, David Morley, et al. (1989). Families, technologies, and consumption. Working Paper, Brunel University, CRICT.
Soja, Edward (1989). *Postmodern Geographies.* London: Verso.
Spigel, Lynn (1990). Television in the family circle. In Patricia Mellencamp (ed.), *Logics of Television.* Bloomington, Ind.: Indiana University Press.
Stam, Robert (1983). TV news and its spectator. In E. Ann Kaplan (ed.), *Regarding Television,* vol. 2, New York: American Film Institute.
Thompson, Edward P. (1967). Time, work-discipline, and industrial capitalism. *Past and Present,* vol. 38.
Wellman, B. (1979). The community question. *Journal of Sociology,* no. 84.
Worpole, Ken (1983). *Dockers and Detectives.* London: Verso.

Satellite Dishes and the Landscapes of Taste

21

Charlotte Brunsdon

I like to like what's better to like—

Billie (Judy Holliday)
in *Born Yesterday*

Rather than offering a theoretical overview of determinations on, and directions for research into, the television audience, I offer a case study that has both implicit and explicit points of theoretical engagement. I would like here to try to show the way in which the personal tastes and preferences experienced and articulated in the domestic context to which ethnography gives us some access, while always being *personal*, are always also profoundly *social*. While this issue is an acknowledged focus of ethnographic concern, it is not usually discussed in the way in which I wish to address it, but more commonly forms a site for a recognition of a certain kind of trouble for the ethnographer.[1]

Ellen Seiter (1990) discusses this lucidly in her reflection on the Tuebingen/Volkswagen project, a "case study of a troubling interview." She points to the way in which the different social statuses of those involved in the interview, as well as their contrasted approach to, and desires for, the occasion, structures the interaction, and indeed could be considered the substance of the interview. Basically, the two interviewees involved do not want to offer detailed textual readings of television programs to two university professors because of their attitudes to television in general and because of their sense of what is appropriate to discuss with professors. These views, to Seiter's discomfort, they are happy to expound. Seiter's account directly addresses and reads this trouble in the interview, this failure to gain the data envisaged in the original research design. It is these failures, these gaps, these pauses—the moments when an interviewee changes tack in the middle of a sentence—with which I am initially concerned, for these seem to me moments in which we can locate the often unconscious recognition and negotiation of cultural power, in that we see here the struggle of an individual to locate him- or herself in relation to already circulating discourses of taste. In Billie's words, "to like what's better to like."

Ang (1985) addresses this directly in her *Dallas* study, and it leads her to her formulation of the "ideology of mass culture." It is thus that she designates the cultural attitudes with which her respondents have to negotiate when recounting the experience of watching *Dallas.* She had actively solicited this self-referential reflection in the formulation of her original advertisement: "I like watching the

TV serial *Dallas* but often get odd reactions to it," and offers a sophisticated analysis of the different discursive strategies whereby her respondents incorporate the recognition that the object of their pleasure is not culturally prestigious. As her respondents sometimes repudiate and disavow any pleasure, Ang also touches on the very complex relation of conscious and unconscious desire, something that has hardly been touched in empirical audience studies.

Janice Radway becomes concerned with the issue of readers' understanding of the cultural value of their pleasure in her 1984 study *Reading the Romance* when she investigates the connotations of reading "to escape" for her readers. Although they recognize that this is what they do, Radway insists: "If given another comparable choice that does not carry the connotations of disparagement, they will choose the more favorable sounding explanation" (Radway, 1984: 89).[2] Later she continues:

> In an effort to combat both the resentment of others and their own feelings of shame about their "hedonist" behavior, the women have worked out a complex rationalization for romance reading that not only asserts their equal right to pleasure but also legitimates the books by linking them with values more widely approved within American culture (p. 90).

Although it is not her central concern, Radway gives clear accounts of the way in which romance reading is legitimated through recourse to less controversial benefits like "learning about other countries."

In what might be taken to be a quite similar case, the reactions of fans to *Gone with the Wind,* we find rather less tortuous work done by the women to justify and explain their pleasures. Helen Taylor (1989) points to the way in which the celebrated "legendary" status and commercial success of *Gone with the Wind* functions partly to legitimate her correspondents' pleasure to themselves, as well as to her. So "what's better to like" is not generically given, although there clearly has been an historical association of feminine genres like the novel, melodrama, soap opera, and romance with the downside of taste. But, as in the case of *Gone with the Wind,* a certain kind of success can lead to a change of category, just as different media—the novel, cinema, television—have all in their time been seen to seduce the (nonmasculine) feeble-minded (Lovell, 1987).

Andrew Ross (1989), in his reading of contemporary U. S. responses to the Rosenberg letters, points to a related process in which the cultural legitimacy—or otherwise—of tastes and vocabularies determines their reading so thoroughly that it can be quite invisible to researchers and commentators. He compares the relative failure of the published letters to gain the Rosenbergs any support from the erstwhile left intelligentsia with the later reception of George Jackson's letters. He argues, most persuasively, that the revelation of the intimate sensibilities of the Rosenbergs, their quotidian middlebrow tastes, and specifically, the fluctuating expressive register of Ethel Rosenberg's style, which mobilizes all her "ordinary" cultural resources to write these (public) private letters, were profoundly embarrassing for the more mandarin tastes of (anti-Stalinist) intellectuals. The letters—indeed the lives of the Rosenbergs—were too centrally formed within a petit bourgeois aesthetic to be readable as authentic by legitimate intellectuals—unlike the authenticity incarnate of Jackson's "otherness." Ross's reading of the letters attempts to address what is specific to these letters, rather than what they are not, and argues for their continuing capacity "to compromise every possible canon of 'legitimate' taste" (1989: 29). Ross concludes this part of his research by arguing that it is the "untidy problematic of lower-middle class culture" which is most neglected in cultural studies.

My case study is the erection and reception in England in 1989–1990 of dish aerials to receive satellite television. Available figures suggest that these dishes have been overwhelmingly bought and rented by those in social classes C1 (19.3%), C2 (35.3%), and D (20.3%).[3] The dishes, which are about two feet across, first became easily available to the private purchaser in 1988, shortly before the launch of Rupert Murdoch's Sky television in February 1989. Cabling is the exception rather than the rule in Britain, and a different dish (the "squarial") is necessary to receive British Satellite Broadcasting (BSB), the more upmarket satellite channel, which started broadcasting in May 1990. There has been a certain amount of public discussion about satellite dishes, and it is on this, mainly as reported in newspapers, that I wish to concentrate. I want to argue that in this one example we see condensed a complex set of issues, including a conflict of taste codes that is illustrative of the history and status of different taste formations in Britain. The argument about who has right to put satellite dishes where provides, if you like, a *mise-en-abyme* of current conflict about broadcasting policy in Britain (Brunsdon 1990b). More germanely for this context, it provides a site within which we can trace the vocabulary and discursive contours of "television tastes," within which individuals experience and articulate their own preferences, which in their turn redefine, extend and reinforce the conflicting fields and their relation to each other.

Like channel selection or program watching, but unlike giving an account of either of these two activities, erecting a satellite dish is not necessarily a verbal activity. Buying or renting a dish can, I think, legitimately be read as an act that signals a desire, a connection with something that these dishes are understood to mean, or connote, or promise.[4] However, unlike channel selection or program watching, which are activities performed in the privacy of the home, erecting a satellite dish is done outside the home. This audience practice is, among other things, a nonverbal signifier of taste and choice—or, as an article in the *London Evening Standard* put it:

> So far in London, take-up of Sky has been slow. If, however, you actually welcome the round-the-clock rubbish being beamed out the problem is that you can't watch it discreetly.
>
> Under normal circumstances if your tastes extend no further than Neighbors, Capital Radio and Dynasty at least you can indulge yourself without the whole street knowing about it (Mark Edmonds, "Fright on the Tiles," 12 July 1989, 23).

This passage uses a structure common in discussion of taste, one that is present in the negotiations with the researchers in work I have cited earlier. This is a distinction between a private and a public taste, an indoors slippers-and-dressing-gown and a Sunday-Best of taste—the invocation of a known hierarchy of "what's better to like," from which ordinary mortals fall away. For this writer, the problem for those who like "round-the-clock rubbish" from space is that they can no longer indulge secretly.

However, this private/public distinction can work more than one way with satellite dishes—people can have quite different attitudes to acknowledged hierarchies of taste. Ondina Fachel Leal (1990) uses the idea of the television "entourage" to discuss the customary decoration of the television set in Brazilian homes. In fact, as her work shows, the notion of the entourage–the doilies, plastic flowers, photographs of loved ones carefully arranged on the television set–is appropriate only in the case of lower-class homes. The upper classes, instead of decorating and celebrating the television, enshrining it as the center of family life, often give it a room of its own, and always leave it unembellished in its techno-austerity. The case of satellite dishes, and the question of their siting, has similarities with the creation (or not) of a television entourage, but also substantial differences. Leal demonstrates that the television in working-class homes must be placed in such a way that passersby can see that the family possesses a television, so the difference between the two is not simply that the entourage is in the home while the satellite dish is outside it, although, at the same time, the public placement of the dish is exactly what is at issue, and this is what I wish to explore. "It's a nice extra, like a jacuzzi, that I'm sure would interest a lot of people" (Mark Goldberg, Hamptons Estate Agents, quoted in *The Evening Standard,* 12 July 1989, 23).

Press coverage of satellite dishes in Britain can be divided into three categories, if we exclude the trade press and advertising features generated by Sky television and BSB themselves. This distinction is not always easy to maintain, as can be seen by investigating the first category, business/industrial coverage of satellite television. This necessarily, and properly, entails the reporting of satellite television within discussions of Rupert Murdoch's communications empire, as well as smaller-scale coverage of employment in dish-making factories and individual entrepreneurs of the dish revolution like Liz Stewart, "a bubbly brunette from Fife," who according to the *Sunday Express* designed a system that "dishes out a blow to satellite giants" (16 October 1988, 25). However, in early 1989 it was noticeable that it was the Murdoch-owned papers, *The Times* and *The Sun,* that carried news reports about the increased demand for satellite dishes. For example, *The Sun* reported in January, shortly before Sky opened, that there were "Thousands in Dash For Satellite Telly Dishes" and quoted a spokesman ostensibly from Dixons, an electrical goods retailer, saying: "The fantastic range of programs being offered by the Sky station has really caused a stir" (5 January 1989, 5) Similarly, in February *The Times* reported that "Sky launch boosts demand for dishes" (4 February 1989, 3), and a couple of days later, "Satellite dish firm expanding" (7 February 1989, 2). The first of these two articles, which includes a statement from the Council for the Protection of Rural England, offers an early formulation of "satellite dishes as a threat to the environment," with which I will be centrally concerned in my third category of coverage. Thus although there is a certain fuzziness, particularly in some newspapers, about this category, we can still legitimately distinguish it from the other two, "consumer guides" and "dish-siting controversy."

The consumer guides are fairly self-explanatory and were mainly a feature of the immediate pre– and post–Sky launch period. For my purposes, what is significant is the way in which the choice of what to buy or rent is presented solely as an individual consumer purchase—a private, domestic matter that is treated appropriately by different newspapers with their different images of the type of consumers their readers are. Thus *Today* makes no mention of BSB and addresses "the big question facing viewers keen to wire up to satellite television" as "whether to rent or buy" (4 February 1989, 22), while *The Independent,* in an article called "The cost to the viewer," consults a range of experts and mentions BSB and WH Smith Television (Astra). Here, the reader is regarded not as a potentially "keen" viewer, but as more distanced—"curious"—as in "For the curious consumer, the message seems to be: tread carefully" (1 February 1989, 13). There are no mentions in any papers of planning restrictions–or indeed of the fact that these dishes will be put on the outside of houses. It is only in the third category of coverage that this private consumer choice is seen to have public consequences. It is the formulations of these consequences that are of interest to me here, articulating as they do a range of oppositions:

- private: public
- consumer: citizen
- entertainment: culture
- supranational: national
- future (innovation): history (conservation)

The controversy about the siting of satellite dishes, peculiarly resonant as it is in 1990 against a decade of "heritage enterprise" in Britain, also reworks and re-presents founding historical conflicts about broadcasting, some of which have simultaneously been articulated in the debates over the quality threshold in the 1900 Broadcasting Bill. Indeed, the BSB/Media Education pack aimed at those taking GCSE Media Studies draws attention to the hostility in some quarters that greeted changes in broadcasting from the 1920s through the juxtaposition of a series of (unattributed) hostile quotations from 1922 to 1982, asking, "Knowing when these comments were made, and the ways in which all of the developments referred to are now part of everyday life, how far do you think people are justified in criticizing satellite television?" (Wall and Chater, n.d.: 12). The false ingenuousness of this question is to some extent redeemed if we consider the most obvious historical parallel to the satellite dish, the television aerial in the 1950s.[5] Oral history and historical ethnographic research such as that of O'Sullivan's (1991) confirm that there was public controversy about the erection of television aerials in the 1950s, although there is no trace of this controversy in the standard histories of the BBC and Independent Television nor in standard textbooks on planning such as Cullingworth (1988) or histories of planning such as Punter (1985) or Cullingworth

(1979). The traces of this history can be found in repetitions, such as the fact that many of the places that have banned satellite dishes, like the Joseph Rowntree Trust village, New Earswick in Yorkshire, also have bans on outdoor television aerials,[6] or the inclusion of television aerials within the strict national restrictions on any alterations to Grade I and II listed buildings. New towns like Milton Keynes, built in the 1960s, were cabled throughout, partly to avoid exterior television aerials. It is noteworthy that the public debate about aerials in the 1950s coincides with the more general debates about the Americanization and commercialization of British culture of the period, which were particularly focused by the opening of commercial television in 1955. Tim O'Sullivan (1991), conducting interviews with people about their memories of first getting television, shows the way in which, to some, the television aerial symbolized a proud stake in modernity.

As Charles Barr (1986) and John Hill (1986) have shown, in British cinema of the 1950s and early 1960s, commercial television functions metaphorically to condense a set of attitudes to the commercial, the American, the mass-produced. Raymond Williams, in his 1960 essay about advertising, dates the battle for the skyline much earlier, to the 1890s, "with 'taste' and 'the needs of commerce' as adversaries" (Williams, 1980: 177). The point about history's repeating itself in this way, now that television aerials have, in the main, become accepted as part of the urban landscape, or as the BSB pamphlet puts it, "part of everyday life," lies partly in the significant differences (e.g., Sky television, rather than television as such). But it also lies in the way in which the similarities of some of the debates and discursive figures, encrusting/constituting a new object, the satellite dish, reveal that new ideas do not drop from the sky, but indeed, as others have argued about the television set itself, are constructed as meaningful within networks of relationships and discourses that antedate the technological innovation (Lull, 1988, 1990; Morley, 1986; Gray, 1987).

Thus much of James Lull's work has been concerned to explore the way in which the television set is used within the familial domestic context. He uses the concept of extension to conceptualize the relationship between the set and the already existing dynamics of interaction within the family. In a 1988 piece he argues that McLuhan's original notion of the mass media's extending the human senses through technological capability can be revamped to allow us to classify extension at three levels, the personal, the familial, and the cultural. I have some reservations about a certain uncontradictory quality in this concept, but would here wish to propose that satellite dishes, as well as being literally extensions, also condense familial and cultural extensions. The several instances of men erecting, and indeed inventing, their own dishes further suggest that this new technology may have a particular place in the gendered division of labor—and personal extension.[7] Recent ethnographic work offers specific instances of the mapping of gender and generation across and through domestic technology (Morley and Silverstone, 1990; Gray, 1991; Seiter et al., 1989).

Thus O'Sullivan (1991) finds that the final decision to buy a television set in the 1950s often rested with the man of the household, who frequently also installed the aerial, and who sometimes retained a residual proprietary power over the on/off switch. Moores (1990) maps the fluctuating gender/generational conflicts around satellite television, in which different family members occupy different positions at different times to different others in relation to the "same" equipment. For example, a woman unhappy with her spouse's purchase of satellite television defends that same purchase to her parents. These fluctuating identifications should make us cautious about ascribing essential qualities to technology and technology use, while still being alert to the patternings of power in specific historical divisions of labor, use, and attitude.

Here it is useful to recall Ang and Hermes' theorization of the interplay of gender and generation in the Meier household recorded by Bausinger (Ang and Hermes, 1991; Bausinger, 1984). Ang and Hermes observe, at the beginning of their article: "Mr. Meier, the male football fan, ends up not watching his favorite team's game on television, while his wife, who doesn't care for sports, finds herself seating herself in front of the TV set the very moment the sports program is on. Gender is obviously not a reliable predictor of viewing behavior here." They proceed to argue for a postmodern feminist understanding of gender, in which the concept of articulation is central. They conclude, "we must accept contingency as posing the utter limit for our understanding, and historical specificity as the only ground on which continuities and discontinuities in the ongoing but unpredictable articulation of gender in media consumption can be traced." Ang and Hermes include within the logic of their argument about gender a similar critique of the way in which class (and race/ethnicity, although this is not developed) can be used within ethnographic accounts as a preconstituted and preemptive explanatory factor.

The public debate in Britain in 1989 and 1990 about the siting of satellite dishes offers us a particular, national, historical example of a conflict of values, which, because it is staged on the skyline, gives us some access to nonverbal audience practice. Working as I do here from one source, national newspapers, will obviously provide only one kind of outline of this debate. I think it offers an account of some of the discourses in play. How individuals position themselves in relation to these differentially available circulating discourses at particular times and in particular contexts cannot be deduced and can only be investigated through particular ethnographies of the type that Ang and Hermes advocate. However, researchers cannot do anything with these particular local knowledges unless there is also an attempt to apprehend a wider discursive field. To "place practices of media consumption firmly within their complex and contradictory social contexts" (Ang and Hermes 1991) requires some mapping or constitution of these contexts.[8]

It is in relation to this argument that I wish to place a discussion of the third category of press coverage of satellite dishes, "controversy about siting." Working from a corrupt and rather random corpus–national, nonspecialist newspaper coverage of rows about the siting of dishes, certain patterns emerge with striking clarity.

First, there is the question of who speaks. Two categories of persons appear in these reports, "anti-dishers" and "dish-erectors." Coverage is overwhelmingly dominated by anti-dishers who are always professional—graphic designers, professors, and so on—and nearly always *representative*—councilors, spokespeople for trusts or estates, residents' associations. Often, of course, they will have initiated the news item as part of their campaign to get a dish removed, but for our purposes what is significant is the way in which they never represent themselves as speaking on their own behalf. Anti-dishers, who in Bourdieu's terms are the possessors of, indeed propagandists for legitimate cultural capital, act and speak at a general social level about a matter of public concern (Bourdieu, 1984).

Dish-erectors, on the other hand, are always particular individuals. For example, the Radford family of Norton-sub-Hamdon, in Somerset, who won a dish in July 1989:

> John Radford, a building worker, pinned it proudly and prominently to the wall of his little cottage, a Grade II listed dwelling and settled back with his wife Jean, a cheese packer, to watch MR Murdoch's old movies dropping in from outer space ("Sky dish is the limit for listed village," *The Independent*, 25 October 1989, 3).

Here we have a construction worker and a cheese packer. In February 1989 we had Steven Davenport, a forty-two-year old unemployed disabled man:

> A disabled TV viewer has been ordered by town hall planners to take down his rooftop satellite dish in a test case which could affect the future of satellite broadcasting in Britain.
>
> Steven Davenport was stunned by the council decision . . . ("Satellite TV dish banned as 'eyesore,'" *Sunday Express*, 19 February 1989, 17).

Most interesting, though, is the case of dish-erector Maggie Brown, who is also the media editor of the national newspaper, *The Independent*. Unlike all other reported dish-erectors she is a well-paid professional with access to the media, a profile more common in anti-dishers. Furthermore, despite her designation of her own house as "undistinguished," given that it is administered by Dulwich College, it must certainly be within reach of the most desirable areas of South London. Reported dish controversies have taken place either within villages, where there may still be local working-class occupancy of desirable cottages, or in areas of terraced urban housing. The erection of dishes on secluded detached houses has aroused no comment.[9] Similarly, the extensive dishing of the river frontage of luxury dockland apartments on the north Thames bank also appears to be uncontroversial. Although, as Muthesius (1982) traces, the terraced house in England was built for, and is occupied by all classes of society, it is not in the very prestigious terraces, for example, of Bath, Brighton, and Leamington Spa, that there has been dish controversy. Nor was there controversy in the inner city, where there is usually a mixture of terraced housing and newer council blocks, despite the fact that the satellite dish has come to signify the conspicuous consumption of a certain kind of poverty, as in this commentary on a Gallup poll for *Moneywise*, which had voted Nottingham the most desirable place to live: "There is relatively little difference between rich and poor in Nottingham; the way to tell the middle-class area from the council estate is that the council houses all have satellite dishes" (William Leith, "Life is not so bad and that's the bottom line," *The Independent*, 26 August 1990, 3). The controversies have taken place in areas that have a more mixed occupancy, where what section 4 of the 1971

Town and Country Planning Act refers to as "the essential character of the area" is a matter of continuous everyday struggle. The character of the docklands development is in some ways perfectly homologous with the character of the satellite dish. Thus although Maggie Brown has many characteristics of the anti-disher, she, for professional reasons—and she is careful to point out the professional necessity of having a dish—is a dish-erector, and her case study, "My dish did not go down well" (*The Independent,* 8 May 1990, 14), conforms to the individuated format of all dish-erectors, which is particularly interesting given that she wrote it herself.

The very contradictoriness of "My dish did not go down well" allows us to outline the contours of anti-dish discourse, which is remarkably uniform. Dish-erectors, in contrast, are normally marooned in the personal, specific, and concrete.

1. Anti-dishers generally make no reference to television programs. Brown, whom we could describe as an anti-disher with a dish, illuminates this point because she wants to have both BSB and Sky. Generally, though, anti-dishers discuss dishes quite formally as alien protuberances perversely attached to the outside of houses by untutored do-it-yourselfers.

2. Reference is always made to architectural provenance—often with some precision: "perfectly good late Victorian terraced house" (*The Independent,* 14 July 1989); "villages built of lovely honey-colored hamstone" (*The Independent,* 25 October 1989); "terraced Edwardian House" (*Daily Telegraph,* 7 October 1989); or, in Maggie Brown's case, "the front of my Victorian house." The contrasted repertoires of knowledge, television versus architecture, are evidently contrasts between less and more culturally legitimate forms.

3. Some knowledge of the relevant environmental regulation and town planning acts is often displayed. Anti-dishers often express regret at the lack of legal restriction, particularly for unlisted buildings. Thus, of Queen's Park London:

> A classic late-Victorian suburb, the district is supposedly a protected area of "special architectural or historic interest," a place where the writ of the home-improvement brigade should not run unchecked; where picture windows, stone cladding and television dishes are outlawed. Does it work? "This may have been declared a conservation area but my answer is 'so what?'" ("Battle against blots on the townscape," *The Independent,* 14 July 1989).

The law, for anti-dishers, is not in these matters strong enough and is not enforced satisfactorily.

4. Certain evaluative phrases recur. The favored written adjective is "unsightly," which is such a normative word that it does not exist in the positive "sightly." "Eyesores," and "Blight/Blot on the landscape" are also favored, often as a headline (see above). This phrase in particular has been offered to me quite unsolicited by many passersby when I have been out photographing houses with dishes, and was also volunteered by the lab that develops my films.

5. Value plays a role here too. Apart from the estate agent quoted earlier who compared satellite dishes to jacuzzis, items of conspicuous consumption of a very particular provenance, seeing dishes as potentially valuable home improvements,

the consensus among the anti-dishers, the worry that underlies the articulated, public-spirited architectural concern for the integrity of the buildings, is one about value. This refers not necessarily simply to house prices, but, for example, in villages has to do with a more generalized "heritage" value. Thus of Norton-sub-Hamdon:

> A council spokesman said: "We'd be absolutely appalled if this became widespread. Somerset's value as a tourist venue relies on the character of its historic buildings and villages"(*The Independent,* 25 October 1989, 3).

In London, though, things are harsher:

> And apart from the effects on the skyline, who knows what will happen to house prices once dishes start sprouting in earnest from our roofs and walls? (Mark Edmonds, "Fright on the tiles," *Evening Standard,* 12 July 1989).

A Mr. Tyler believes that most householders appreciate the benefits of following the rules. House prices in a conservation area can be significantly raised:

> "People who buy houses round here tend to have regard to the ambience of the area as a whole" (Michael Durham, "Battle against . . . ," *The Independent,* 14 July 1989).
>
> The dishes have to face south, and a view of a south-facing facade with satellite receivers along part of it is not going to make a house on the opposite side of the street easier to sell (Tim Rowland, "The blight of the satellite dish," *Daily Telegraph,* 1 October 1989).

It is, in the end, not what the dishes look like that matters—it is what they mean. And what they mean is both very simple and very complex. It is not necessary to be reductive to say that it is all about house prices really, because this move is made so spontaneously by the anti-dishers themselves, it is not made by all of them, and I am sure most would protest if this was offered, say, by an audience researcher, as a primary motivation. Sometimes, indeed, mention of house prices figures as a rhetorical last resort, an attempt to speak the language of philistines–or to recognize the values most endorsed in Britain in the last twelve years. Furthermore, financial gain does not have to be a primary motivation, if we follow Bourdieu's arguments about the disinterestedness of the inheritors of legitimate culture, for financial reward to result. There is a vision inspiring many of the anti-dishers, a vision of a particular England, as Patrick Wright (1985) puts it, of an old country. This harmonious, orderly community is self-policing because of its shared values and assumptions, vigilant against the autodidacts of the environment. To continue the quotation I began above:

> "People who buy houses round here tend to have regard to the ambience of the area as a whole," he said. "By and large they are not the kind of do-it-yourselfers who think it's the bee's knees to put in a mock Georgian frontage, or go down to a DIY superstore to pick up a timber cladding front porch. They tend to stick to traditional styles, though we still have to be vigilant—there are exceptions (Mr. Tyler, quoted in Michael Durham, "Battle against . . . ," *The Independent,* 14 July 1989).

The shifts in this speech—"People who buy houses round here. . . " "They. . . " "We . . ." trace the fragility of this community of natural taste, the way in which it is made through vigilance as well as in which it is born. There is here a particular characterizing of the relationship between the public and the private, which, as I argued earlier, is a significant division within discourses of taste. Judy Attfield (1989) provides a fascinating example of conflict over this distinction in her account of the net curtain war between tenants and architects in Harlow New Town in the 1950s. In this, in some ways analogous, public conflict the architects repeatedly complain that tenants ruin their open plan picture window dwellings in a quest for coziness pursued through the obstructive placing of furniture, heavy curtaining, and nets. Attfield observes, "Through the appropriation of privacy by the concealment of the interior from the uninvited gaze, people took control of their own interior space and at the same time made a public declaration of their variance from the architects' design" (1989: 228). This example provides another indication of the way in which public and private are constructed spaces, perceived differentially and differently accessible to different persons. Mr. Tyler's speech above is the discourse of the unselfconscious inheritors of public space who accept external uniformity in the absolute confidence of internal, private uniqueness. This vision of Britain has to be set against the aggressively downmarket image of Sky. As Mark Edmonds put it in 1989:

> But unless Sky changes tack and goes for a more upmarket audience, a satellite dish protruding from the front wall will do about as much for your standing in the neighborhood as a visit from a rat-catcher (Edmonds, "Fright on the tiles," *Evening Standard,* 12 July 1989).

Dish erectors defend themselves with considerable resignation, in vocabularies of the personal, which hints at other meanings for dishes. Thus Mr. Bolton, of New Earswick:

> The village is beautifully kept and much more desirable place to live than most council estates, but the people who have a big say in our lives tend to be old folk with old-fashioned ideas. Like it or not, satellite TV is here to stay and it's frustrating to be denied it (Daily Mail, "TV dish ban in 'Quaker' village," 4 December 1989, 15).

Or Mrs. Kidd of the same village, quoted in another newspaper:

> We saved up £130 to buy the dish. We have three kiddies who love the films ("Village bans TV satellite dishes," *The Guardian,* 4 December 1989).

Or the Radfords' son, Colin:

> I know all my friends would like one. I also know what they're saying about spoiling the village. For me it's rather *boring* ("Sky dish is the limit . . . ," *Independent,* 25 October 1989).

The dissatisfactions that leak out of these plain statements, the half-expressed desire for another order, one more modern, or more fun for the kiddies, less *boring*, are of a quite different type to the confident, regulatory, public-spirited complaints of the anti-dishers. Dishes are a do-it-yourself chance of a better environment of satisfaction. In the classic privatized consumer transaction, you pays your money and you takes your choice. The fact that the choice itself may be less than anticipated, or that the quality of what is available may be disappointing, is another matter. The discursive context within which this choice is made, the double jeopardy of satellite television in the terms of legitimate culture, militates against elaborated defense or critique. The *Daily Telegraph* (13 April 1989, 23) epitomized this point in heading an article about a man who made a dish from a dustbin lid, "The man who cannot complain about trash on TV." For the anti-dishers, though, each 60-centimeter platter, "scarring rows of houses at exactly the level where their uniformity remains most intact" (Tim Rowland, "The blight of the satellite dish," *Daily Telegraph*, 1 October 1989), signals an opting out of, an impediment to a certain public vision. And it is this, as I have argued, that is the hegemonic taste code, that constitutes dish-discourse, that frames the terms and reporting of dish conflict, and that will therefore also provide one of the contexts in which dish-erectors articulate their defenses. It is thus that the nonverbal aspects of the practice become significant. We could say actions speak louder than words. Certainly at a general level the dishes can be approached as conspicuous consumption, and classically, conspicuous leisure consumption. But the erection of a dish is also historically specific, a particular act, a concrete and visible sign of a consumer who has bought into the supranational entertainment space, who will not necessarily be available for the ritual, citizen-making moments of national broadcasting (Scannell, 1988; Chaney, 1979)–who is abandoning the local citizenry and the national landscape of heritage and preservation (Morley and Robins, 1989). Erecting a satellite dish on the front of your house is partly a declaration of not being bothered "to like what's better to like."

Notes

1. My argument should be understood, if I may phrase it this way, as "post–Clifford and Marcus" (1986). That is, I am not pointing out that power is always inscribed within the ethnographic enterprise (and the rest), and would here follow Geertz (1988) in his response to postcolonial and epistemological critiques: "The moral asymmetries across which ethnography works and the discursive complexity within which it works make any attempt to portray it as anything more than the representation of one sort of life in the categories of another impossible to defend" (1988: 144). Within this argument, he does still defend the enterprise. I am concerned with how, if you like, cultural power is spoken and circumscribes speech.

2. See also the review of this book, Ien Ang (1988) "Feminist Desire and Female Pleasure."

3. BARB figures, "Socio Economic Breakdown of Satellite Viewers," week ending 10 June 1990. Published in *Broadcast*, 11 August 1990, 10–11.

4. Thorstein Veblen's classic notion of "conspicuous consumption" provides one obvious possible approach to the acquisition of satellite dishes (Veblen, 1899). Another is offered by Alfred Gell, who discusses the (hard-earned) purchase of television sets by Catholic Sri-Lankan fishermen. He argues that for these poor, hardworking people the (electrically unconnected) television set functions like a work of art to negate/transcend the real world. He argues that this is "adventurous consumerism," "which struggles against the limits of the known world," rather than dull, unimaginative consumerism, which reiterates the class habitus. Attractive as this argument is, and it would obviously form one direction for future research, I think it would be more tenable in the case of satellite dishes if they did not in fact receive satellite television (Alfred Gell, 1988, "Newcomers to the world of goods: consumption among the Muria Gonds.") The Algerian film, *From Hollywood to Tamanrasset* (directed by Mahmoud Zemmouri, 1990), a comedy of dish-passion in Algiers, posits the anarchy of satellite reception against the staid state channel.

5. Television ownership and rental increased very rapidly in Britain in the 1950s. Calculated from the number of sound and television licenses issued, the increase is from 763,941 in 1951 to 9,255,422 in 1959. (Figures from John Montgomery, 1965.)

6. "The elected representatives of the villagers say they do not want these unsightly dishes everywhere," he said. "It is not that they dislike satellite television. We already have a strict rule of indoor aerials only." (Mr. Cedric Dennis, director of housing, New Earswick, Yorkshire, *Daily Telegraph,* 4 December 1989, 5.

7. A *Daily Telegraph* story, "Camouflage designed to hide eyesore dishes," featured Mr. Peter Plaskett, photographed painting a dish in a leafy pattern. Mr. Plaskett is marketing individually designed dishes and "said he was being kept busy by people who are concerned about their neighborhoods" (8 December 1989, 9). Another *Daily Telegraph* story, with a guest writer from *Electronics Weekly,* Leon Clifford, featured a similarly posed "man and dish" photograph of Mr. Stan Bacon, who has made a satellite receiver from a dustbin lid. The Sky spokesperson commented that this method was used for reception of Superchannel in Poland (13 April 1989, 23).

8. It is this that I understand John Fiske (1990) to be specifying in his use of the *langue/parole* distinction in a discussion of taste, or Lull (1988) in his notion of cultural extension.

9. John Wyver made a feature for the Channel Four arts program, *Without Walls* (17 October 1990), which had several wonderful examples of satellite dishes in secluded spots.

References

Ang, Ien (1985). *Watching "Dallas."* London: Methuen.

———. (1988). Feminist desire and female pleasure: On Janice Radway's *Reading the Romance. Camera Obscura* 16 (January 1988): 179–192.

Ang, Ien, and Joke Hermes. (1991). Gender and/in media consumption. In James Curran and Michael Gurevitch (eds.), *Mass Communication and Society.* London: Edward Arnold.

Attfield, Judy (1989). Inside Pram Town: A case study of Harlow House Interiors, 1951–1961. In Judy Attfield and Pat Kirkham (eds.), *A View from the Interior,* 215–238.

Attfield, Judy, and Pat Kirkam (1989). *A View from the Interior.* London: Virago.

Barr, Charles (1986). Broadcasting and cinema: 2: Screens within screens. In Charles Barr (ed.), *All Our Yesterdays.* London: British Film Institute.

Bausinger, Hermann (1984). Media, technology, and daily life. *Media, Culture, and Society* 6(4): 343–351.

Bourdieu, Pierre (1984). *Distinction.* Translated by Richard Nice. London: Routledge & Kegan Paul.

Brunsdon, Charlotte (1990a). Television: Aesthetics and audiences. In Patricia Mellencamp (ed.), *Logics of Television,* pp. 59–72. Bloomington, Ind., and London: Indiana University Press and the British Film Institute.

———. (1990b). Problems with Quality. *Screen* 31.(1): 67–90.

Corner, John (1991). *The Homely Image: Essays in Television History.* London: British Film Institute.

Chaney, David (1979). *Fictions and Ceremonies.* London: Edward Arnold.

Clifford, James, and George E. Marcus (1986). *Writing Culture: The Poetics and Politics of Ethnography.* Berkeley: University of California Press.

Cullingworth, J. B. (1988). *Town and Country Planning in Britain.* 10th ed. London: Unwin. First edition, 1964.

———. (1979). *Peacetime History of Environmental Planning,* vol. 3, *1939–1969, New Towns Policy.* London: HMSO.

Fiske, John (1990). Ethnosemiotics: Some personal and theoretical reflections. *Cultural Studies* 4(1): 85–99.

Geertz, Clifford (1988). *Works and Lives: The Anthropologist as Author.* Cambridge, Eng.: Polity Press.

Gell, Alfred (1988). Newcomers to the world of goods: Consumption among the Muria Gonds. In A. Appachurai (ed.), *The Social Life of Things* Cambridge, Eng.: Cambridge University Press.

Gray, Ann (1987). Behind closed doors: Video recorders in the home. in Helen Baehr and Gillian Dyer (eds.), *Boxed In Women and Television,* pp. 38–54. London: Pandora.

———. (1991). *Video Playtime: The Gendering of a Communications Technology.* London: Comedia/Routledge.

Hill, John (1986). *Sex, Class, and Realism.* London: British Film Institute.

Leal, Ondina Fachel (1990). Popular taste and erudite repertoire: The place and space of television in Brazil. *Cultural Studies* 4(1): 19–29.

Lovell, Terry (1987). *Consuming Fiction.* London: Verso.

Lull, James (1988). Constructing rituals of extension through family television viewing. In James Lull (ed.), *World Families Watch Television.* Newbury Park, Calif.: Sage.

———. (1990). *Inside Family Viewing: Ethnographic Research on Television's Audiences.* London: Routledge/Comedia.

Montgomery, John (1965). *The Fifties.* London: Allen and Unwin.

Moores, Shaun (1990). Dishes and domestic cultures: Satellite TV as household technology. Unpublished paper, Polytechnic of Wales.

Morley, David (1986). *Family Television: Cultural Power and Domestic Leisure.* London: Comedia.

Morley, David, and Kevin Robins (1989). Spaces of identity: Communications technologies and the reconfiguration of europe. *Screen* 30(4): 10–34.

Morley, David, and Roger Silverstone (1990). Domestic communications: Technologies and meanings. *Media, Culture, and Society* 12.(1): 31–55.

Muthesius, Stefan (1982). *The English Terraced House.* New Haven, Conn., and London: Yale University Press.

O'Sullivan, Tim (1991). Television memories and cultures of viewing, 1950–1960. In John Corner (ed.), *The Homely Image: Essays in Television History.* London: British Film Institute.

Punter, John (1984–1985). *A History of Aesthetic Control I: The Control of the External Appearance of Development in England and Wales,* 1909–1947 (1984), 1947–1985 (1985). Reading: Department of Land Management, University of Reading.

Radway, Janice (1984). *Reading the Romance.* Chapel Hill, N.C.: University of North Carolina Press.

Ross, Andrew (1989). *No Respect: Intellectuals and Popular Culture.* New York: Routledge.

Scannell, Paddy (1988). Radio Times. In P. Drummond and R. Paterson (eds.), *Television and Its Audience.* London: British Film Institute.

Seiter, Ellen (1990). Making distinctions in TV audience research: Case study of a troubling interview. *Cultural Studies* 4(1): 61–84.

Seiter, Ellen, et al. (eds.) (1989). *Remote Control.* London: Routledge.

Taylor, Helen (1989). *Scarlett's Women: Gone with the Wind and Its Female Fans.* London: Virago.

Veblen, Thorstein (1899). *The Theory of the Leisure Class.* New York: Macmillan.

Wall, Ian, and Louise Chater (N.d.[1990?]) *British Satellite Broadcasting: Study Material.* London: BSB and Media Education.

Williams, Raymond (1980). Advertising: The magic system. In *Problems of Materialism and Culture.* London: Verso. (Essay originally published 1960s.)

Wright, Patrick (1985). *On Living in an Old Country.* London: Verso.

Afterword: The Place of the Audience: Beyond Audience Studies

James Hay

A prologue and postscript. July 24, 1988. One of the last entries in Eric Michaels' *Unbecoming*, an entry dated a few weeks before his death, recounts an encounter he had with the television series *Dallas* (1990: 178–179). He remarks that that evening there was something familiar about the bizarre twists of the soap, and something bizarre about the episode's redoubling of his own, all too familiar and increasingly immobile, position: Pam in a hospital bed, wrapped in sticky tape. But his sense of identification and estrangement, he continues, had to do with more than seeing his own deteriorating condition (and his own autonarrative) as soap opera. He had, as we learn earlier in the book, moved to Australia after living for some time in Texas. The "backstory" becomes significant on this evening of television viewing because he is also compelled by the difference between watching the small-screen sagas of Texas's Ewings in the United States and in Australia. He notes that in Australia, unlike in the United States, episodes of *Dallas* are broadcast out of sequence and on varying nights of the week on different networks in different regions of the country. How he asks (sardonically, but also out of frustration) can a viewer in Australia ever really feel part of the television narrative, much less its family: "How can you call Australia a country if everyone is at a different point in the story?" He concludes by wishing he were somehow reconstituted as a part of an imaginary audience with friends (Horace, Jackie, Jim, and others) whom he had left in Texas and who might under different conditions have been watching the same episode as the one on his screen. But he acknowledges the impossibility of such an audience. This "afterword" contemplates the possibilities and impossibilities of that night (and of another night some years later when I read his book). I wish I could say that we were watching, Eric, somewhere else.

"Audience," as semiotics would remind us, is a subject and concept that is the product of convention. (And as some have pointed out, the term may already by losing a technical, if not a commonsense, acceptance.) A lesson of semiotics that has remained at the margins of debates over the definition of audience, however, is that the sign ("audiences")—as both what gets designated and how it gets designated—is a *marker* or, more specifically, as way of marking a point in space and in a network of positions. "Gaining definition," in this sense, refers as much to the production of meaning as to a spatial production, as process of marking, demarcating, tracing, and connection points. While the world we inhabit is linguistically and discursively organized, it is also spatially organized and mapped.

Discussions about audiences contribute to the spatial organization of places we inhabit and are in turn conditioned by a world that is spatially organized and practiced. Audiences get defined from particular sites, in relation to particular places, and within particular contexts. To the extent that "audience" became the object of a struggle over definitions, this struggle is implicated in a realignment of the sites where audiences get studied and from which audience research gets produced. To see audience study otherwise assumes that the audience exists nowhere, as a completely relativized concept, because it is so readily "filled with meaning" by various modes of inquiry.

Given the various debates and areas of consensus that run across the essays in this collection, it seems particularly important to ask, as a rather immodest afterthought, what is the point of foregrounding *audience* in communication and media studies, or, for that matter, in any kind of analysis? As a response to this question, I want to consider a passage from Michael de Certeau about *landscape* and the politics of "everyday life":

> The *imaginary landscape* of an inquiry is not without value, even if it is without rigor. It restores what was earlier called "popular culture," but it does so in order to transform what was represented as a matrix-force of history into a mobile infinity of tactics. It thus keeps before our eyes the structure of a social imagination in which the problem constantly takes different forms and begins anew. It also wards off the effects of an analysis which necessarily grasps these practices only on the margins of a technical apparatus, at the point where they alter or defeat its instruments. It is the study itself which is marginal with respect to the phenomena studied. The landscape that represents phenomena in an imaginary mode thus has an overall corrective and therapeutic value in resisting their reduction by a lateral examination. It at least assures their presence as ghosts. This return to *another scene* thus reminds us of the relation between the experience of these practices and what remains of them in analysis. It is evidence, evidence which can only be fantastic and not scientific, of the disproportion between everyday tactics and a strategic elucidation. Of all the things everyone does, how much gets written down? Between the two, the image, the phantom of the expert but mute body, preservers the difference (1988: 41–41; emphasis added).

What exactly might de Certeau's use of the expression "landscape" have to do with the current theorization and analysis of audience? (I pose the question less as a strategic elucidation of de Certeau's work than as a means of reframing the terms of audience work.) Landscape, as it is used in this passage, is a "corrective and therapeutic" *technique* and thus the basis for a kind of analytic *ethics*. It refers analysts to a field of activity and multiform practice ("popular culture" and *everyday life*) that surround analysts and their object of analysis. For de Certeau, this field can never be circumscribed, completely mapped or measured, but, by virtue of its being recognized, underscores the provisional and modeled status of any map and the "strategic" implications of the repertoire of surveying instruments comprising the analyst's "technical apparatus".

Landscape also refers to the play of foregrounding and backgrounding that accompanies an inquiry. Again, in his linking of "landscape" and "everyday life," he notes that "of all the things one does, how much gets written down?" Of all the practices that organize a social formation, which ones become the ones that matter? And just as important, which become the preferred modes of inquiry and explanation? Against the background of everyday life and a society's multiple, common ways of sense making, what kinds of knowledge gets "*institutionalized*"? Which social practices and knowledge practices, through the apparatus and normative institutions that sustain and drive any kind of "research," become privileged (foregrounded) and accepted in representing a social structure—a sense of the way things are? As part of an ethic of inquiry, landscape thus serves as a reminder that the production of knowledge—its merit, veracity, and acceptance—depends upon the privileging of certain social practices out of everyday life and certain techniques for explaining them. Landscape, in this sense, concerns the reservoir of innumerable "minor" practices or ones that do not seem to matter because they lack the structural coherence imparted through a discursive formation to some of these practices.

Importantly, de Certeau's use of the term landscape to characterize how some modes of inquiry reduce the social to a dominant set of procedures and single structure raises the issue of representation, specifically the observed and "imaginary" status of landscape: "it keeps before our eyes the structure of a social imagination in which the problem takes different forms and begins anew." (The French terms *paysage*, use by de Certeau here, also refers to *scenery*, whether natural or depicted.) Landscape designates something that is not recognized or noticed because it is "common," familiar, and ordinary as well as too robust and multiform. Landscape is therefore something that can only be imagined but that always exceeds attempts to fix it through imaging or modeling. Only specific features of everyday life—characterized here as landscape—get framed through a technical apparatus of observation, imagining, (i.e., filmed, photographed, painted, drawn, written, narrated). Knowledge rests upon certain preferred observational techniques and technologies and upon the power of these techniques, and of the artifacts they produce, to *represent* (e.g., as snapshots) social, economic, or political structure. The materiality of landscape (something that de Certeau leaves vague in this passage and elsewhere) is a consequence of making the "discrete" practices of everyday life visible, of constructing a coherent way of seeing the social as a structure, of assuming that there is a position from which one can observe the social as coherent, of objectifying, in other words, the fluidity of everyday life as a proper, stable set of relations.

Given that "landscape" is as ambiguous a term as "audience," there is a risk of replacing one slippery expression with another. On the other hand, the range of connotations surrounding de Certeau's use of this term offers a way of reconsidering objectives and methods of inquiry about audiences. From many of the essays in this collections, it should be clear that landscape, as I have discussed it thus far, is what audience studies—or, for that matter, media studies—have come to grapple with. Recognizing a relation between audiences and *landscape* underscores that there is a continuous, though richly variegated, surface of activity both surrounding and constituting the subjects being examined in audience analysis. Audience then, is a way of recognizing, of studying, and (for some) of explaining a particular activity amidst this landscape. The degree to which audience activity is seen as complex has to do with the willingness of those studying them to recognize and address an audience's connection to this field of activity. Ultimately, audience has a double role: audience serves as a reminder of the contingencies and fragility of any system, pattern, structure, institution, conclusion, or map, but audience also potentially becomes a way of privileging a set of practices and fixing a set of relations in everyday life.

This double role is most critical in media and communication studies, where audiences became a privileged object of research. Since the 1940s, audiences have become the point of efforts to bolster as well as to qualify or challenge the seeping generalizations about the role or effects of media in organizing (homogenizing) social, economic, and political structure. Just as important, however, is that the very *disciplinarity* of audience research in media and communication studies (as "discursive formations" with institutionalized/departmentalized procedures and technologies for producing knowledge) has bracketed the audience's complex connection to everyday life, even as it has attempted to discuss audiences as a complex set of processes. Thus, not only has discussion about audiences been inflected by the disciplinization of media studies and the institutional status of its research, but also assumptions about the audience's relation to everyday life and culture have been "disciplined," shaped by the privileged statues of its discourse—as "research"—in everyday life. And one consequence of this has been that pronouncements and findings about audiences through this research and its dissemination in everyday life has continually (though not without debate) attribute to everyday life certain patterns of behavior and effects. Many audience studies have assumed that they can explain the familiar and the ordinary, and they set off to locate a stable set of relations upon which the familiar and the ordinary rest.

Indeed, "everyday life" or the equally inflated term "popular culture," which de Certeau associates with landscape, have been phantom-like concerns pervading media and audience studies since their emergence. And it is quite important to understand when and how these subjects/terms gave provided a king of lexical and rhetorical backdrop—an implied objective—in media and audience studies. But communication and media studies' claim on audience study—the authority of its techniques and technologies—make "audience" a specific, through arbitrary and relatively discreet, object of analysis because the object appears to be shaped wholly or primarily by media and communication processes. An attention to "landscape," however, discourages seeing audience as a subject literally relegated to these or any single "discipline". It should lead those studying audiences to ask not only what audience analysis has to do with the study of (among other things) art, history, politics, economics, science, education, and geography, but also how every mode of inquiry selectively observes the complex heterogeneity of everyday life.

In 1988 Janice Radway came to a somewhat similar conclusion when she noted that "instead of segmenting a social formation automatically by construing it precisely as a set of audiences for specific media and/or genres, I have been wondering whether it might not be more fruitful to start with the habits and practices of everyday life"(1988). Radway would redirect the traditional object of audience analysis toward the more "endlessly shifting, ever-evolving kaleidoscope of daily life and the way in which the media are integrated and implicated in it"—an object of analysis, she argued, that was less easy to fix.

Audience analysis may lead to the recognition of everyday life, but there are also a number of problems about the term "audience" that an attention to everyday life makes acute. Being an audience is only one activity in everyday life, even though the analysis may obscure this fact. Still, it is hard to envisage a discussion about contemporary everyday life that does not address the issue of being an audience. Under what conditions is one not an audience? Given the ambiguities of audience, much of everyday life (driving a car, riding a subway, shopping) could be described as an audience *anywhere*, the key issue becomes locating the socially sanctioned and culturally produced sites where "audience" assumes a significance and meaning.

Radway proposed accomplishing this through an "ethnography" that would bring together multiple disciplinary perspectives and could more effectively consider the practices of everyday life "as they are actively, discontinuously, even contradictorily pieced together by historical subjects themselves *as they move nomadically via disparate associations and locations through day-to-day existence* [italics mine]." In many respects, her proposal for a multidisciplinary project offered a useful way of beginning to rethink assumptions underpinning communication and media studies that had set the parameters of audience analysis for decades. And her valorization of ethnographic methods has also been central to debates reshaping audience study since the 1980s. But ethnography, as Lawrence Grossberg (1988) noted in his response to her essays and as others have argued since, has not been particularly sensitive to the *mobility* of subjects through everyday life. What remains implicit in Radway's proposal (and even in Grossberg's response to it) is spatial problematic—a more full-fledged consideration of the issue of everyday life as a spatially-constituted field of practice.

I want to suggest that this spatial problematic has always pervaded media, audience, and ethnography study but has seldom been (and only very recently begun to be) an issue attracting substantial attention. And in order to do so, I want to consider yet one more implication of de Certeau's use of the term "landscape."

By associating landscape with "popular culture" and *everyday life*, de Certeau confers a spatial significance to these latter terms. And while the spatial significance is only suggested by his use of "landscape" in this passage, it becomes more pronounced elsewhere in his discussion of the city as both a structured place and a "lived, disquietingly familiar space." Everyday life does not just refer, then, to the "kaleidoscope" of social practice and lived experience that exceed attempts to fix them as social structure, but also to the transience of social subjects amidst the spatial organization of the social world. For de Certeau, the visible city is a planned, geometric environment/landscape—the total grid of places, buildings, streets, and son on, whose identity and relations seem relatively secured through proper names and maps. Amidst this city, de Certeau argues, there is also a less visible, "migrational" and mobile city—a "metaphoric" city—comprised of the multiple, vernacular passages, producing new spaces, across and through the institutionalized and marked/visible relations among places. The landscape produced by this mobility (a metaphoric city) is both that which, as the continual creation of invisible spaces, exceeds attempts to structure the environment of social life and simultaneously that which regulates and disciplines by "secretly structuring the determining conditions of social life" (1988).

While Radway's proposal rightly understands the importance of considering the audience's relation to a broad field of social practice, it stops short of recognizing how this landscape is spatially organized and continually, discreetly reorganized. While it acknowledges that audience study has been fundamentally shaped through media and ("mass") communication studies' particular ways of disciplining knowledge about this landscape, it only begins to consider how audience research has occurred through particular sites, and it stops short of addressing how audience research has deployed and perpetuated certain spatial models and how audience work has taken place within changing spatial frames of reference. de Certeau's observation about "landscapes of inquiry," seems crucial as an afterthought about audience studies because it serves as a reminder that "defining" audiences is a spatial project. It underscores that listening, reading, and viewing occur in and around particular sites and through a social world organized geographically. But it acknowledges that these activities simultaneously produce "paths"—from one site to another—through everyday life, and that these paths elude modes of inquiry that privilege certain sites in everyday life. "Landscape" is the margin of any audience formation, identity, or site, and it is also the margin of audience study.

Thus while audience study may be one way to recognize that television is a site, audience study may just as likely to understand "television audience" only from that site rather in terms of television viewers' movement among other sites in their everyday lives. Furthermore, in many studies about audiences, television becomes a preferred site for understanding "audience" and for formulating generalizations about everyday life without reflecting upon its constitution as a *site* within a *landscape* of other sites and spatial references. As Lynn Spigel (1990) has noted, television viewing during the 1950s was organized through a complex transformation of domestic space (e.g., interior design, artchitectonics, landscaping, and suburban development). Television watching, therefore, contributed to and is circumscribed by an environment of spatial models and references. But focusing on a "television audience" offers a very particular understanding about the landscape of that audience.

To speak of a "landscape" of audience study therefore suggests that "defining" audiences is a spatial project, but that audience research as seldom seen its project overtly as a spatial problematic, that is, that the issue of space has seldom been foregrounded. And even though recent critiques and theories of audience research have problematized the disciplinarity of audience study, by emphasizing the audience's relation to "everyday life," they too have been slow to discuss its project as a spatial problematic. But it is one thing to suggest that audience research has never considered its spatial designation of audience as an issue, or that only recent debates about audience study are making space an important issue. It is quite another to assume that audience research has *never* been a spatial project.

To consider how the study of audience has relied upon and perpetuated spatial models, references, and rhetorics is a task that lies beyond the scope of this essay. One could, however, simply take stock of assumptions about space underpinning studies that conceive of the audience in terms of a communicative process. Undoubtedly, this is the most commonsense and disciplinary way of understanding the audience, and one whose limits the concepts of *landscape* calls into question. Not only does the concept of landscape emphasize that being an audience involved occupying a site, but it also alluded to a field of practice within which communication and audience are caught up and of which they are only partially constitutive.

As James Carey (1989) has suggested, the traditional model of communication, as sender-message-receiver, is deeply rooted in a dominant cultural and religious idea of transmission and distribution through space for the control of distance and people. Although Carey counterpoises this "transmission" or "transportation" model of communication with what he calls a "ritual" model (the "maintenance of a society" and its sense of communality "in time"), he implies that each strategically illuminates otherwise interrelated features about the sociality of communication. Carey's observation, which he attempted to elaborate through his reading of Harold Innis's work, affirms that the social study of communication has been fundamentally a spatial problematic even though little about (mass) communication research has seen its project explicitly in those terms. And again, while current audience studies have discounted the notion that a coherent message is simply transported, they have only begun to consider how media subjects' activity is a question of mobility within a social landscape of media sites and networks.

Carey's efforts to acknowledge the spatial features of the most commonly accepted model of communication also calls attention to an intellectual tradition from which audience research emerged. Sociologist Robert Park, Ernest Burgess, and R. D. McKenzie—all associated with the Chicago School in the early twentieth century—are generally considered the forerunners of mass communication, media, and audience research (see Park and Burgess, 1967). Scant attention, however, is given to how their views of community, social relations, and mobility were tightly bound up with their understanding of modern, urban life as an "ecological" system, that is, as constellations of communal life separated and bounded by "communal institutions" such as churches, theaters, playgrounds. Or, to turn this around, their conception of community, social relations, and mobility as an environmental issue was based upon their having used particular Chicago neighborhoods as reference points and their having privileged the city as material site and metaphor for research into modern life. Their "research" (whether one sees its connection to Progressivist politics) was implicated in an effort to map, and thus to make visible, the newly emerging, often unrecognized zones, sites, and paths of social life in the city.

Beyond these examples, there are other ways that audience study has adopted or accepted certain landscape genres. A concept such as "the family" is predicated upon spatial models of domesticity. A "1950s audience, "however obliquely, refers to a complex historical set of spatial relations and frames of reference that include, but are by no means confined to, the transformation of everyday life amidst a suburban environment. An audiences' relation to economic, political, or social milieus and concepts (e.g., totalitarianism, postindustrialism, imperialism) is constructed through/as spatial models. And the concept of postmodernism conjures an array of landscape metaphors.

Seen in terms of the "Landscape of inquiry" surrounding various intellectual traditions and dispositions, audience study can be said to occur from (moving among) particular sites and through a world organized and reorganized spatially. It has drawn upon and reproduced spatial models, rhetorics, and frames of reference—whether lived or imagined. As a partial and selective way of fixing social relations and of explaining everyday life, its operation is continually *resituated* yet *territorializing*.

As a place-binding and space-producing activity and as a form of inquiry that privileges certain sites of engagement and certain relations ("networks") among sites and subjects, then, audience study might also be understood as "strategic" or "tactical"—expressions de Certeau uses to emphasize a spatial politics of knowledge production. For de Certeau, strategic inquiry refers to a relatively institutionalized set of procedures for managing and predicting, through the production of maps, the places and the terrain where social relations are formed. The tactical recognized the multiform practices in everyday life as the basis for imagining, negotiating, and transforming an already mapped terrain. Strategic analysis is "geometrical," "geographical," visual, and panoptic. Tactical analysis attempts to locate "another spatiality," and "anthropological," poetic, and mythic experience of space.

There is something seductive about dividing audience analysis this way, of seeing it has having developed strategically and, more recently, tactically. Certainly, some forms of audience analysis have assumed that their activity can be panoptic and geographic, that messages are transported intact and are uniformly consumed or understood, that "audience" designates structure and coherence, that various media uses by audiences can be charted and predicted with certainty, and that the landscape of audiences is stable, real, and natural—an assumption supported tautologically by audience research's realist techniques for representing audiences. Other, generally more recent, ways of conceptualizing audience and of doing audience analysis could be described as tactical. They focus on the mythic and symbolic ways of conceptualizing social relations and structure, see television and television viewing in relation to "everyday life," underscore the constitutive features of "audiencing" in social relations and structure, use audience to demonstrate the instability of media messages, ideology, and power, and deconstruct the panopticism and realism of corporate and institutionalized audience research.

At worst, however, these terms decontextualize and generalize the methodology, definitions, theoretical orientations, and politics of audience studies. At best, they rehearse the rejection, by critical media and cultural studies, of mass communication research's empirical, realist method and its faith in the stability of social and communicative structure over process. Understood as a binarism, the terms also affirm the belief, in and outside cultural studies, that de Certeau is too often deployed by television criticism to conceptualize how television audiences "evade" the socially structuring role of media, and thus how television (as a dominant form of contemporary popular culture) does not necessarily control its audiences or uniformly reproduce a dominant ideology. Historically, the terms do describe conflicting audience study projects before the late 1980s, since audience study (as "qualitative" research or as "cultural study") came to be so closely aligned with ethnographic accounts of the everyday lives of audiences and with debates about revising traditional ethnography in a landscape where television had come to occupy such a central position. And at best, the terms serve as a reminder that what one studies and how one studies it is part of spatial politics.

It is this last point that I particularly want to retain and emphasize, in part because cultural studies of media and audiences can (indeed have at times) focused so much upon the contingencies, instabilities, and the hegemonically resistant activities of audience that issues of space seem trivial, irrelevant, or anathema (i.e., space equals structure equal that which is politically resisted). In recent audience-related work, temporality and spatiality have become somewhat more pronounced as something to be described, but it has remained primarily a methodological rather than a conceptual and political issue. Audience ethnography still tends to emphasize television (or another medium) as a privileged site. It is prone to understand its project as one of interpreting subjects' narratives rather than considering how a subject's engagements with television and access to and investment in other sites produce spaces and temporalities (and are part of a field of spatial reference). And, as James Clifford (1992) has noted, ethnography too often overlooks the spatial politics of "participant observation," that "being in the field" is an issue of an ethnographer's access to and mobility among particular sites and of the ethnographic subject's transience.

To say, therefore, that the object of audience study and audience ethnography is a landscape is to play upon an ambiguity. Landscape is that which already is spatially organized but which is continually traversed and gradually reconfigured. And as such, it discourages a simple opposition between the "geographic" (i.e., the spatial structuring of the social) and the nomadic or temporal (i.e., that which resists or undoes the spatial). Landscape is an expression of our faith that the world around us is both stable and plastic; it is a consequence and a condition of social structure and mobility.

What, then, is the materiality of this "imaginary landscape"? The question is both crucial and difficult for a variety of reasons. The reason I want to emphasize first concerns the Marxist intonation of de Certeau's argument. In one sense, landscape is the material condition and impression (the "scape") of social structure that we inhabit. On the other hand, its materiality is contingent upon its being practiced. The "spaces" produced across everyday life are also attributes of this landscape; they make a landscape seem both plastic and structured. The issue of a landscape's materiality is also complicated by de Certeau in his seeing language (i.e., names and words) and its practice (reading) as spatial (i.e., walking as the production of a passage through streets with names). "Society" is spatially structured through names and words, just as the city is divided, its various points organized, through language. But if that is the case (as deconstructionists have always contended it is), then that structure is always contingent and deferred, while reading (audiencing) is *productive*—in a materialist sense of spaces.

From here we confront an issue that has conjoined some of the concerns of Marxism and textual theory, as well as having underscored the limitations of each. Landscape is the material condition and consequence of a discursive economy, and it is the linguistic condition and consequence of a material economy. The important variation here, however, is that landscape poses this as a spatial issue. And what the debates shaping audience study today have done is to conjoin *and* call into question the limits of a materialist or discursive analysis.

Audience study informed by semiotic and structuralist theory rejected the classical Marxist notion that audiences are simply consumers of media industries an their ideologies. Rather, semiotic and structuralist theory have become one basis for describing the "reader's" productivity. But the productivity of viewing, for semitic and structuralist theory, involves the production of meaning and consciousness. In this sense, it becomes an interpretive rather than a materialist issue.

Audience study informed by semiotic and structuralist theory has also challenged the empiricism and realist techniques of traditional social and behavioral research about (mass) communication and the audience. Language and empirical reality, these studies have argued, cannot by readily disentangled, since reality is always mediated by language. British cultural studies also convincingly maintained that the meaning of media texts are contingent upon audiences, compentencies to "decode" these texts. But while British cultural studies was interested in rethinking the audience as a site of cultural production (or expanding, if you will, the traditional Marxist conception of production and the "transmission" model of communication), it tended toward a textual explanation of that site and it understood an audience's "positionality" in historical and class terms rather than in spatial terms. Thus, while an audience study driven by textual theory may have identified audiences through their "competence" with particular codes, they seldom considered the spatial significance of those competencies as determined through and as determining material sites of production. It matters, therefore, how competencies are established in particular places in relation to particular technologies, but moreover how competencies of particular technologies enable or constrain one's ability to negotiate, cope with, or transform those places—to make one's way amidst established avenues. And when this becomes and issue, then space itself becomes more of a problematic. If everyday life is comprised of various procedures for negotiating places and producing paths and spaces, then "audience" becomes a way of considering how, when, and where certain technologies (television, radio, books) and techniques (viewing, listening, reading) are deployed to produce contexts, relations among sites, and networks.

To the extent that audience became a subject for rethinking and expanding the site of ideological and cultural production, it reformulated the relation between texts and their contexts—in a way consonant with Volosinov's belief that the production of ideology and discourse occurs through a social and discursive terrain that is always already organized. Audience as a discursive or textual subject suggested that contexts are also discursively produced and organized. But again, "context" was often a fuzzy term that did more to underscore the (historical) situated-ness of meaning, ideology, and power than to address the complexity of the site of production (or rather, the complex relations among sited of production). It was more about the continual formation of discourse and that contexts are discursively produced and reproduced than about the limits, the situated-ness, the multiple and complex trajectories, the colonizing and territorializing of discursive formation. In short, context, was more a term rooted in an Historical materialism of texts and interpretation than in a spatial materialism of access, mobility, networks. Precisely for this reason, the issue of "contextualization"—something central to current audience research—generalizes that which "landscapes" accentuates somewhat more effectively.

Achieving a more "expanded" sense of the site of production is in some respects what Radway is suggesting when she proposes a more interdisciplinary kind of ethnographic audience research. But, as Grossberg noted in his response to her proposed project, ethnography too often understands the conditions of its subjects in discursive rather than material terms. I would add that in the whole the traditional subjects of ethnography are positioned within a landscape it tends to emphasize the speaking subject over its landscape of material and discursive practice.

Engaging with the spirit of Radway's proposal would involve what Doreen Massey (1994) has termed a "power geometry." Massey, a geographer, uses the term to describe how the relative mobility of different social groups and different individuals is part of a terrain of other kinds of flows and interconnections. She illustrates how a geography of her own everyday life and her own sense of place would involve recognizing these multiple, taken-for-granted trajectories that pass and intersect through her neighborhood in London and, in so doing, connect her sense of neighborhood to other sites. The residents of her neighborhood who have migrated there from other regions and nations, the overhead jets, the trucks, cars, and trains, the store windows displaying items assembled from all over, vendors of magazines and newspapers, telephone and fax lines, television antennae, all constitute a nexus—a "landscape" so to speak—of circulation (not simply reception). For Massey, her allegory becomes a means of demonstrating that neither social subjects nor the places they inhabit are static, and that localities, as points of regional, national, and global flows, are "extroverted" and thus part of non-Euclidean "geometries of social relations and power."

Massey's notion of a "power geometry" is instructive for audience study because it offers a way of considering how to expand the notion of a "site of production." "Power geometry," while linking people and social groups to particular places, decenters them and turns attention to its landscape.

Her writing about geography does provide and important perspective for the disciplinary forces in media and the audience study that seldom see their project as overtly "geographic". Neither Massey nor other critical, Marxist geographers, however, have discussed media or the issue of audience a particular, and in most instances complex, processes. While I am not suggesting necessarily that they should have, it does indicate a long-standing area of neglect and a point for future interdisciplinary work. Such a project would not as much entail the kind of macroanalysis of spatiality (across twenty centuries no less) supplied by David Harvey's (1989) attempt to bridge materialist and "culturalist" theory.

It would, I believe, be closer to Arjun Appadurai's (1990) conception of production as occurring at the conjunctures and disjunctures of various kinds of "-scapes." For Appadurai, these would include "ethnoscapes" ("the landscape of persons"), "technoscapes" ("global flows of technology"), "finanscapes" ("the disposition of global capital"), "mediascapes" ("the distribution of the electronic capabilities to produce and disseminate information . . . and the images of the world created by these media"), and "ideoscapes" ("concatenations of images . . . and ideas"). Like Massey's notion of a "power geometry," Appadurai's conception of production decenters, though never excludes, human subjects (e.g., his notion of the ethnoscape). An "ethnoscape" of individuals, families, and other social groups is itself part of other "-scapes", and each generates and is determined by the others. Nor does his conception of production privilege media and their effects, preferring to understand the issue of media effects more as a question of media subjects' relations to the multiple "-scapes," conjunctures, disjunctures, and effectivities. And in this way, he also sees the issue of landscape in both materialist and ideological terms. Finally, Appadurai makes explicit something only implied by de Certeau or Massey's example, namely that production/consumption occur within and generate multiple, fractured, interplaying "-scapes" and that these "-scapes" are the conditions of and consequences of flows, (of people and social groups in relation to capital, media images, and ideologies in space).

In a way that more directly engages the tradition of audience research, the work of David Morley and Roger Silverstone has attempted to examine domestic television viewing as a spatial issue.[1] Their work calls attention to the viewer's relative mobility among various sites in the home (rather than just in front of the television set) and thus to the way that domestic space is a socially organized set of relations among sites. Relying upon Bourdieu's concept of the "habitus," they propose that broader social relations are played out (*produced*) domestically through media audiences' competence with and access to particular technologies that align them with particular places in the home. Importantly, this kind of study decenters the activity of television viewing by placing viewing in relation to other media sites and technologies organizing the home. Some of their work has also considered how individuals and families form identities through their conceptualization of their own domestic space, through "mental maps."[2]

The importance of Morley and Silverstone's approaches to domestic space lies in its attempt to expand the site of production while remaining sensitive to the complexities of media as they discuss the practices surrounding media. One could argue that their work has remained too focused on television as a privileged site in domestic life, though to have done otherwise would ignore the status accorded television by those whose domestic lives they have observed. If anything, their conception of space may be too easily explained as a condition and consequence of practices surrounding television and media. How, for instance, are viewers' concepts of domestic space informed by and continually renegotiated through a field of spatial reference that only partly results from their investment with television? Furthermore, if domestic space (as "habitus") is a site for reproducing a broader set of social relations, then it becomes crucial to situate domestic space within—or to broaden the discussion of domestic space by acknowledging—a field of sites and of spatial references beyond domestic space.

Considering television viewing in the home in relation to other sites, however, would involve taking to heart Appadurai's contention that production/consumption occurs not so much in a static "field" as amidst "flows". As his argument affirms, Bourdieu's notion that the "habitus" is a particular site of social production is made somewhat problematic if one understands the "field" of social production as comprised of interurban, interregional, and transnational circulation. These flows become important to recognize in order to avoid seeing the spatial problematic of audience study as simply a matter of mapping fixed points through a Euclidean geometry. Analyzing engagements with particular media technologies (the common subject of audience study) therefore involves understanding how production, identity, and power operate through multiple an particular *locations* (or networks of sites that territorialize production), as well as through multiple and converging *flows*.

Charlotte Brundson's essay in this collection, for instance, concerns the transformation of British landscape (itself a point of rich generic reference) in the age of satellite dishes. It illustrates how changing practices involving television are played out across sites such as home, neighborhood, city, and nation, and how architectonics, landscape, and urban planning become subjects in a national discourse and policy about television consumption and viewing. by concentrating on the material technology of reception (its acquisition, display, use in accessing), however, Brundson also points to a way of thinking about domestic viewing as a site traversed and "scaped" by various flows. In order to elaborate on this point, one might use her observations as a basis for considering what the emergence of dish reception in England and Scotland had to do, among other things, with the scattered successes of dish and cable companies attempting to import foreign and transnational channels of programming and with the relative mobility in England (or with the movement to England) of social and ethnic groups for whom watching television is but one way of negotiating new and former senses of place.

The project of considering as spatial practices the technological engagements and practices in the everyday lives of social subjects thus involves examining how individuals or social groups have access to, come to occupy, become invested in, transform, move among, and are traversed by an connected to other sites through various other flows. In the spirit of audience ethnography, such a project is interested in the concrete specificity of particular sites whiles recognizing (in Massey's sense) their "extroverted" nature. While John Hartely and Ien Ang are right in concluding that the audience is a fiction, it is important to add that its ontological status is conditioned by, and a consequence of a "landscape" (or in Appadurai's sense, multiple "-scapes"). Understood in relation to landscape, the subjects of audience research are always decentered (i.e., in and out of the privileged media sites of audience research), yet always situated and navigating in relation to a complexly "scaped" and spatially organized environment where media often serve as a basis for one's sense of place.

This way of thinking about (or rethinking) the subject of audience research is informed by a "spatial materialism": Audience becomes an issue of how social relations are spatially organized and how the production of meaning, texts, discourse, ideology, and identity—all accepted objects of recent audience study—always occurs from and amidst concrete sites. Film studies, for instance, devoted considerable attention to the "positioning" of spectators by a cinematic apparatus, but it generally ignored the movie theater, the home, the arcade, the city, the nation, the piazza, and so on as concrete sites of engagement by social subjects. But to examine how social relations are spatially organized or how the production of meaning is always situated entails more than a political economy of that production. Places are designated and mediated discursively. They become signified and signifying frames of reference for social subjects. Models and myths of places do become a basis upon which social relations are imagined, fetishized, and refashioned (albeit from particular sites and through particular technologies and modes of circulation).

One problem, in seeing the concept of audience and the project of audience research through a spatial materialism however, is that the term "spatial" may suggest a retreat from or rejection of historicism. This problem is somewhat compounded by de Certeau contending that the spatial structuring of social relations is endlessly remade through everyday life, implying that landscape is both a spatial and historical horizon. Yet his theorization of space lacks concrete examples whose historicity is an issue. His most central and fully elaborated analogies for discussing everyday life (and being an audience/reader) are the activity of walking through a city or watching a landscape from a train. Analogies such as these may loosely define a modern landscape and may broaden the sites or activities that have come to be privileged in audience studies, such as book reading or film or television watching, but clearly they serve more as metaphors than as specifically contextualized practices.

What is lacking, in other words, is a sense of how spatial relations and frames of reference are produced historically. Although the subjects of audience study (or social relations more generally) are spatially organized, in everyday life these relations are never fixed. As de Certeau implies and as Massey has argued more directly, space need not be understood as stasis or "coexistence within a structure that establishes the positive nature of all its terms." Space may be regulated, but this regulation occurs through historical flows, convergences, and divergences. Regulation also involved ritualization (or a repetition over time) and continual realignments and renegotiation of spatial frames of reference. What gets emphasized, however, by understanding audience as a spatial problematic, is how sites of consumption (as production) are both the condition and consequence of history.

By redirecting attention from audience to the shifting contours of landscape, the study of audiences, through a pedigree or intellectual tradition of "audience research," becomes arbitrary. Or rather, one would have to look to a variety of histories in order to begin reconstructing and reimagining landscapes through which audience has been defined. Such a project would involve considering new relations between places associated in the past with audiences and other sites seldom associated with audience. And it would entail reconsidering how subjects in audience study are situated within emerging and residual landscapes.

Audience study has been one of the many techniques of surveying a modern landscape, particularly the city, the suburb, the nation, and their organization through media technologies transformed everyday life and one's sense of place (or displacement). For many Western artists and intellectuals at the beginning of the twentieth century (predecessors of de Certeau), the city street became the metaphor and "stage" of modern everyday life. It was a stage *en plein air*, but constituted of an urban architectonics and geography. It became the point of reflections about realigning or blurring an earlier relation between audience and performance and between the observer and the observed.[3]

As theorist of postmodernity are quick to remind us, however, we have come to live less within a set of modernist distinctions (between urban and rural, private and public space) than in a networked society—a profusion of media vectors interconnecting virtual sites or "terminals." Here, the concept of audiences, understood as a question of whether media engagement constitutes "activity" or "passivity" (a fundamental issue in audience studies since the 1950s), is gradually replaced by the concept of "interfacing" or "interactivity." The most dire pronouncements about "postmodernity" emphasize the cybernetic features of individuals living in a networked society where media technology apparatus. In order to underscore the simulate state of things, theories of postmodernism have characterized its landscape as empty, as "hyperspace." Rehearsing Adorno and Horkeimer's argument about the uniformity of mass culture, but in decidedly more spatial terms, those describing the condition of postmodernity have gone so far as to suggest that, because of media vector and flow are everywhere, this environment is constituted wholly of virtual spatial references.

In some respects, this kind or privileging of media as the constitutive force and dynamic in social and cultural formation is consonant with audience study's own penchant for emphasizing media as objects of study. In so doing, both assume or leave the impression that there is a stable and homologous relation between media and its landscape. Where, more recently, audience study has come to emphasize the complex and uneven ways that media sites have been engaged in everyday life, they have evinced a potential to avoid making these kinds of homologies. Only in recognizing that "audience study" is a spatial problematic, however, can this potential be developed in a way that might envisage the "electronic city" as but one kind of "scape" against which the general issue of consumption might be examined and as a "scape" wherein it becomes crucial to figure out how media vectors, "webs," and "networks" get mapped and what their mapping has to do with the territorialization and reterritorialization of consumption and production of social relations.

Mapping is an important concept and practice that merits some clarification here. I take it to be, in the most general sense, the practice of recording or remember coordinates. As a formalized image of a place (or places), a map is ideological, always having the potential to blur its difference from the places or territory it signifies. Maps are also ideological in their function as spatial recordings and instructions. But beyond their ideological features, maps have become administrative, instruments of spatial control, constraining and enabling flows and travel and regulating the relation of people or social groups to places. Since the earliest European explorations of New Worlds, cartography has been implicated with imperialistic and scientific projects.[4] And as such, it has been as means and the object of struggles and claims over territory.

Historically, mapping has taken various forms and genres, some forms becoming more conventionalized, institutionalized, and "scientific" than others. Traditional audience research has operated through a kind of "geographic" apparatus that has objectified spatial practices into places occupied by individuals or groups with a stable, coherent (and thus measurable) identity. Its empirical methodology has assumed that its subjects are real (not ideological) and that its representation of them is realist. Its techniques and technologies for surveying landscape thus aspire to the authority and scientificity of topographs. The Nielsen company's method of audience survey has operated through procedures that map the television "home" and the relation of viewers to places in this way. Like empirical geography, its project involved tracing a set of pre-existing and thus stable relations whose coherence is preeminent. One of the reasons for discussing audience research as a spatial problematic is to call attention to its investment in the surveillance and territorialization of everyday life.

But because mapping takes many forms, one cannot assume that it is simply the specialized labor of professionals or scientists. If one looks beyond the most formal kinds of maps, mapping can be seen as a much more widespread and common practice of navigating and recollecting places and paths in everyday life. This is not to say that one is free to design any map one wants. New maps are constrained and enabled by cultural and economic capital, and some maps do become more institutionalized than others. Furthermore, preferences for or investments in certain kinds of maps will obscure alternative ways of mapping. This applies as much to the most socially institutionalized maps as to the least formalized in everyday life. In both senses, mapping helps secure an individual or group's relation to place, in part by excluding or marginalizing other ways of imaging/imagining a relation to place.

The question then becomes that of how to discuss the way social subjects become invested in and navigate an environment that is constituted in part of media sites, without the techniques of mapping perpetuated by traditional audience research. In part this would involve decentering the human subjects found in traditional audience research, charting their mobility among sites, their investment in or access to sites, and the way any particular site is traversed and organized by media and other flows. The empirical nature of such an analytic practice lies in its recognition of material "networks" of reception, that is, the technological intersecting of media vectors that traverse and connect specific sites of engagement by social subjects. But such a project would also need to recognize the limits of an empirical explanation that presumes a site to be discrete and stable. Otherwise it would overlook the play of emerging and residual features of landscape and features that remain invisible to it own method of mapping. And it would need to avoid an overly economistic explanation of sites and networks of production/consumption.

Furthermore, it would need to acknowledge the textually constructed features of mapping, while avoiding a textual determinism. Landscapes are like palimpsests—material surfaces that can be etched and etched over, however gradually. Places are imaged and imagined (and thus textual and ideological). But places or landscapes are not simply the textual projections of maps or the result of the flow of images and texts. A variety of flows contribute to the spatial formation of social relations as territory. What is needed, in this sense, is a cartography of sites and flows.

Mapping could also serve as a response to the interpretive practices characteristic of ethnographic audience study. Ethnography may attend to audiences in everyday life, but it too often emphasizes its subjects' accounts of their everyday lives, rather than their landscape. Consequently everyday life is made to seem largely an issue of consciousness rather than of material sites, networks, and mobility. At worst, an ethnographic discussion of an audience's environment is purely descriptive or assumes that the relation between a subjects's accounts and his or her environment is self-evident and thus unnecessary to conceptualize. At bets, it privileges a media site but fails to acknowledge the spatial problematic surrounding the formation of audience subjects.

One consequence of ethnography in audience study has been the growing preoccupation with the analyst's relation to the subjects being analyzed. The attractiveness of ethnography for those critical of traditional empirical analysis lay, in part, in their recognition that media or audiences could not be understood without the researcher's participating in that experience and culture. Seldom, however, is the issue of participation understood in overtly spatial terms. The challenge for audience ethnography is in considering how the site of media engagement is transformed by ethnographers and ethnography, rather than assuming that an analysts' temporary occupation of the same sites as her/his subjects makes it unnecessary to consider the analyst's and subject's relative mobility to and investment in that site.

To the extent that audience study is an attempt to chart the everyday lives of media subjects (or social subjects in a media environment), it should be aware of differential maps, differential modes of mapping, and what power particular maps do or do not wield over an environment. This task of *mapping everyday life* may in fact oxymoronic unless one recognizes that in order to delineate the structure through which everyday life occurs, one must also recognize the contingencies of mapping structures. And in order to discuss the multiform practices and flows of everyday life, one must also consider how they (and the cartographer) reproduce structures and rely upon maps. As a practice of marking, recording, and deciphering landscapes, mapping is therefore both empirical and allegorical. Its empiricism assumes that there are relatively stable and tactile spatial references that one experiences in everyday life and that become coordinates for describing any subject. As allegory, it concerns the transient, symbolic, arbitrary, and generic (conventionalized) disposition of observer and observed in and to particular places.

That maps are both empirical and allegorical suggests that the landscape being mapped is both stable and malleable (material but continually "scaped"). Its method and subject are spatial first, but with an understanding that mapping and the territory being mapped are the basis for and consequence of historical practices and flows, that spatial organization is produced by and produces history. The audience's "definition" (in spatial terms) is a process of becoming, yet always (in spatial terms) is inadequate or "unbecoming." Precisely for this reason, Eric Michaels' narrative about his solitude before the television set in a hospital room, dying, and his desire for an imagined audience that was geographically dispersed can be read as a confession and an allegory of possibility and impossibility. His story affirms that the need for "audience" rests upon a sense of historical rupture in one's landscape—of nostalgia for and a desire to remember, reinvent, or reestablish, in spatial terms, a mythic family and community.[5]

I would like to thank Lawrence Grossberg and Jane Juffer for their suggestions about the issues I address in this essay.

Notes

1. See Morely, 1992, and Silverstone, Hirsch, and Morely, 1992.

2. See Morely, 1988. The concept of mental or cognitive mapping has been developed in somewhat different ways by Kevin Lynch and later Fredrich Jameson. Morley's strategy, however, differs from Lynch's reliance upon gestalt psychology and from Jameson's effort to discusses this process largely as a challenge to artists and critics.

3. One could consider the poetry of Guillaume Apollinaire (particularly "Phantom of the Clouds"), the cinema of Dziga Vertov, and Walter Benjamin's "Berlin Chronicle," to cite just a few references.

4. Derek Gregory (1994) has discussed the historical relation between European exploration, cartography, and other sciences that organized knowledge about the New Worlds and their inhabitants.

5. I invoke Michaels' story also because his writing about television practices among Aboriginal communities in Australia (particularly his *For a Cultural Future*) offers a compelling "ethnographic" account of the politics of differential mapping in contemporary media culture. His insights would have made an invaluable contribution to this book.

References

Appadurai, Arjun (1990). Disjuncture and difference in the global cultural economy. *Public Culture* 2(2) (Spring 1990): 1—24.

Carey, James (1989). *Communication as Culture: Essays on Media and Society*. Boston: Unwin & Hyman

Certeau, Michael de (1988). *The Practice of Everyday Life*. Berkeley: University of California Press.

Clifford, James (1992). Traveling Culture. In Lawrence Grossberg, Cary Nelson, and Paula Treichler (eds.), *Cultural Studies*, pp. 96–112. New York: Routledge.

Gregory, Derkek (1994). *The Geographical Imagination*. Cambridge, Eng.:Blackwell.

Grossberg, Lawrence (1988). Wandering audiences, nomadic critics. *Cultural Studies* 2(3): 377–391.

Harvey, David (1989). The Condition of Postmodernity. Oxford: Blackwell.

Massey, Doreen (1994). *Space, Place, and Gender*. Minneapolis: University of Minnesota Press.

Michaels, Eric (1990). *Unbecoming: An AIDS Diary*. New South Wales, Australia: emPress Publishing.

Michaels, Eric (1987). *For a Cultural Future*. Sydney: Artspace.

Morley, David (1988). Chapter in James Lull (ed.), *World Families Watch Television*. Newbury Park, Calif.: Sage.

———. (1992). Television, technology, and consumption. In *Television, Audiences, and Cultural Studies*. New York: Routledge.

Park, Robert, and Ernest W. Burgess (1967). *The City: Suggestions for the Investigation of Human Behavior in the Urban Environment.* Chicago: University of Chicago Press (originally published 1925).

Radway, Janice (1988). Reception study: Ethnography and the problems of dispersed audiences and nomadic subjects. *Cultural Studies* 2(3):359–376.

Silverstone, Roger, Eric Hirsch, and David Morley (1992). Information and communication technologies and the moral economy of the household. In Roger Silverstone and Eric Hirsch (eds.), *Consuming Technologies: Media and Information in Domestic Spaces*, pp. 15–31. New York: Routledge.

Spigel, Lynn (1990). *Make Room for TV*. Chicago: University of Chicago Press.

About the Book and Editors

This book offers a major reconceptualization of the term "audience," including the landscape of a given audience—the situated and territorializing features of any way of seeing and defining the world. Given de Certeau's hypothesis that listening, watching, and reading all occur in places and result in produce transformed paths or spaces, the contributors to this landmark volume have provided innovative essays analyzing the transformations that take place in the geography between sender and receiver. The book acknowledges, in the face of conventional "discourse analysis," the contextual features of discourse, to produce a complex and textured understanding of the concept of audience.

The Audience and Its Landscape presents the work of a vital cross-section of international scholars including Sweden's Karl Erik Rosengren, the UK's Jay G. Blumler and Roger Silverstone, Australia's Tony Bennett, Israel's Elihu Katz, Canada's Martin Allor, and the United States's Janice Radway, Byron Reeves, and John Fisk, to name a few. This book is truly groundbreaking in its depth and scope, and will speak to students of rhetoric, mass communication, cultural studies, anthropology, and sociology alike.

James Hay is an associate professor in the Department of Media and Cultural Studies at the University of Illinois at Urbana/Champaign. **Lawrence Grossberg** is the Morris Davis Professor of Communication Studies at the University of North Carolina at Chapel Hill. **Ellen Wartella** is dean and Walter Cronkite Regents Chair in Communication Studies at the University of Texas–Austin College of Communication.

Index

For Product Safety Concerns and Information please contact our EU
representative GPSR@taylorandfrancis.com
Taylor & Francis Verlag GmbH, Kaufingerstraße 24, 80331 München, Germany

www.ingramcontent.com/pod-product-compliance
Ingram Content Group UK Ltd.
Pitfield, Milton Keynes, MK11 3LW, UK
UKHW022024061126
477531UK00002B/17
* 9 7 8 0 3 6 7 3 1 8 3 0 7 *